High Integrity Systems and Safety Management in Hazardous Industries

High Integrity Systems and Safety Management in Hazardous Industries

J R Thomson

ELSEVIER

AMSTERDAM • BOSTON • HEIDELBERG • LONDON
NEW YORK • OXFORD • PARIS • SAN DIEGO
SAN FRANCISCO • SINGAPORE • SYDNEY • TOKYO

Butterworth-Heinemann is an Imprint of Elsevier

Butterworth-Heinemann is an imprint of Elsevier
The Boulevard, Langford Lane, Kidlington, Oxford OX5 1GB, UK
225 Wyman Street, Waltham, MA 02451, USA

Notices
Knowledge and best practice in this field are constantly changing. As new research and experience broaden our understanding, changes in research methods, professional practices, or medical treatment may become necessary.

Practitioners and researchers must always rely on their own experience and knowledge in evaluating and using any information, methods, compounds, or experiments described herein. In using such information or methods they should be mindful of their own safety and the safety of others, including parties for whom they have a professional responsibility.

To the fullest extent of the law, neither the Publisher nor the authors, contributors, or editors, assume any liability for any injury and/or damage to persons or property as a matter of products liability, negligence or otherwise, or from any use or operation of any methods, products, instructions, or ideas contained in the material herein.

British Library Cataloguing-in-Publication Data
A catalogue record for this book is available from the British Library

Library of Congress Cataloging-in-Publication Data
A catalog record for this book is available from the Library of Congress

ISBN: 978-0-12-801996-2

For information on all Butterworth-Heinemann publications
visit our website at http://store.elsevier.com/

Typeset by Thomson Digital

Printed and bound in United States of America

Working together
to grow libraries in
developing countries

www.elsevier.com • www.bookaid.org

Dedicated to Ahmet Erdem, Colin Seaton, and others who were killed in accidents where they received some of the blame, yet where accountability really lay elsewhere.

Contents

PART 2   HISTORICAL OVERVIEWS OF HIGH-INTEGRITY TECHNOLOGIES

PART 3   SAFETY MANAGEMENT

About the Author

Jim Thomson's career has included nuclear plant operations, research and development, academia, design engineering, safety management, engineering management, and safety consultancy. After completing his PhD in process engineering at Aberdeen University, he started his career in 1979 at the Prototype Fast Reactor power station, Dounreay, latterly as a shift manager. He was a lecturer at Edinburgh University before joining NNC (now part of the AMEC Group) in Cheshire to develop safety cases for Heysham 2 and Torness nuclear power stations. In 1989 he moved to the South of Scotland Electricity Board (later Scottish Nuclear Limited) doing design/project management and safety case management, before becoming Nuclear Safety Manager and then Electrical and Control Systems Manager for British Energy (now EdF Energy). Since 2004 he has worked as a safety consultant, and has been a director of two major international safety consultancies. He now runs his own company www.safetyinengineering.com where his recent clients have included AMEC, Areva, Invensys, Rolls-Royce, and Ultra Electronics.

He has worked as a consultant in Belgium, Brazil, Canada, Finland, France, Hungary, Germany, India, Japan, Kazakhstan, Qatar, Sudan, Romania, UAE, UK and USA. He has worked on nuclear power stations in USA, UK, France, Germany, Finland and Hungary, offshore oil platforms in the North Sea, the Black Sea and the Indian Ocean, and refineries in Europe and Africa.

He is a Fellow of the Institution of Engineering and Technology, a Fellow of the Institution of Mechanical Engineers, and a Fellow of the Nuclear Institute. He has completed the Advanced Management Programme at Oxford University and the Senior Nuclear Plant Management Program at the Institute of Nuclear Power Operations, Atlanta, Georgia.

He was the author of "Engineering Safety Assessment" (Longman 1987) and co-author of "Elements of Nuclear Power" 3rd edition (Longman 1989). He has chaired two international conferences on nuclear engineering matters. As well as being a lecturer at Edinburgh University in the mid-1980s, he has been a visiting lecturer at Imperial College, Manchester University and Strathclyde University. He has also been a member of the degree accreditation panel for the Institution of Engineering and Technology. He was awarded the 2013 Pinkerton Prize by the Nuclear Institute.

He is married with two grown-up daughters.

Preface

This book is about the engineering management of hazardous industries, such as oil and gas production, hydrocarbon refining, nuclear power and the manufacture of chemicals and pharmaceuticals. It is also about how large enterprises must exercise their duties of corporate social responsibility. Its scope includes:

- An overview of the design processes and design considerations for high-integrity systems.
- The history and technologies of some aspects of hazardous process plant, and how we have learned (and must keep learning) to make things safer.
- Systems engineering – in its widest sense – and how multidisciplinary engineering efforts are necessary to make complex, hazardous plant operate safely.
- An overview of safety management processes as applied to hazardous industries.
- How these complex multidisciplinary enterprises, to design and operate hazardous plant, can sometimes fail (despite our best intentions).
- The subtlety and fragility of the robust safety culture that is required to operate hazardous plant safely.

It is aimed at professional engineers – those who design, build and operate these hazardous plants. When major accidents happen and attract large headlines, there are sometimes instant judgments and ill-informed opinions presented in the news media. This book presents an informed engineer's view of how we try to manage safety, and also how, in some selected major accidents case studies, we have got it badly wrong.

This book is also written for business schools and university engineering departments to help in engineering management education. It seeks to use plain language and no mathematics, while striving to maintain a rigorous approach.

The book is also written for engineering students trying to decide which way their career should go. The choice between design engineering and plant operations can be a big decision: I have managed to enjoy both.

My intent is to describe, in a few hundred pages, the wide range of challenges which face engineers when we try to make hazardous industries safe. I have previously written a book on technical safety – the detailed assessment of aspects such as system reliabilities, accident consequence assessments and plant risk – and this present book does not attempt to cover any more of that area. Instead this book is about the challenges of engineering management in hazardous industries.

I have worked in hazardous industries – nuclear and oil and gas – for over 35 years. There is no doubt that the vast majority of the issues that surround safety management in hazardous industries are common to all countries, and to all hazardous industries. We have to learn from each other. The parochial attitude of "Well, that could never happen here" is misguided and pernicious. As George Santayana said, "Those who cannot remember the past are condemned to repeat it".

I am grateful to many erstwhile colleagues who have taught me so much. I am also grateful to my wife Jane for her patience and her proofreading. Any residual errors are entirely my fault.

The selection of material for this book has been prolonged, and there are many other things I would have liked to include. I have previously used a significant portion of the material in this book in various presentations or in journal articles, and some parts have already been published on my website www.safetyinengineering. com. Any feedback would be gratefully received – I can be contacted via my website.

Finally, I am mindful of the many hundreds of people who were killed or injured in accidents that have been used here as case studies. The objective is to learn from past mistakes.

Jim Thomson, Lanarkshire, UK
www.safetyinengineering.com

Introduction

HAZARDOUS INDUSTRIES, HIGH-INTEGRITY SYSTEMS AND MANAGEMENT PROCESSES

This book is about hazardous industries, and how the right processes and people are required to prevent major accidents.

Hazardous industries are those where there are dangerous materials that have the potential to injure or kill large numbers of people, or to do serious environmental damage. These include the obvious, such as nuclear energy, offshore oil, refineries, and chemical and pharmaceutical plants. However, there are also some not-so-obvious industries and industrial artifacts where, arguably, we have become inured to the hazards, or else we have been so successful in managing the risks that the hazards are no longer self-evident. We must not let familiarity with a risk lessen the attention it receives. (Risk is a function of the size of the hazard and the likelihood that there will be an accident involving the hazard. For example, dam failure may be a major hazard, but the risk might be acceptable because the likelihood of dam failure is remote.)

Examples of "not-so-obvious" hazardous industries include any that use pressure vessels, where there can always be an explosion hazard. Another example is alcohol distilleries - remembering of course that alcohol was the original rocket fuel, for the V2 missile in World War 2. Once-every-so-often these "not-so-obvious" hazards remind us of their existence; the huge explosion at the Buncefield, UK, petrol (or gasoline) storage facility on December 11, 2005 was such a case where, if the accident had not happened on a Sunday morning, there would have been many deaths. Prior to the event, no one really expected that a gasoline leak in a storage facility could lead to a massive vapor cloud detonation with a destructive shockwave, although a large fire would have been expected of course.

Figure 1.1 presents a very high-level timeline of some major industrial accidents that have happened in the last 60 or so years. Several of these accidents will be discussed in some detail later in this book.

Hazardous industries require high-integrity systems and safety management processes, of various sorts, to ensure safety. High-integrity systems are those parts of the hazardous plant where failure could lead to an accident, and for which high reliabilities are claimed. In particular, this refers to the pressure boundary of pressurized systems, and to instrumentation and control (I&C) systems that are used, for example, for functions like emergency shutdown or fire and gas leak detection and mitigation. The objective of this book is to provide an overview of the necessary management

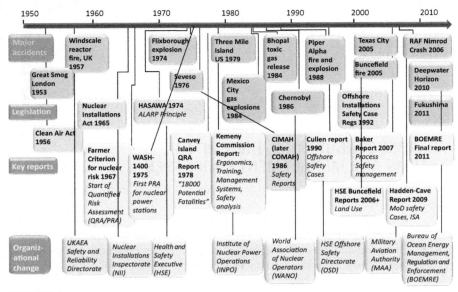

FIGURE 1.1

This diagram shows a timeline of significant industrial accidents between 1950 and 2012 which have had a major impact on the way we manage safety. Only one aerospace accident, the RAF Nimrod crash in Afghanistan in 2006, is included, because it has particular relevance to safety management. Other aerospace accidents that might have been included are the space shuttle accidents (Challenger 1986 and Columbia 2003) and the 2009 loss of Air France AF447.

processes to ensure that high-integrity systems are designed, implemented, operated and maintained successfully. In addition, this book will present some key case studies of incidents and accidents to illustrate how things can go wrong, thereby showing why these processes are needed.

At least four groups of people are crucial to the safe design and operation of hazardous installations.

- The people who assess the safety of industrial plants of all sorts (i.e., the consideration and assessment of potential hazards and accident sequences).
- The people who design the high-integrity systems that protect against accidents.
- The operators and maintenance personnel of the hazardous installations.
- The industry safety regulators.

Yet another group is academia, where the understanding of industrial safety can be more difficult because of language issues (jargon) and also physical and experiential distance. A common problem is that *all* these groups of people do not understand each other's perspectives very well. This is not a criticism of any group, it is merely a reflection of reality – someone who is an expert in the design of high-integrity

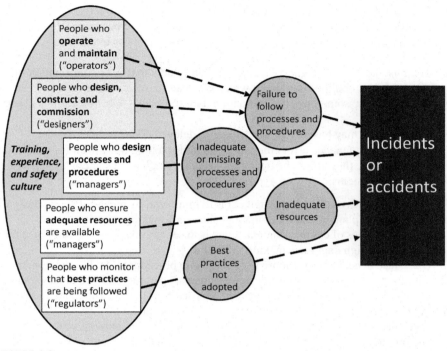

FIGURE 1.2

Compliance failure can be a cause of accidents - either people do not follow the procedures (due to pressures of time, budget, novel situations, too many responsibilities, inexperience or lack of training, or some combination of these, or conscious violation, or just carelessness) or the procedures may be inadequate or missing. This diagram presents a simplified overview of the safety roles and responsibilities of operators, designers, managers, and regulators. In practice, the roles overlap and can be blurred, and other players are also involved, such as research and development (which leads to new technologies and new standards). Managers are also responsible to ensure adequate resources, and regulators must monitor that best practices are adopted in all aspects that affect safe operation. Also, it is wrong to expect "blind compliance" from operators - they need to be alert at all times to the possibility that their procedures are inappropriate or contain errors.

systems cannot also be an expert in, say, the assessment of explosion consequences, or the operation of hazardous plant, or the regulatory framework for safety. This book is about all these groups of people responsible for the safety of hazardous plant.

Accidents happen when people, under pressure, make mistakes. The pressure may be time, budget, novel situations, too many responsibilities, inexperience or lack of training, or some combination of these, or else conscious violation – or there may be no pressure at all, in which case we call it "carelessness". The mistakes made by people – human errors - may be of various types:

- The mistakes may be of commission (doing something incorrectly, without intent) or omission (not doing something that should have been done).
- Errors of omission may be genuine (unintentional) errors or intentional violations.
- The mistakes can be cognitive errors ("I didn't know that"), or confirmation bias ("I misunderstood the situation"), which can lead to overconfidence.
- Operators may sometimes lose "situational awareness" ("What on earth is going on?").
- The mistakes may have either immediate or delayed consequences. Latent design or maintenance errors may lie dormant for a long time, even for years, if for example they affect the operation of rarely challenged standby systems or protection systems.

One of the roles of the design engineer is to minimize, by means of sound design, the likelihood of plant operators or maintenance staff making mistakes.

One of the roles of engineering managers is to set up processes that minimize the likelihood of design engineers making mistakes.

One cause of accidents is failure to comply with procedures. All the above-named groups of people – safety analysts, design engineers, operators and maintenance staff, regulators – have procedures that they are supposed to follow. Compliance failure is when procedures are not followed. Figure 1.2 illustrates this.

There also exist organizational failures, where subtle combinations of things such as time pressure, cost pressure, staffing levels and recruitment issues, unclear or misleading expectations from senior management, contract and subcontractor management, supply chain issues, and training or experience issues, can somehow all come together at one particular time and place – and an accident ensues.

This book is *not* about the detailed requirements arising from international (or national, or industry) standards – although some standards and their requirements will be mentioned. Instead, this book seeks to present "what good looks like" in the management of hazardous industries, together with examples of incidents and accidents where good practice has not been followed. The emphasis within this book will not be a cookbook description of how to follow major standards; instead, the emphasis will be on the general principles that are behind the major standards.

Case studies are often the clearest way to illustrate these general principles. The case studies presented include some very well known ones, and also some less well-known examples. These are presented primarily with the intent of drawing out the key lessons for others.

Although this book is specifically about hazardous industries, and not about transportation, I will nevertheless include a number of case studies relating to aviation accidents. Although aviation safety is explicitly outside the scope of this book, there are many important lessons for the safety of hazardous installations that can be learned from aviation accidents, for various reasons.

- First, aviation accidents are extremely well documented, and cockpit voice and data recorders mean we usually know exactly what happened immediately prior to the accident. Hence aviation accident reports can offer unparalleled insights into human behavior under extreme pressure. They can also offer insights into how the operator (pilot) can be confused by the control systems in emergencies.

- Second, modern terrestrial industries are following the trend, begun in aviation, of the extensive use of computer control. Hence, hazardous industries must be assiduous about learning from aviation experiences. Already, there have been several major air crashes where the causes of the accidents are at least partly due to issues arising from their computer systems.

An objective in this book is to write in plain English, and not to fall into the common science and engineering habit of excessive use of jargon and mathematics. If something is important, as I believe the matters in this book most certainly are, they deserve to be written about in a way that can be clearly understood. Most of engineering effort in high-technology industries actually is not about mathematics and advanced technologies – it is usually about orchestrated, collective effort to achieve a successful outcome.

The contents of this book look both forwards and backwards: we must remember mistakes from the past but we must also look toward the technologies of the future. From the past, boiler explosions may now be a rarity but we have to be mindful of the lessons of the past to avoid any complacency. At the other extreme, digital control systems are the present and the future, but engineers and scientists must strive to continue to find better ways of making them safe. Also for the future, a relatively new risk, cyber security, became world headline news in 2010 when the world's first industrial cyber weapon, Stuxnet, was discovered.

The safe design and operation of hazardous plant requires multidisciplinary teams each with many collective decades of experience. These teams will include all the engineering disciplines, plus detailed knowledge of the workings of each specific type of plant. It is beyond any one book to review all of these areas. The intent of this book is therefore to summarize the design and management processes that are needed for the safe design and safe operation of hazardous plants that use high-integrity systems.

THE STRUCTURE OF THIS BOOK

This book is divided into three parts. The first part, High-integrity Safety Instrumented Systems (Chapter 2), deals with the following aspects:

- Designing for high reliability, including:
  - Logic elements for high-integrity instrumentation and control systems – microprocessors, analog systems, gated arrays, and others.
  - The processes for producing high-integrity software.
  - The architecture of I&C systems (where "architecture" means the highest level of design).
- Cyber security of industrial plant control systems.

- Some cautionary tales relating to the design of the human-machine interface (HMI - in simple terms, the control room) for digital systems.
- Managing aging analog and digital instrumentation and control systems.

The second part of the book consists of two historical reviews of different areas of safety-critical technology: the history of pressure vessels integrity and failures, and a brief history of computer development. In the nineteenth century, "boiler explosions" (catastrophic pressure vessel failures where the energy comes from the compressed gas or steam inside the vessels) were fairly commonplace, largely because the ability to manufacture pressure vessels had proceeded far ahead of any understanding of what causes pressure vessels to fail. By the mid-twentieth century, pressure vessel failures had become very rare indeed, to the point where the term "boiler explosions" had taken on a different meaning, usually referring to gas-fired boilers where there had been a gas leak that had ignited. Earlier concerns about the potential for devastating pressure vessel failures have, to all but the dedicated cognoscenti, almost been forgotten about – the engineers and scientists have done their jobs too well. By contrast, computers are still in their infancy and the potential for industrial accidents arising from the use of computer systems in control and protection systems is real. These two chapters are positioned adjacent to one another in order to invite the comparison of the history of the safety of these two, completely different, technologies. It took a long time for pressure vessels to become relatively safe – we must try to accelerate that process with computers.

The first and second parts of this book are therefore about the basic technologies – high-integrity control systems and pressure vessels - that all (or almost all) hazardous industrial plants rely on.

The third and final part of the book, safety management, covers safety management processes and a range of case studies. The processes that are needed for the safe design, operation and maintenance of hazardous plants are common to various industries including oil and gas, nuclear, chemical industries, and pharmaceuticals (with inevitably some local differences). They include:

- Safe working arrangements on hazardous plant
- Technical safety and risk assessment methodologies
- Safety justification and safety regulation
- Accident and incident investigation
- Emergency planning

Case studies will include reviews of some major accidents including Chernobyl and Fukushima, Deepwater Horizon, Piper Alpha, Texas City and other less infamous but nonetheless tragic accidents.

- The element of the human factor in some major accidents will be dwelt upon – in some accidents, the behavior of one or two key individuals may sometimes lead to appalling losses, and it is informative to try to "get inside their heads." Really extreme pressure may sometimes lead to strange behavior, which seems to defy logic. Case studies will discuss examples.

- Organizational failings and safety culture will be a recurring theme – and, indeed, these aspects pervade almost all accidents.
- Also, brief consideration will be given to "Tragedies of the Commons" – environmental disasters where everyone and no one are to blame: the biggest of these may yet face us in the twenty-first century, in the form of global warming.

High-integrity Safety Instrumented Systems

The Design of High-integrity Instrumentation and Control (I&C) Systems for Hazardous Plant Control and Protection

The design of high-integrity I&C systems for hazardous plant is an area that has seen truly enormous changes in the last 30 years or so with the widespread introduction of digital (computer-based) systems. Before the 1980s or 1990s, all plant control systems and control rooms used analog sensors, analog logic based on discrete electronic components, and simple control systems, with alarm annunciator panels consisting of rows of lamps lit by incandescent bulbs. By comparison, modern computer-based plant control systems now have intelligent ("smart") sensors sending digital signals to distributed control computers which connect back to an all-digital control room consisting of a few flat screen displays, where plant mimic diagrams are shown, alarms are displayed, and the operator can make plant changes using touch screens.

These changes are now irrevocable, since the supply chain for I&C systems and components has moved with the times, and few manufacturers now supply older analog control system equipment.

This revolution in plant control has been led largely by the aviation industry, which was ahead of process plant in the adoption of digital control systems. For that reason, the design of digital plant control systems for hazardous process plant can learn a great deal from the experiences, incidents and accidents in the aviation sector as it changed to digital systems, as we shall see.

Particular attention must be given to the design of digital equipment where the early conceptual design (or front end engineering design, FEED) has identified the need for high reliability (or "high-integrity" systems) to protect against major hazards.

This chapter provides an overview of the design considerations for high-integrity I&C systems including the following aspects.

- The safety lifecycle for I&C equipment
- Reliability requirements for high-integrity systems
- Software quality management
- Functional specifications and traceability
- Setting up a high-integrity software project
- Common-mode failure
- I&C architecture

High Integrity Systems and Safety Management in Hazardous Industries. 978-0-12-801996-2

- The selection of logic elements and vendors
- The quality management of software suppliers.

THE SAFETY LIFECYCLE FOR THE DEVELOPMENT OF I&C SYSTEMS

The "safety development lifecycle" concept is enshrined in an international standard called IEC 61508 [1]. This is intended as a "standard of standards", for use across all process industries, the energy sector, and rail, automobile and aviation. Other standards have then been written which put the IEC 61508 requirements into an industry-specific framework. These include IEC 61511 (process industries), IEC 61513 (nuclear industries), and Do-178 (aircraft), although their scopes may vary.

IEC 61508 "Functional Safety of Electrical/Electronic/Programmable Electronic Safety-Related Systems" is a very large and detailed standard. IEC 61508 aims to ensure that, in any project involving I&C systems for protection against hazards (i.e., accidents), the functional and safety requirements are correctly identified at the outset, and then implemented properly in the final realization of the design.

- *Functional requirements* mean both the logical requirements of what the I&C system must do (such as "only permit drive X to operate if conditions Y and Z are satisfied"), and any other physical requirements such as screen formats, voltages, etc.
- *Safety requirements* mean the reliability requirements of safety-related functions, e.g., "the rate of failure of a given function must be better than 10^{-2} per annum." Systems response time and processor loadings are also safety requirements.

IEC 61508 tries to achieve this by:

- mandating a *project safety lifecycle* to ensure that safety issues are properly identified before design begins, and are then tested properly after manufacture, coding and system integration,
- recommending methodologies for determining the required reliabilities, (i.e., the safety integrity levels or SILs) for the safety functions in the E/E/PES,
- recommending techniques to ensure that the required software SIL levels are achieved, and
- recommending techniques for assessing hardware reliabilities.

The project safety lifecycle for the design, operation and eventual decommissioning of a hazardous plant is summarized in Fig. 2.1. The most important purposes of the safety lifecycle are to ensure that (a) design work is properly planned, and (b) safety requirements are traceable from beginning to end.

The first step in a new major project is overall concept design: what do we expect the plant to look like? A front end engineering design (FEED) project stage then develops an overall concept, including the definition of the *plant hazards* (i.e., what accidents are conceivable) and their necessary prevention and mitigation measures

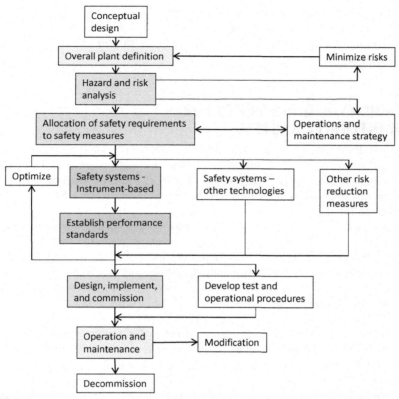

FIGURE 2.1

The safety lifecycle for instrumentation and control systems. From an overall plant definition (the conceptual plant design), safety requirements can be assigned to safety-instrumented systems, other safety systems, and other measures such as procedural controls. The safety-instrumented systems must then be designed, implemented, tested and commissioned, while maintaining strict design change controls and ensuring traceability between functional requirements and testing requirements. Finally, once operational, any modifications must be subject to strict controls to ensure that any changes are made with an equivalent level of consideration as the original design.

(which are sometimes confusingly called "controls"). Overall safety risk criteria should have been defined by the client organization (perhaps indirectly from the safety regulator) and, from these criteria, technical safety specialists can then define the functional and safety requirements for the high-integrity safety systems. (We shall return to the FEED process and risk criteria in Part 3 (Chapter 10).)

A key requirement is that, having defined a schedule of functional and safety requirements, these requirements must remain *traceable* throughout the construction project, to make sure that eventual commissioning tests actually do test the right things. Also, the schedule of functional and safety requirements must itself remain subject to rigorous *change control*; i.e., elements in the schedule can only be changed subject to careful consideration, e.g., a revision of the original safety analysis done in the FEED stage.

RELIABILITY REQUIREMENTS FOR HIGH-INTEGRITY SYSTEMS

IEC 61508 deals with both low-integrity ("non-safety" or "safety-related") and high-integrity ("safety systems") applications. Reliability requirements are defined in a series of bands called Safety Integrity Levels (SILs). Here we shall be focusing on high-integrity safety systems, which in this book is taken to mean SIL 3 and SIL 4 systems (see Table 2.1).

Hence *high-integrity systems* are defined as those with reliability requirements more onerous than 10^{-3} on demand (for low demand systems) or 10^{-3} per year (for high demand systems). These are normally referred to as SIL 3 and SIL 4 systems. The exact definitions of these terms may vary from industry to industry, and from country to country. For example, the aviation industry tends to refer to failure rates per hour, whereas the process plant industries tend to refer to failure rates per year. Since there are $(24 \times 365) = 8760$ h in a year – or roughly 10,000 – this means that 10^{-3} failures per year (fpy) equates to approximately 10^{-7} failures per hour, etc. Also, the European nuclear industry has imposed a more restrictive reliability limit on the SIL band definitions, as shown in the right hand column of Table 2.1.

So, what do these numbers actually mean? This simple question actually gets to the core issue. In older, analog systems with discrete components it was possible, in principle, to take failure rate data for each component and thereby determine an overall system reliability, using basic arithmetic. Now that most new systems are digital (microprocessor-based), although it is still possible to do a failure rate calculation for the hardware parts of the system, the reliability of the software cannot be determined by numerical means. And, furthermore, the reliability of the software is likely to dominate the overall system reliability.

Table 2.1 Safety integrity levels

Safety integrity level (SIL)	Low demand mode of operation (probability of failure on demand, pfd)	High demand mode of operation (failures per year, fpy)	Aviation industry – high demand mode of operation (failures per hour)	European nuclear industry (pfd or fpy)
1	10^{-1} to 10^{-2}	10^{-1} to 10^{-2}	10^{-5} to 10^{-6}	Not better than 10^{-1}
2	10^{-2} to 10^{-3}	10^{-2} to 10^{-3}	10^{-6} to 10^{-7}	Not better than 10^{-2}
3	10^{-3} to 10^{-4}	10^{-3} to 10^{-4}	10^{-7} to 10^{-8}	Not better than 10^{-3}
4	10^{-4} to 10^{-5}	10^{-4} to 10^{-5}	10^{-8} to 10^{-9}	Not better than 10^{-4}

To solve this problem – that we cannot determine numerical reliabilities for software – software reliability is assumed to be some sort of function of the quality control measures adopted during the software specification, production, and testing processes. In some way that cannot be readily measured, therefore, software reliability is a function of all the quality management arrangements related to software specification, coding, and testing. Hence, each SIL reliability band in the international standard IEC 61508 corresponds to a range of techniques and measures to be applied in the development of the system hardware and software. Thus, if these techniques and measures are applied, the system is assessed to have the reliability of the relevant SIL band. This aspect will be considered in more detail in later sections.

However, it is worth emphasizing at this stage that these quality criteria for software reliability and SILs (i.e., the quality assurance controls and measures necessary to achieve each SIL), as presented in IEC 61508, are really just intelligent judgments. Since the software failure rates will always be low (even SIL 1 would mean only one failure every 10 years for a high demand mode system), and the application software will be different in each plant system (so we cannot use measured software failure rate data from one system and compare them with any other system), there can be no meaningful dataset that can be used to compare actual failure rate data with the assessed SIL levels. Hence, we have SIL bands that are assigned according to the judgments of experts but without any really solid foundation. SILs are the best means we have of assessing software reliability, even though they are really only intelligent guesses.

The numerical equivalence between the SIL limits for low demand rate (pfd) and high demand rate (fpy) systems sometimes causes confusion. If a system has a low demand frequency, such as a fire alarm system, then the approach is straightforward: Periodic tests are done to confirm that the system is working properly. High demand systems such as control systems are effectively in continuous demand – and yet if (say) they are both SIL 2 systems, designed to broadly similar quality management standards, then the numerical reliability of both is taken to be 10^{-2}, either probability of failure on demand *or* failures per year.

Finally, these SIL levels are applied to the whole *safety loop*, from sensor, through logic solver, to actuator. The SIL level is also generally applied to the whole *system*, where the system may contain a large number of safety loops.

HARDWARE AND SOFTWARE SYSTEMS DESIGN

IEC 61508 defines I&C hardware systems as either "Type A" or "Type B". Type A systems have well-defined failure modes, and behavior in fault conditions that can be completely determined, and there are dependable failure data from field experience. Type B systems are systems that do not satisfy the criteria for Type A. Type A equipment is normally composed of simple components that can be fully analyzed, due to a knowledge of their reliability and failure modes, in a failure mode and effects analysis (FMEA). This allows dangerous failures of the system to be identified and hence a failure probability value calculated. In practice, this almost always means that Type A hardware is analog logic, with "old-fashioned" electro-mechanical input sensors and output transducers, and no software.

Hence a system employing software is almost by definition Type B, since its failure modes are hard to define, its behavior in fault conditions may also be hard to define, and there may be no dependable failure data. Because of this additional complexity and the non-deterministic calculation of failure probability, IEC 61508 has stronger architectural constraints for type B systems than for type A. Typically the pfd range claimable for type B systems is an order of magnitude less than for an otherwise architecturally similar Type A system.

These classifications within IEC 61508 lead to two different ways to design a very safe system. A Type B digital (software) system requires a greater degree of redundancy and self-diagnostics to achieve the same SIL rating, whereas a Type A system can be designed as a simple redundant analog system which is primarily fail-safe. In essence, a digital system can make up for its relative complexity and undefined failure modes by having thorough online diagnostic systems for detecting dangerous faults.

The use of software-based diagnostic systems to monitor the behavior and operation of high-integrity software-based systems leads to interesting philosophical questions, like the Roman poet Juvenal who asked "Quis custodiet ipsos custodes?" ("Who guards the guardians?"). Designers have to design systems where unrevealed dangerous failures of the software-based diagnostics system cannot occur under any foreseeable circumstances. They must, at the same time, ensure that spurious warnings from the diagnostics systems, which would reduce operator confidence in the systems, are minimized.

THE "SOFTWARE PROBLEM" AND SOFTWARE QUALITY MANAGEMENT

We must remember that "high-integrity" should really mean in practice that "we have done everything reasonably practicable and we do not believe this system should ever fail dangerously". Since software is an essential element for most high-integrity systems in the twenty-first century, we need to ask ourselves why and how software presents special problems.

The first and most important fundamental issue that makes software different is as follows.

Issue 1
Software systems are not readily amenable to inspection and testing in the same way as older analog equipment.

One cannot inspect software in the same way that one could with old-fashioned analog equipment. This may seem an obvious, simplistic statement, but it has many ramifications, as we shall see below.

Another fundamental difference between microprocessor-based systems and other logic solvers is as follows.

Issue 2

In microprocessors, all signals (input and output) go through a single element – the microprocessor central processing unit (CPU).

In microprocessors, all functions are dealt with serially by the same CPU. The processor cycles round each function in turn at extremely high speed (MHz). There is no unique "signal flow path", as would be the case in analog systems that use discrete components. Hence, in principle, it is conceivable that there may be unknown dependencies between separate safety functions that have, somehow, been incorrectly written into the software. In short, you cannot readily prove that the software will not do *extra* things you do not want it to.

With *analog* systems using discrete components, in order to produce a high-integrity product we have to carry out the following steps thoroughly and carefully.

1. specify the required functions accurately (specification),
2. verify the design (verification),
3. identify and analyze possible failure modes (failure modes and effects analysis or FMEA), and
4. test the product (validation).

These four steps are also required for software systems, but each step becomes more complicated – and there are additional areas of concern also:

1. Ensuring accurate software specification (the specification problem).
   a. The specification of software includes the specification of the operating system as well as the application software. For example, Microsoft Windows has tens of millions of lines of code. (For this reason, Windows is not used for high-integrity systems.)
   b. With application software, it is relatively straightforward to add as much functionality as anyone can want, so almost inevitably functionality rapidly grows. Hence, extra functions are added and the software grows in size and complexity. It is not unusual for the application software for an industrial control system to have millions of lines of code. Even high-integrity protection (or safety) systems – which should, almost by definition, be simple – may have hundreds of thousands of lines of code.
2. Complexity of software (the verification problem). If it is difficult to *specify* large amounts of code, it is equally difficult to *verify* them in any meaningful way. To ensure software code can be verified, it must be written in a clear manner, and verification planning must be done with care to ensure that verification is comprehensive and auditable.
3. Complicated failure modes of components that use microprocessors (the FMEA problem). In high-integrity systems, we need to be able to carry out detailed and thorough analyses of all the potential failure modes. This is known as failure modes and effects analysis or FMEA, sometimes called failure modes,

effects and criticality analysis or FMECA. The problem of complex software is that it can become very difficult to predict all the potential failure modes. This can be especially important where third-party-supplied microprocessor-based components can interact with other systems.

4. Microprocessor "physics of failure"(the aging failure modes problem). The microprocessor hardware on which the software operates may have indeterminate aging behavior which may cause additional failure modes as the equipment ages.

5. Proof testing of the installed system. There are far too many possible system states so full negative testing is impossible (the validation problem). "Positive testing" means "testing to show that what you expect will happen does in fact happen". In contrast, "negative testing" means "testing to show that *only* the things you expect to happen do in fact happen", i.e., testing that no possible combination of inputs produces an incorrect output. Because all the inputs signals go through the same CPU, to really proof test the software you would need to do full negative testing for every possible plant system state. As we shall see, the amount of negative testing required for anything but the very simplest system becomes impossible. Only positive testing and (at best) some very limited negative testing is practicable. In the absence of full negative testing, we have to rely on rigorous quality control measures to ensure that unwanted functionality does not exist.

6. Cyber-attack (the security problem). The requirement to ensure that software-based systems are protected from malicious attack has become a significant new area of work in the development of high-integrity systems.

7. Demonstration of the claimed reliability (the reliability problem). As already discussed in pages 14 and 15, software-based systems are assigned reliabilities on the basis of the quality management techniques and measures employed during the system design. If we now seek to substantiate these reliabilities by statistical testing, we run into the difficulty created by the validation problem (issue 5) above. In principle, to demonstrate high reliabilities, each function would have to be tested, in each and every plant state, thousands of times. With very many thousands of plant states and possibly several hundred input signals in a typical industrial plant, the number of tests required becomes astronomical and impossible. So, we come back to the position described in Section "Reliability Requirements for High-integrity Systems". We have SIL (reliability) bands with defined quality management techniques and measures for each band, and each reliability band is assigned according to the judgments of experts, but nevertheless without any really solid basis. As stated in Section "Reliability Requirements for High-integrity Systems", SILs are the best practical means we have of assessing software reliability, even though they are really only intelligent guesses.

8. The "**diverse software**" problem. In extremely high-reliability systems such as nuclear reactor shutdown systems, as we shall see, two diverse high-integrity protection (safety) systems may be required. The question is, is it alright for

both these systems to be software-based? Given the areas of difficulty described above, can we rule out common-mode software failures?

9. Components sourced from other companies that use Previously Developed Software (PDS): This issue most often arises due to the purchase of sensors and actuators which have embedded software – the so-called "Smart Sensor" problem. The difficulty here is that the pedigree of the PDS may be unknown, or the quality control system used by the supplier company may be considered inadequate, or else the supplier company may be reluctant to divulge information about the software because they consider it threatens their Intellectual Property (IP). (*Acronym alert*: Software-based devices can also be known as COTS devices – since they use Commercial Off The Shelf software, as opposed to bespoke software. Software whose origins are unknown can be called SOUP – Software Of Unknown Provenance. All PDS, COTS and SOUP devices require thorough investigation and testing before they can be used in a safety-critical application. See page 41.

Specification (item 1), verification (item 2), and testing (item 5) all require rigorous management of the software design process throughout the system's lifecycle. In particular, *traceability* – the tracing of all aspects of the functional and safety specification through to the design realization and its testing, both forward and backward – is an absolutely essential part of the process. The means of doing this is called a *traceability matrix*, and software tools such as DOORS are used to help with this aspect. Traceability also requires rigorous *change control*, i.e., any design changes that arise (as they surely will) must be considered and specified very carefully before they are implemented, and the test requirements changed to reflect the new functions.

The above are the main areas where high-integrity software – and the justification of its use for the operation of hazardous plant – runs into difficulties. Figure 2.2 presents a summary. We shall consider aspects of them in more detail, with examples of relevant accidents or incidents, in the rest of Chapter 2 and the succeeding Chapters 3–6.

All of the above can be considered to be aspects of *software quality management*.

FUNCTIONAL SPECIFICATIONS, TRACEABILITY, AND THE V-MODEL

There is a lot of experience to show that the most critical aspect of designing high-integrity systems is to make sure the specification is correct. Figure 2.3 presents UK Health and Safety Executive data which shows that 44% of control system failures in industrial control systems can be attributed to poor specification.

For software in industrial control systems, the V-model process in Fig. 2.4 shows that there will typically be a series of nested specifications, from the overall functional requirements specification, through intermediate system specifications, to detailed

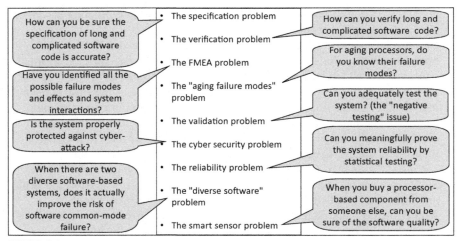

FIGURE 2.2

The "software problem" – some major sources of problems in software-based high-integrity safety systems.

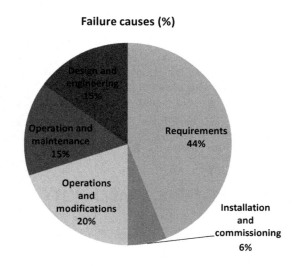

FIGURE 2.3

The causes of systematic and human failures in the industrial plant control systems. Almost half (44%) of such failures can be attributed to errors in requirement specifications. (Data from Health and Safety Executive, *Out of Control – Why control systems go wrong and how to prevent failure*, HSG 238, 2nd Ed, 2003.)

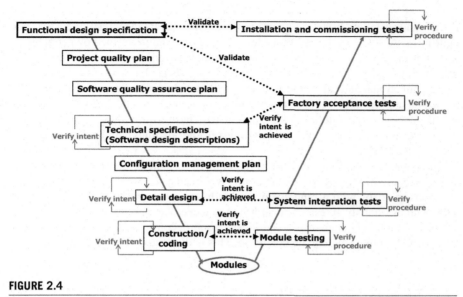

FIGURE 2.4

The software development "V" model. High-integrity software follows a development process from a functional and safety design specification, through intermediate stages of more detailed specification, to software coding. Modules of software are then tested in accordance with the specifications. System integration tests are followed by factory acceptance tests (FATs) and installation and commissioning tests. Installation tests are sometimes known as site acceptance tests (SATs).

module specifications and eventual software coding. This means that there is a real potential for the specification at each stage to be misunderstood or garbled. Also, the people who do the software coding at the end of this specification process will probably have had no involvement in the functional and system specifications. They will simply be following the coding specification they have been given. The top-level specifier therefore cannot rely on the people doing software coding to have an understanding of what the overall *design intent* of the system is. The software coders will simply follow the software specifications precisely. Hence the specifications must be precise, accurate and unambiguous. One must therefore proceed with great care when managing the interfaces in a complicated specification process from functional requirements through to software code.

At the start, a detailed schedule of functional requirements – especially safety functional requirements – must be drawn up from the specifications. These safety functional requirements (SFRs) must then be traced through the various intermediate stages to ensure proper implementation.

Each level of specification must be subject to careful verification at each stage, by knowledgeable people who have not been directly involved in its preparation.

The SFRs are used as the basis for module testing, factory acceptance testing (FATs) and installation testing (also known as Site Acceptance Testing SATs). For high-integrity systems, the test procedures are written and performed by people who were not involved in the system specification and design, to ensure that anomalies and ambiguities are identified and resolved. These tests are known as *validation* tests.

The functional and safety requirements will include environmental requirements, for which appropriate testing in the installed configuration will be required (i.e., inside the cabinets that will house the equipment). These requirements include humidity, temperature, shock loadings (including seismic loadings), and electromagnetic interference (EMI).

Validation testing may sometimes be required to include "statistical testing" as a confidence-building technique. This involves repeat testing of a range of safety functions, perhaps with a range of plant conditions (or input states). However, the value of these statistical tests is nugatory: in practice, the number of tests required to produce a genuinely statistically significant result becomes astronomical for even a single function in a single plant condition. This is illustrated in Figure 2.5.

No. of demands (or operating time) No. of observed failures N	10		100		1000		10000		100000	
Confidence	50%	80%	50%	80%	50%	80%	50%	80%	50%	80%
0	1.6 E-1	2.3 E-1	1.6 E-2	2.3 E-2	1.6 E-3	2.3 E-3	1.6 E-4	2.3 E-4	1.6 E-5	2.3 E-5
1	2.69 E-1	3.89 E-1	2.69 E-2	3.89 E-2	2.69 E-3	3.89 E-3	2.69 E-4	3.89 E-4	2.69 E-5	3.89 E-5
2	3.92 E-1	5.32 E-1	3.92 E-2	5.32 E-2	3.92 E-3	5.32 E-3	3.92 E-4	5.32 E-4	3.92 E-5	5.32 E-5
5	7.42 E-1	9.27 E-1	7.42 E-2	9.27 E-2	7.42 E-3	9.27 E-3	7.42 E-4	9.27 E-4	7.42 E-5	9.27 E-5
10	13.0 E-1	15.4 E-1	13.0 E-2	15.4 E-2	13.0 E-3	15.4 E-3	13.0 E-4	15.4 E-4	13.0 E-5	15.4 E-5
20	23.9 E-1	27.0 E-1	23.9 E-2	27.0 E-2	23.9 E-3	27.0 E-3	23.9 E-4	27.0 E-4	23.9 E-5	27.0 E-5
30	34.6 E-1	38.3 E-1	34.6 E-2	38.3 E-2	34.6 E-3	38.3 E-3	34.6 E-4	38.3 E-4	34.6 E-5	38.3 E-5

Indicative SIL levels | >10⁻¹ | SIL 1 | SIL 2 | SIL 3 | SIL 4 |

FIGURE 2.5

Statistical testing requirements using Bayesian inference. This table illustrates failure probabilities (and hence claimable SIL levels), after a number of observed failures *N*, with a given number of operating demands or operating time *M*, at 50% and 80% confidence. Hence, e.g., in order to demonstrate statistically the achievement of SIL 3 for a single safety function in a single plant state at 80% confidence, 10,000 repeat tests would be required with not more than a handful of failures. When there are large numbers of inputs and safety functions, with many plant states (as would be typical for process plant), any meaningful statistical test regime becomes hopelessly impracticable.

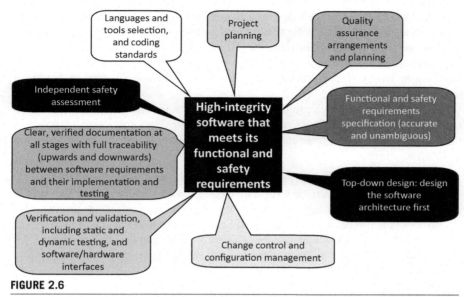

FIGURE 2.6

The principal elements of a high-integrity software project.

HIGH-INTEGRITY SOFTWARE: SETTING UP A SOFTWARE PROJECT

High-integrity software projects must be managed carefully, resourced properly, follow carefully thought-out and verified procedures, and have "gates" at suitable key points in the project where management reviews take place before proceeding to the next stage. The key elements in setting up a successful project are illustrated in Fig. 2.6 and comprise:

- project planning;
- quality assurance (QA) arrangements and QA planning;
- the selection of software languages and software tools;
- Functional and safety requirements specification (as discussed elsewhere);
- "top-down" design (which means thinking and planning the software architecture – its structure – before any coding work starts);
- change control and configuration management arrangements;
- the approach to verification and validation;
- a clear documentation structure; and
- arrangements for independent assessment

Having all these elements in place will not guarantee a successful project, but not having them in place will make project failure much more likely.

There is a history of large safety-related software/I&C projects going significantly over budget and suffering significant cost overruns. In particular, high-integrity software projects should *not* be treated as IT projects. They must be treated as safety-related projects led and managed from the engineering side of the business – they must not be led by IT departments, although some IT personnels may be seconded into the project teams.

There are three key high profile roles in large software/I&C projects: the systems architect, the project manager, and the project sponsor. Their roles, and a number of other key learning points from earlier large high-integrity software/I&C projects, are as follows.

1. Large software/I&C projects are complex multidisciplinary projects.
   There should be very senior level sponsorship, and the senior *project sponsor* should be able and willing to devote a significant amount of his/her time to assisting the project toward a successful conclusion. He/she is required to mentor the project manager and his team through the project, and to ensure that all relevant parts of the organization are aligned for successfully carrying out the project.
2. The site operators must support the project and must be well represented on the project team to ensure that the end product gives them what they want.
3. A lead technical role in large high-integrity software projects is the *systems architect*. He or she is key to producing a robust design.
4. There needs to be careful and early consideration and planning to minimize potential plant downtime.
5. The *project manager* needs a broad set of skills which include leadership, commercial awareness, and communication skills.

TECHNIQUES AND MEASURES TO PRODUCE HIGH-INTEGRITY SOFTWARE

For a high-integrity software project, it is necessary also to define the techniques and measures that will be adopted to achieve the desired SIL level – i.e., the software reliability. IEC 61508 and other standards offer, in some cases, different approaches that will yield equivalent SIL levels, so there are design choices to be made. The recommended techniques and measures for high-integrity software are complicated – and the detail belongs in a software engineering textbook which is well beyond the scope of this book – but can be grouped together under the following headings.

1. Functional safety assessment: checklists, truth tables, failure analysis, common-cause (or common-mode) failure analysis, reliability block diagrams
2. Software requirements specification: formal or semi-formal methods, traceability, specification tools
3. Fault detection, error detecting codes

4. Diverse monitoring techniques, diverse programing
5. Recovery mechanisms or graceful degradation (mostly for hardware faults)
6. Modular design, simplicity, stateless design
7. Trusted/verified software elements (but with caveats on claims for "proven-in-use")
8. Forward/backward traceability at all stages
9. Structured or semi-formal or formal methods, auto-code generation
10. Use of proven software development tools
11. Guaranteed maximum cycle time, time-triggered architecture, maximum response time
12. Static memory allocation, avoidance of interrupts, synchronization
13. Language selection, language subsets, suitable tools
14. Defensive programing, modular approach, coding standards, structured programing
15. Testing: dynamic, functional, black box, performance, model-based, interface, probabilistic
16. Process simulation, modeling
17. Modification/change control: impact analysis, re-verification, revalidation, regression testing, configuration management, data recording and analysis
18. Verification: formal proof, static analysis, dynamic analysis, numerical analysis

The degree to which each technique or measure has to be implemented depends on the SIL level required for the equipment. Not all techniques and measures are required for all SIL levels. Other techniques and measures may also be applicable – the above is intended merely as a checklist of some of the more common methods used. Detailed definitions of the terms used above are given in IEC 61508 part 7.

I&C FAILURE ANALYSIS TECHNIQUES

This section briefly outlines some of the main techniques used in I&C failure analysis, which can be carried out for hardware and subsystems: Failure modes and effects analysis, fault tree analysis and sneak circuit analysis.

Failure Modes and Effects Analysis (FMEA) is an "inductive" engineering analysis to determine what can fail, how it can fail, how frequently it will fail, what the effects of failure are, and how important the effects of failure are. (Another related approach, failure modes effects and criticality analysis (FMECA) focuses more on the significance of each failure.) FMEA is used to aid the design concept, and to prompt the designer to "design out" failure modes that involve severe penalties. It is also useful for considering the operational uses of the system, and for focusing design effort on the most significant safety-related failure modes.

FMEA is carried out at subsystem level, using functional diagrams and schematics of each subsystem. A complete component list, with the specific function of each component is then prepared for each subsystem. Failure mechanisms that could occur are then determined. The failure modes of each component are then considered. (For example, a circuit component might fail open circuit or short circuit.)

FMEA is "inductive" because it relies on the skill and knowledge of the engineers doing the analysis to identify the failures and their significance.

Fault tree analysis (FTA) is a deductive logical approach to study a known undesired event. FTA systematically considers all known events and faults which could cause or contribute to the undesired event. FTA is mainly a qualitative technique used to identify areas of detailed design that require careful consideration because of their known potential to cause accidents.

Sneak circuit analysis (SCA) is used for evaluating electrical circuits, with the intention of identifying latent (sneak) circuits and conditions that inhibit desired functions without component failure having occurred. Sneak circuits can occur in all types of electrical and electronic systems, so SCA is required for high-reliability systems. Sneak circuits can create sneak paths (which allow current to flow along an unexpected route) or sneak timing (which causes functions to be inhibited or to occur unexpectedly).

Common-mode failure (CMF) is considered in more detail in the next section "Common-mode Failure".

An important point here is that failure analysis needs to be done early in the design process and product lifecycle if it is to have value. The identification of important failure modes when a subsystem or system is already in manufacture, or even operational, causes stress in organizations and they may then have to react rapidly. Late identification of important failure modes also undermines the client's confidence in the capabilities of the design organization and may lead to wider reviews being started.

COMMON-MODE FAILURE

This section describes the potential causes of common mode failure (CMF), how we can design I&C systems to prevent CMF, and how we can take possible CMF into account when doing plant risk assessments. It addresses both software and hardware systems and components. Common mode failure can also be referred to as common-cause failure or dependent failure. Some people may say that there are subtle differences between these terms, but let us keep it simple.

Much of the basic thinking around CMF issues dates back to the 1970s and 1980s. It was recognized that the avoidance of CMF was to a great extent an issue of quality assurance and quality control throughout the lifecycle of the I&C system. As already discussed, IEC 61508 is based on the assumptions that (i) the realistically achievable reliabilities from redundant C&I systems are in some ways proportional to the degree of quality assurance and quality control employed throughout their lifecycles, and (ii) there are limits to the achievable reliabilities in non-diverse systems.

One precise definition of CMF is: "a common mode failure is the result of an event(s) which because of dependencies, causes a coincidence of failure states of components in two or more separate channels of a redundancy system, leading to the defined systems failing to perform its intended function".

A list of possible causes of common-mode failure is given in Table 2.2.

Table 2.2 Possible causes of common-mode failure

	Source of CMF	Some possible causes of dependent failure
1	Specification or design failure	Failure to recognize within the specification the full range of circumstances in which the plant must operate
		Wrong/inadequate standards used
		Inadequate management of change (control of plant modifications)
		Common aging processes on redundant channels
2	Construction/installation/ inspection/commissioning failure	Poor quality control of components and subsystems during manufacture
		Lack of physical/electrical separation during installation
		Improper installation
		Commissioning testing: failure to test adequately all credible circumstances
3	Maintenance or operations failure	Failure to repair defective equipment in a timely manner
		Maladjustment of set-points, limit switches, etc.
		Improper or inadequate maintenance or test procedures
		Failure to follow maintenance procedures
		Poor control of overrides or interlock defeats
		Poor housekeeping
		Poor quality spare components
4	Environmental aspects	Temperature
		Humidity
		Vibration
		Stress
		Corrosion
		Contamination (abrasive material, chemical agent, etc.)
		Radio frequency interference (RFI)
		Radiation
		Static charge
		Extreme weather (rain, snow, hail, ice and wind)
		Seismic event, tsunami
5	Other external and internal hazards	Fire
		Flood
		Explosion
		Air crash
		Terrorism

Although the scope of these failure mechanisms is very large, there is no guarantee that the list of CMF causes given above is comprehensive, since CMF is a catch-all, i.e., it includes "any other possible dependent failure mechanisms you have not considered". Also, most of the above issues are underpinned by the employment of personnel with suitable qualifications, training and experience – ultimately it is the people who count. They have to be professional, disciplined, knowledgeable and engaged. They must also be able to differentiate between important and non-important issues.

Because the causes of CMF must include "dependent failure mechanisms you have not considered", the defences against CMF must address both real, identifiable potential causes of CMF (e.g., fire, or inadequate routine testing), and also more abstract, philosophical concerns. A very short list of some important defences against CMF is presented below:

1. Clear, robust quality assurance and quality control arrangements
2. Clear functional specifications (logic, environment, ergonomics)
3. Fail-safe design
4. Independent verification and validation (IV&V)
5. Testing at component, module, subsystem and system level
6. Clear traceability between functional requirements and testing, in both directions
7. Separation of control and protection
8. Physical separation between channels
9. Electrical separation between channels
10. Protection against fire and explosion
11. Flood protection
12. Seismic design as appropriate
13. Functional and equipment diversity *(in very high-integrity applications)*
14. A routine testing regime which effectively tests the functional requirements
15. Staggered testing so that channels are tested at different times

CMF concerns in I&C systems apply in particular to several key areas of design.

1. First, separation between the (lower integrity) control system and the (higher integrity) protection (or safety) systems must be thorough and absolute, in order to prevent the risk that failures in the control system could also cause failures of the protection system. This applies in particular to sensors and actuators, where shared devices or devices of the same design should be eliminated wherever possible by design.
2. The AC electrical supplies to separate and diverse systems must also be separated, ideally up to transformer level, with appropriate overvoltage and undervoltage protection arrangements is place.
3. Separation of systems should pay particular attention to the risks of local fire and flood which could potentially affect both the separated systems at the same time.

For the very highest integrity applications, such as nuclear reactor protection systems, there may be a need for a second, diverse and separate system of detecting fault conditions and initiating a reactor shutdown. To be unambiguously diverse and separate, the following requirements would have to be satisfied for an ideal, fully independent design of two systems.

1. The diverse reactor protection system (RPS) design should be developed by a different team, using independently derived safety functional requirements;
2. the diverse RPS should be electrically and physically separated, and should be protected from common hazards such as fire or flooding;
3. it should use different input sensors measuring diverse operating parameters;
4. its signals should pass via separate routes and be processed by diverse types of logic solver;
5. its final actuating devices (usually electrical breakers) should be from a different manufacturer; and
6. its means of shutdown should use different physical principles (e.g., boron injection vs. control rods).

In addition, there remains a non-disprovable concern that there may be a weakness to common-mode failure (CMF) if both sets of logic solvers in two nominally diverse systems are software-based. This concern is related to the complexity of software systems, and the associated difficulties of verification and validation (V&V). In some undefined way, similarities in the software code design and production processes may yield the possibility of CMF – even if different software languages and operating systems are employed in the two systems. Because of this concern, it has become common in some countries to specify that the diverse protection system should be hard-wired.

Conventional reliability assessments of random failure rates for hard-wired systems are based on measured failure rates for all the system's components and an assumption of "perfect" routine testing (i.e., the routine testing detects all latent faults). This assessment approach can often yield unrealistically low predictions for actual systems failure rates. The system failure rates will in practice be dominated by common-mode failures and not by random failures.

However, the application of probabilistic (or quantitative) risk assessment techniques means that there is a need to make judgments about the reliabilities of redundant systems, even when their failure rates are dominated by CMF. Two main approaches have been developed for modeling the effects of CMF in probabilistic (or quantitative) risk assessment:

i. The **SIL (or "reliability cut-off") method** assumes that the reliabilities of a redundant multiple-channel system can be represented by a CMF "cut-off" value, which is judged according to the perceived or assessed quality of the equipment, its design, manufacture, installation, operation and maintenance. The best achievable system reliability (the "cut-off" level) is usually worse than (i.e., less reliable than) the value determined by reliability calculations. The SIL method has already been described in "The Safety Lifecycle for the Development of I&C Systems" page 12.

ii. The **Beta Factor method** assumes that the frequency of dependent failures between redundant channels is proportional to the assessed random failure rates of each redundant channel. The constant of proportionality is called the Beta factor. This method is discussed below.

A measure of the effects of CMF on system reliability can be obtained by using the Beta factor, which is defined as "the probability that, if a failure occurs in one channel of a redundant system, other channels will also fail due to a common cause". Hence, if the reliability of a single channel can be determined by calculation or otherwise, the CMF of the redundant system can be determined by multiplying the assessed single-channel reliability by the Beta factor.

IEC 61508 quotes data suggesting Beta factors typically lie in the range 0.07–0.4. A detailed discussion of the Beta factor methodology can be seen in IEC 61508 part 6, Annex D.

There are philosophical difficulties with Beta factors, however:

- In principle, there is no reason why single-channel failure rates *should* be proportional to common-mode failure rates. They are different things altogether.
- Beta factor calculations can be used to produce spurious levels of accuracy for the reliability of redundant systems. Two or three significant figures have sometimes been claimed, quite inappropriately. With simple SIL "orders of magnitude" numbers you at least know that the figures are really just intelligent judgments, and nothing more precise.

Hence both the SIL and Beta factor methods have underlying philosophical concerns. The SIL approach uses unverifiable judgments about the relationship between system reliability and quality control techniques and measures, whereas the Beta factor approach draws unverifiable correlations between single-channel failure rates and common-mode failure rates.

If two diverse and separate protection systems have been installed in accordance with the principles set out above, then any failure of one system *should* always be completely independent from a failure of the other system. Hence the probability of simultaneous failure of both systems can be modeled in probabilistic (or quantitative) risk assessments as the multiple of the two wholly independent systems' failure probabilities.

Any failure to meet the independence requirements described above will to some extent compromise this conclusion.

The following two case studies (Ariane 5 launch failure in 1996, and the Forsmark power failure in 2006) illustrate the care and attention to detail that is required in the design phases to ensure the risk of common-mode failure can be minimized.

CMF CASE STUDY: ARIANE 5 LAUNCH FAILURE, FRENCH GUIANA, 1996

A well-known example of CMF is the Ariane 5 space launcher failure in 1996. This launch was the first of ESAs new Ariane 5 rocket launcher, which was a much bigger vehicle than its predecessor Ariane 4. Like its predecessor, Ariane 5 used dual-redundant Inertial Reference Systems (IRSs) working in a duty/standby arrangement. Thirty-seven seconds after launch from French Guiana on June 4, 1996, both the duty and the backup IRSs failed, which led to the launcher going off-course and its disintegration.

The cause was a software fault in equipment which was unchanged from Ariane 4, but which was unsuitable for the changed flight trajectory of Ariane 5. There was inadequate analysis and simulation of the systems in Ariane 5. The dual-redundant IRSs used identical hardware and software, with one active and one on hot standby. Because the flight trajectory was different from Ariane 4, the active IRS declared a failure due to a software exception (i.e., an error message). The standby IRS then failed for the same reason.

Actually, the affected software module only performed alignment of the inertial platform before launch – it had no function after launch. So, the error message was inappropriately handled, although the supplier was only following the contract specification.

CMF CASE STUDY: FORSMARK, SWEDEN, 2006

Forsmark 1 nuclear power plant was working normally at full power on July 25, 2006. Routine maintenance was being carried out in the 400kV switchyard, outside the plant. The work was being done by the grid system operator, Svenska Kraftnat, and was not under the control of the power plant operator. A short circuit occurred in the switchyard during this work, which resulted in a severe disturbance to the electrical systems of the nuclear power plant. (Svenska Kraftnat had misjudged the need to interlock an earth fault protection. If this had been done correctly, the size of the resulting electrical disturbance would have been much smaller.) The voltage fluctuation led to an automatic reactor shutdown, and the plant was disconnected from the electricity grid. However, the size of the voltage fluctuation caused part of the battery-backed AC internal distribution network to fail, and only two of the four diesel generators started automatically. Some control room equipment also failed so, at first, the operators were unable to understand fully what was happening.

After 22 minutes, power was restored and the other two diesel generators were started up. Reactor core cooling was never impaired.

This relatively minor incident, which harmed no one, is of interest because it illustrates loss of defence-in-depth. Several safety systems which were supposed to operate independently in fact failed simultaneously because of the external fault – the voltage disturbance caused by the short circuit in the 400kV switchyard. The principal that diverse systems should be resistant to common-cause failure was not achieved in this instance.

The large voltage fluctuations fed through the transformers which supplied local power systems throughout the plant. All four diesel generators started up. However, the overvoltage caused failures in two of the four uninterruptible power supplies (battery-backed systems). These failures in turn prevented two of the four diesel generators from connecting to the busbars which supply their electricity within the power plant – battery power was needed to effect the connections. Thus there was a common-mode failure between the batteries and the diesel generators, two systems which should have been wholly independent. In addition, the control room screens failed so the operators had only a limited view of what was happening. However, after 22 minutes, the operators manually closed the breakers (high-voltage switches) which connected the two failed diesels to their busbars, and full power was restored.

Failure modes and effects analysis (FMEA) should have considered the effects of all conceivable voltage fluctuations within the battery-backed systems. It appears that various electrical system modifications which had been carried out in 2005 had not been adequately modeled and tested.

There were still significant margins in this incident – even if three diesel generators had failed, the reactor core would still have received adequate cooling, and the operators would still have been able to manually close the breakers even though the battery supplies had failed. However, the incident is an uncomfortable foretaste of the much worse events at Fukushima in 2011, when total loss of backup power supplies occurred, and which will be discussed later.

(Ref: Analysgruppen Bakgrund, no 1, vol 20, Feb 2007, "The Forsmark incident, July 25, 2006", published by KSU Nuclear Training and Safety Center, www.analys.se).

I&C ARCHITECTURE

This section aims to set out the basis for good I&C safety systems architecture. In particular, the I&C architecture of nuclear power stations of nuclear power stations will be considered since these architectures are more complex than any other process plants. Here, "architecture" means the highest level of systems design.

First, why are the architectures of safety systems different in nuclear, oil and gas, and aviation? This may seem a stupid question, but it is worth thinking about. Some fundamental differences that affect I&C architectures in nuclear power plants, oil and gas (O&G) facilities such as oil platforms and refineries, and civil aircraft, are as follows:

1. The hazard magnitudes may be significantly different. The potential hazards to the general public from nuclear power plants – especially in terms of the risk of having to evacuate significant areas of land for many years – are in general greater than those for any other potential industrial hazards. (There are certain exceptions, e.g., the 1984 Bhopal accident in Gujarat, India, and also some potential dam failures, where the hazards are as bad, or worse than, nuclear power hazards.)

2. Also, in civil aviation, the persons at risk (the passengers) are accepting that they are taking on the risk by buying their tickets – we each do some sort of (probably subconscious) risk/benefit assessment. (The same might be said for employees on, say, offshore oil and gas platforms.) There is therefore a difference between voluntary and involuntary acceptance of risk, and between risks where there is also benefit (e.g., salary) and where there is none. These

factors – and others – ultimately mean that the reliability requirements are different for the I&C systems for nuclear plants, oil and gas plants, and aircraft.

3. Aircraft inevitably have to mix up control systems and protection systems, at least to some extent, whereas in both NPPs and in oil and gas facilities it is possible (and desirable) to separate control and protection. Modern digital aircraft systems have tended to become more "integrated" – which means the separation between control systems, pilot display systems, and protection systems (which act to limit the flight envelope to safe areas) has become more diminished.

For large process plants, the overall I&C architecture has to be frozen early in the design process. This is because:

i. the I&C architecture includes the major protection systems and therefore defines much of the safety case; and

ii. the I&C architecture leads to definitions of space requirements and system separation requirements (for cubicles, switchgear, and cable routes) so that the civil structures can be designed.

Hence the program for major process plant construction needs to address I&C architecture at an early stage. The typical order for I&C system specification, design and implementation would be as follows:

1. Define the overall I&C architecture, including space and separation requirements
2. Prepare system functional specifications for I&C systems
3. Prepare detailed technical specifications for I&C systems
4. Place contracts
5. Detailed design and manufacture
6. Works testing
7. Site testing
8. Commissioning testing

A fundamental principle for good I&C systems design in process plants is that *control systems and protection systems shall be separated.*

Control systems, in their broadest sense, include sensors, plant status indications, logic systems, alarm systems, and auto-control systems. With process plant digital systems, the "control systems" are usually referred to as the digital or distributed control system (DCS). The DCS includes the human–machine interface (HMI) – the computer screens in the main control room – for the normal operation of the plant. The functions of the DCS are self-evidently to control the plant parameters within normal limits, and to advise the operators of plant status, and to raise alarms when normal parameter ranges are exceeded.

Protection systems include sensors, logic, actuators, and the dedicated HMI for the protection systems. These systems should play no active part in the normal operation of the plant whatsoever – their only role is to detect anomalous behavior and initiate protection/mitigating actions, and keep the operator informed of what they are doing.

The control and protection systems should be electrically and physically separated to try to eliminate common-mode failures between control systems and protection systems. At no point should there be a direct electrical connection between a control system and a protection system. Any required electrical signal connection (e.g., communications) should be via buffered links, e.g., opto-isolators. Electrical separation includes also the power supplies for the systems. In general, the power supplies should be sourced from different transformers, and the RPS should use guaranteed supplies. The systems shall also be physically separate, ideally with physical/fire barriers in between. These measures are necessary to prevent common-mode or common-cause failures.

All protection systems will employ redundancy to some extent, and usually control systems also have redundancy. The extent of redundancy will depend on various factors including reliability requirements, maintenance requirements, and the *single failure criterion*. (The single failure criterion should always be applied to protection systems – it means that no single failure of a system should cause a dangerous system failure.) A high-integrity protection system will usually use two-out-of-four (2oo4) logic. Control systems are often "duplex" (1oo2) where this is dictated by plant availability requirements.

The HMI (human machine interface) encompasses displays, alarms and manual controls. Evidence shows that human reliability is not good in high-pressure situations with serious time constraints. For this reason, normally only weak reliability claims (typically of the order of 10^{-1} pfd) are made for operators in fault situations, and it is normally assumed they do not have to react quickly. Consequently there is little purpose in designing an extremely high reliability HMI, so it is normally just a SIL 1 system.

One key concern, however, is that the HMI should not mislead operators into making a bad situation worse – for this reason, the *ergonomics* of plant displays deservedly gets a lot of attention.

Also, indication systems that show the status of safety-related systems (such as are used for monitoring the plant in accidents) need to have high integrity. Hence these systems often feature in special high-integrity displays in control rooms with SIL 2 or even SIL 3 reliabilities.

Auto-control systems can initiate faults if they fail. This can lead to plant downtime, and also to challenges on the protection systems, i.e., they can cause "initiating events" in fault sequences. It is therefore common to implement duplex control systems which can achieve SIL 2 reliability (10^{-2} pfd or pa).

I&C architectures are probably at their most complex in nuclear power stations. Here, the levels of hazard and the accordingly high reliability requirements mean that diverse high-integrity protection systems are required. Furthermore, the need for extremely high-reliability removal of decay power from the reactor core postshutdown means that there has to be very high-reliability backup electricity supplies. A simplified schematic of a modern I&C architecture for a nuclear power station is presented in Fig. 2.7.

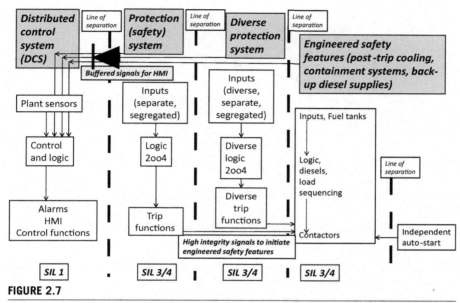

FIGURE 2.7

The most complex process plant I&C architecture – a simplified schematic diagram for nuclear power plant I&C systems.

I&C ARCHITECTURE CASE STUDY: THE I&C ARCHITECTURE FOR THE EUROPEAN PRESSURIZED WATER REACTOR (EPR)

At time of writing, Electricite de France (EDF) is about to make a decision to finally commit fully to the construction of a large new twin pressurized water reactor (PWR) at Hinkley Point, Somerset, in the UK. The plant will be two European Pressurized water Reactors (EPRs), designed by Areva and built by EDF. Similar plants are already in construction at Olkiluoto (Finland), Flamanville (Normandy, France), and Taishan (China).

The process to build these reactors in the UK began some years previously, and detailed proposals were submitted to the UK's nuclear safety regulator, the Office of Nuclear Regulation (ONR), in 2009, for the Generic Design Assessment (GDA). The original proposal for the I&C architecture is shown in simplified form in Fig. 2.8.

Notable features of this architecture are as follows.

1. The EPR I&C consisted entirely of computer-based systems, Areva Teleperm XS (for the high-integrity protection functions) and Siemens SPPA-T2000 (for lower integrity control functions). Furthermore, both these systems were originally developed by Siemens, so any claim for design diversity seemed weak.
2. The Siemens SPPA-T2000 system, which is a relatively low-integrity system designed to SIL 2 or equivalent standards, was being used to change parameters on the high-integrity Teleperm XS system. In general, there was a high level of

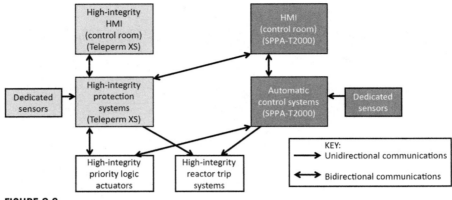

FIGURE 2.8

A simplified schematic diagram of the EPR I&C architecture as originally proposed.

connectivity between two complex systems, although these systems were being claimed to be wholly independent of each other.

3. The Teleperm XS system was being claimed by the designers to have a reliability of 10^{-5} failures on demand. The SPPA-T2000 system was being claimed to have a reliability of 10^{-3}. Both of these reliability claims are at the very top end of what might be realistically achievable (see Table 2.1).

As a result of the Generic Design Assessment, the UK regulator requested significant changes in the I&C architecture [4,5]. In particular,

- It was agreed that a non-computerized backup system would be implemented in order to provide protection and controls in case of total loss of C&I functions from the Teleperm XS and SPPA-T2000 platforms.
- It was agreed that one-way only communication would be implemented from the Teleperm XS system to the SPPA-T2000 system.
- It was agreed that the reliability claims would be reduced for the Teleperm XS (10^{-5} pfd to 10^{-4} pfd) and SPPA-T2000 (10^{-3} pfd to 10^{-2} pfd) systems.

A notable outcome of this review process in the UK is that EPR designs in UK, Finland and France will now each have different I&C architectures. All three countries are, of course, members of the European Union, so one might have expected common standards of regulation to apply. However, in each country the safety regulator has taken a different view of what constitutes "best practice". The UK position is as described above. In Finland, the safety regulator (STUK) has also requested a non-computerized backup system but in that country a FPGA-based system has been adopted (see "The Selection of Logic Elements and Vendors for High-integrity Industrial Safety Systems"). In France, however, the safety regulator (ASN) has more-or-less accepted the original architecture proposal with the proviso that more safety-related functions, which were originally to be in SPPA-T2000 systems, will now be in the high-reliability Teleperm XS system. The Chinese design will be the same as the French.

The key issue here is that it shows there is no international consensus on best practice in I&C architecture. This is an issue that should be resolved, but unfortunately it is also an issue where national interests and pride can work against the achievement of consensus. A final point of note is that there exists a forum for nuclear safety regulators called the Western European Nuclear Regulators Association where these issues could and should have been resolved. However, the French regulator is not presently a member.

THE SELECTION OF LOGIC ELEMENTS AND VENDORS FOR HIGH-INTEGRITY INDUSTRIAL SAFETY SYSTEMS

At present, the marketplace for high-integrity voting logic elements, for use in (primarily) the oil and gas, petrochemical, pharmaceutical and nuclear industries, is overwhelmingly dominated by microprocessor-based systems. These systems are used to implement logic algorithms in particular for emergency shutdown systems and fire and gas detection systems. Some principal vendors of such equipment are listed in Table 2.3. Each of these systems is claimed to be SIL 3 or SIL 4 capable.

Table 2.3 Some principal vendors of high-integrity I&C logic solvers

Vendor	System brand name	Logic elements
ABB	System 800	Microprocessor
Areva	Teleperm XS	Microprocessor
Areva	Unicorn	Magnetic
CTEC	Firmsys	Microprocessor
Emerson	DeltaV SIS	Microprocessor
HFC-Doosan	HFC-6000	Microprocessor
Honeywell	Safety Manager	Microprocessor
Invensys-Schneider	Tricon	Microprocessor
Mitsubishi	MELTAC	Microprocessor
RADIY		FPGA
Rockwell	Triplex	Microprocessor
Rolls-Royce	Spinline	Microprocessor
Westinghouse	Common Q	Microprocessor
Westinghouse	CSI	FPGA
Yokogawa	ProSafe SLS	Magnetic

One key vendor selection criteria is that of obsolescence. Plant operators look for their equipment to be supported by the original equipment manufacturer (OEM) for as long as possible. In particular, the following aspects are important.

1. The vendor should offer through-life support.
2. The vendor should offer backward-compatible upgrades.
3. The system design should facilitate mid-life renewal.
4. The vendor should demonstrate arrangements to maintain its skills and capabilities.
5. The vendor-claimed safety integrity levels (SILs) may need careful validation.

(Obsolescence issues are discussed further in Chapter 6.)

Microprocessor-based systems dominate the market. Each manufacturer of microprocessor-based high-integrity logic solvers guards its intellectual property very carefully, since the design, production and marketing of these systems represents a large investment. It is not at all easy to reach any clear conclusions about whether any system is "better" than any other. The market leaders in high-integrity microprocessor-based logic solvers are (arguably) Areva, Invensys-Schneider, Rolls-Royce and Westinghouse. In particular, all four have been approved for use by the US Nuclear Regulatory Commission. Their main claims for the strengths of their systems are as follows.

- Areva Teleperm XS has strictly cyclic operation, with no process-controlled interrupts. There is no dynamic memory allocation, so each variable in the application software has a permanent dedicated location in the memory. It uses a standardized, simple software structure, and automatic code generation from function diagrams.
- Invensys-Schneider Tricon uses triple modular redundant (TMR) architecture in which three parallel control paths are integrated into a single overall system. This gives high fault tolerance, with no process-controlled interrupts and extensive self-test and diagnostics.
- Westinghouse Common Q uses ABB AC160 modular controllers with self-test and diagnostics, and has watchdogs to detect interrupts.
- Rolls-Royce Spinline was developed in the Esterel SCADE environment, has extensive self-test and diagnostics, and is said to be fail-safe and fault-tolerant.

Even for people who dedicate themselves to the analysis of such systems, it is difficult to reach any obvious conclusions about which is "best".

Other systems based on magnetic logic or field-programmable gated arrays (FPGAs) are also available. Older, mostly obsolete, analog systems used discrete electronic logic (TTL – total transistor logic, etc.), or else relay logic.

Magnetic logic consists of small magnetic cores, mounted on printed circuit boards, that have OR and AND gates configured on them using primary and secondary coils wound around the cores. Yokogawa claim their ProSafe SLS system is inherently fail-safe and has no unrevealed dangerous failure modes. Magnetic systems are therefore discrete element logic systems which are best suited to simple logic only.

FPGAs are a type of programmable logic device (PLD) which consists of digital integrated circuits containing blocks of AND, OR and NOR gates that can be configured to produce individual logic algorithms by burning in the connections between the appropriate logic blocks. FPGAs have been in widespread use for some time now

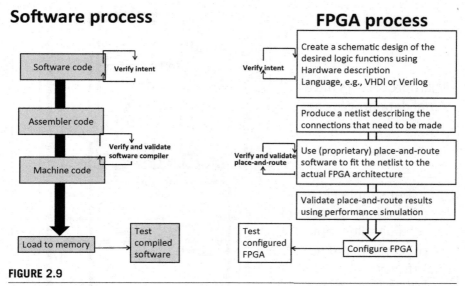

FIGURE 2.9

A comparison of the software/microprocessor design process with the FPGA process for high-integrity systems.

in other fields of electronics, but their use is now growing in high-integrity logic solvers.

FPGAs may have memory functions and basic maths functions, or even embedded processors. Validation then gets more difficult due to the large number of possible system states. For high-integrity applications, only simple FPGAs have been used to date.

The design process for FPGAs is highly software-intensive. Validation and certification of the FPGA design tools are critical. Figure 2.9 presents a comparison of the software (microprocessor) design process compared to the FPGA design process. In essence, an FPGA itself is software-free, but the high-integrity software is contained in the design process before the FPGA is configured.

There are three types of FPGA: SRAM which are manufactured by Altera, Atmel and Xylinx; Flash manufactured by Microsemi; and Antifuse manufactured by Microsemi. Antifuse is non-rewritable, whereas the others are rewritable. It has been suggested that SRAM FPGAs could be susceptible to cyber-attack at the production stage.

Very simple FPGAs using "flat logic", i.e., a simple chain of logic elements going from inputs to outputs across the FPGA chip where each safety function is separate from every other function, seem to offer the best of all possible worlds for high-integrity systems. They have the convenience of manufacture that a microprocessor system has, and in addition it should be possible to subject "flat logic" FPGAs to full negative testing.

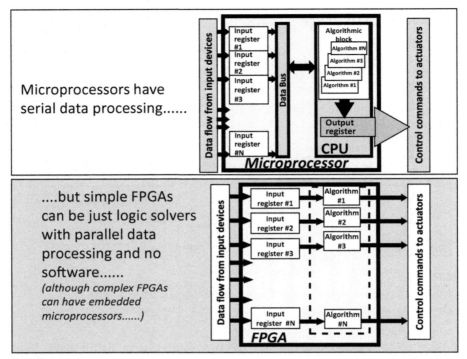

FIGURE 2.10

Microprocessors compared to FPGAs – serial versus parallel data processing.

A comparison between the operation of a "flat logic" FPGA and a microprocessor is presented in Fig. 2.10. In essence, a microprocessor operates in a serial mode at very high frequency, with each piece of logic occurring in series in the same physical location, namely the CPU. In comparison, FPGA uses parallel logic so each safety function can be completely separate on the FPGA integrated circuit.

If we now consider the desirable attributes for high-integrity logic systems in general, the ideal system would have all the following.

- Can handle complex algorithms
- Pre-service full testability
- Verifiable
- Validatable
- Licensable
- Cyber-attack immune
- Resistant to age-related failure modes
- Ease of maintenance
- Ease of configuration management
- Low obsolescence risk
- Low cost

Figure 2.11 presents the author's judgments about the relative merits for use in high-integrity systems of microprocessors, FPGAs, magnetic logic, discrete electronic logic, and relay logic.

The general conclusion is that there is no "perfect" technology for high-integrity logic solvers. Although microprocessors are the most versatile, there are issues about their complexity and hence the potential for unrecognized failure modes. There are also issues about cyber security and short lifecycle leading to obsolescence.

Magnetic logic systems are extremely robust but can only handle simple logic.

FPGAs seem to have a promising future in high-integrity applications. In particular, the potential for "flat-logic" FPGAs lies in their testability – it should be possible to carry out full negative testing, which may then reduce the extent of absolutely rigorous quality control that is currently required during design, coding, and manufacture for microprocessor-based systems.

Nevertheless, for some time into the future microprocessor-based high-integrity systems will dominate the market, because these are the systems into which vendors have mostly placed their efforts.

QUALITY MANAGEMENT OF SOFTWARE SUPPLIERS

One area of ongoing difficulty is ensuring the quality of software systems in components that have been bought from third-party suppliers. In particular this issue is important for so-called "smart devices" – primarily sensors and actuators – which have embedded software that may carry out important safety-related functions.

This issue has become particularly important as more and more devices have embedded processors. Simple old-fashioned analog devices are becoming quite rare.

The quandary is that, whereas the purchasers of smart devices may want to understand fully the nature of the software in the devices they are buying, the suppliers of those devices may want to guard their intellectual property carefully. The software represents a considerable investment and often contains commercially valuable information.

Hence, purchasers need to enter legally binding agreements with the suppliers not to disclose any information.

One approach to this difficulty has been the EMPHASIS tool used in the UK nuclear industry [2]. This software tool (a spreadsheet) asks standardized questions of equipment suppliers with the objective of identifying whether their quality management arrangements are adequate for safety-related applications of their software and equipment.

EMPHASIS asks questions about the supplier's quality management systems, such as:

- The type of quality management arrangements employed
- Staff competency and defined areas of responsibility
- Field performance of their products
- Failure modes assessments, especially for dangerous failure modes
- Change control arrangements

Type of logic elements	Can handle complex functions and algorithms?	Pre-service testability?	V&V	Licensing and safety?	Cyber-attack?	Single Event Upset (SEU) and other age-related failure modes such as electro-migration	Maintenance aspects	Configuration management and change control	Obsolescence risk for operator	Cost
Micro-processor	Can handle complex functions	Full negative testing cannot be achieved because of large number of inputs going into a common logic-solving element	V-model approach well-defined but regulators can always ask for more, e.g., dynamic and statistical testing	Ultimately depends on robust QA, comprehensive documentation, and full traceability from functional requirements, via implementation, to testing	Potentially susceptible	Susceptible (especially for smaller feature size <100 nm)	On-line monitoring and test arrangements can reduce workload	Configuration management and change control need to be extremely thorough	High (short lifecycle)	Cost dominated by Engineering costs, i.e., hardware costs are not significant.
FPGA or PLD	Cannot handle complex functions unless embedded processors are used (in which case advantages over processors are lost....)	Full negative testing could be achieved if (i)logic functions are simple and (ii) functions are segregated on FPGA chip and (iii) it could be proven by inspection that functions are segregated	V-model approach with full traceability. No OS but VHDL and place-and-route software need full V&V	VHDL (and other) software used in design is complex and safety-critical. Current standards treat FPGAs like microprocessors but, if full negative testing could be carried out, then regulators would be more relaxed	Probably immune	Susceptible (especially for smaller feature size <100 nm)	Straight-forward (like hard-wired logic)	May require configuration management and change control similar to microprocessor systems (although in principle it is fixed at installation)	Said to be low	As above
Magnetic logic	Cannot handle complex functions or algorithms	Full negative testing can be achieved	V-model approach well-defined. Full traceability required	Some types of magnetic logic have been licensed in UK for RPS applications	Immune	No	Straight-forward	Fixed at installation	Low	As above + bigger space requirements
Analogue electronic logic	Cannot handle complex functions or algorithms	Full negative testing can be achieved	V-model approach well-defined. Full traceability required	Licensable (because unreliability if individual elements is known)	Immune	No but other potential unrevealed failure modes	Straight-forward	Fixed at installation	Low	As above + bigger space requirements
Relays	Cannot handle complex functions or algorithms	Full negative testing can be achieved	V-model approach well-defined. Full traceability required	Licensable (because unreliability of individual elements is known)	Immune	No but other failure modes such as contact welding	Straight-forward but maintenance burden can be high	Fixed at installation	Low	As above + large space requirements

FIGURE 2.11

Comparative attributes of various types of logic solver for high-integrity protection (safety) systems (author's judgments).

- Record keeping and documentation
- Problem reporting and corrective actions
- Verification arrangements and levels of independence
- Use of subcontractors and use of bought-in components
- Willingness to disclose error reports and change control records
- Use of prescribed design lifecycles
- Use of software analysis tools and techniques
- Environmental testing

The objective is to provide confidence that the supplier understands the challenges of making high-integrity safety-related equipment, and has appropriate arrangements in place for the production of such equipment.

CASE STUDY ON SMART DEVICES: A SMART DEVICE WITHIN A HIDDEN SURPRISE

A good illustration of the sort of issue that can face plant operators is the case of a "paperless" chart recorder in use at Sellafield reprocessing plant, Cumbria, UK, in the early 2000s [3]. Industrial plant all over the world has for many decades used chart recorders, where a parameter of interest – say a temperature in a particular part of process plant – is monitored using a pen on a roll of paper. The difficulty with pen recorders was that, whenever you wanted to investigate something, it always seemed that the ink had run out, or the paper had run out. So, with the advent of micropro-cessors and memory chips, suppliers started to provide "paperless" chart recorders, where the section of "paper" that would normally be visible was replaced by a small LCD display, and the memory inside the instrument stored the signals for later replay if required. They were made to be the same size as old paper chart recorders so they were "like-for-like" replacements of the older units. They have now replaced the older paper chart recorders almost everywhere.

However, in about 2005, operators at Sellafield nuclear reprocessing plant began to notice problems with one particular make of paperless chart recorder. They began to "go to sleep" or require constant re-booting.

Eventually, it was discovered that the operating system software in the paperless chart recorder contained what programers call an "Easter egg". In the factory-installed software, there was a game installed called "Cave Fly". If a particular combination of buttons on its front panel were operated, the chart recorder changed to a game console. The problem was that the game caused the memory to overflow, and dumped old stored data, and also caused the whole chart recorder to start operating slowly and temperamentally.

This "Easter egg" is one of numerous examples. An old version of Microsoft Excel (Excel 95) had a similar hidden computer game, if one knew where to find it, called "The Hall of Tortured Souls". Normally, this sort of thing should not matter (although it is certainly not good practice by the programers.)

The recorders were swapped for spares, returned to the manufacturer and recon-figured over a period of about 18 months, with no real improvement in reliability.

REFERENCES

[1] IEC 61508, Functional safety of electrical/electronic/programmable electronic safety-related systems, second ed., 2010.

[2] R. Stockham, Emphasis on safety, engineering and technology, February 2009, pp. 47–48.

[3] T. Nobes, Conference paper at sixth international conference on control and instrumentation, University of Manchester, September 2007.

[4] EPR system modifications satisfy UK regulator, World Nuclear News, 16 November 2010.

[5] www.onr.org.uk/new-reactors/gda-issue-close-out-uk-epr.htm#close-out-reports.

FURTHER READING

DOE Fundamentals Handbook, instrumentation and control (two volumes), DOE-HD-BK-1013, US Department of Energy, 1992.

Center for Chemical Process Safety, Guidelines for safe and reliable instrumented protective systems, Wiley-Interscience, 2007.

US Nuclear Regulatory Commission, Instrumentation and Controls in Nuclear Power Plants: An Emerging Technologies Update, NUREG/CR-6992, 2009.

C. Maxfield, FPGAs Instant Access, Newnes, 2008.

M. Lyu, Handbook of Software Reliability engineering, IEEE/McGraw-Hill, 1996.

N. Storey, Safety-Critical Computer Systems, Addison Wesley Longman, 1996.

Cyber Security, Cyber-attack and Cyber-espionage

3

"Digital weapons work…they don't put forces in harm's way, produce less collateral damage, can be deployed stealthily, and are dirt cheap. (They have) changed global military strategy in the 21st century... (Stuxnet) will be remembered as the opening act of cyber-warfare."

Ralph Langner

"Cyber offense is well-funded and implemented straightforward within a military chain of command. At the same time, cyber defense of critical national infrastructure is expected to be implemented voluntarily by a dispersed private sector that feels little desire to address matters of national security by ill-coordinated risk management exercises that negatively affect the bottom line."

Ralph Langner

STUXNET

In June 2010, the Stuxnet computer worm was discovered by a Belarus-based antivirus software vendor called Virus BlokAda. The Stuxnet worm had been used to attack the Natanz, Iran, uranium enrichment plant – a gas centrifuge process – since 2008. It caused disruption and delays to Iranian enriched uranium production, by introducing malware into Siemens S7 controllers used in the Centrifuge Drive System.

The Stuxnet worm is "50 times as big as a typical computer worm", according to Symantec. Its filename is s7otbxsx.dll. Stuxnet was developed by the US Department of Energy, US National Security Agency, CIA and Mossad in a project code-named "Olympic Games". It is believed to have been spread by Mossad agents, at first possibly by leaving memory sticks on the ground in a car park, or possibly by infecting laptops owned by contractors with access to Natanz. It was a massive project that required huge resources, including the construction of a mock-up of the Natanz enrichment plant in USA [1].

The Stuxnet attack strategy was to, first, identify physical vulnerabilities (e.g. centrifuge rotors) by, say, review of plant HAZOPs. Then the project team had to understand the control and protection systems of the enrichment plant, and finally to identify software vulnerabilities.

The Siemens S7 controllers at Natanz used Microsoft Windows software. Windows has between 30 and 50 million lines of code (depending on version), and weaknesses in this code are extremely valuable to those who seek to carry out cyber-attacks. Stuxnet exploited four "zero-day" weaknesses, i.e. weaknesses which had not previously been identified and therefore defences were not in place.

High Integrity Systems and Safety Management in Hazardous Industries. 978-0-12-801996-2

The Stuxnet worm was very focused in targeting the control systems of the Iranian enrichment plant. It worked by causing the centrifuge rotors to operate at the wrong speed while simultaneously misleading the operators by displaying that the centrifuge rotors were operating at the right speed. This caused the Iranian uranium enrichment plant to fail to operate properly possibly throughout the period 2008–2010.

The use of Stuxnet to attack the Iranian plant was allegedly agreed by President Obama [2]. In doing so, a Pandora's Box may have been opened where cyber-attacks by governments upon other nation states become commonplace. In addition to Stuxnet, there have been attacks by the Duqu and Flame worms (both possibly related to Stuxnet), and there have been attacks on various US military development projects possibly coming from APT1 (see below). As an article in PC World (David Jeffers, June 1, 2012) put it, "Stuxnet may have achieved the goals it was developed for. Regardless of whether we agree that the mission was admirable or necessary, though, we now have to deal with the Internet equivalent of a mustard gas or Agent Orange leak that has the potential to affect us all". In case this sounds like hyperbole, consider the attack on Saudi Aramco – the biggest oil company in the world – on August 15, 2012. Saudi Aramco was attacked by the Shamoon virus which infected and disabled some 30,000 Windows-based machines within the company. According to the International Institute for Strategic Studies ("The Cyber Attack on Saudi Aramco", April 1, 2013), Shamoon caused significant disruption to Saudi Aramco and also spread to other companies. Even a partial disruption of oil production from Saudi Aramco could have a significant effect on world oil prices. There was speculation that Iran could have been the source of Shamoon.

Until Stuxnet became known publicly, and despite the warnings of experts in cyber security, most people in the field of industrial control and automation did not worry too much about cyber security. Stuxnet therefore had huge implications and became a "game-changer" in industrial control and automation. It was clearly no longer the case that the only thing to worry about was a lone hacker working in his bedroom. Industry now needed to worry about concerted criminal- or government-led attacks upon business software systems and industrial control and automation systems.

Stopping Stuxnet-style attacks against industrial control and automation systems requires special attention. Antivirus software will not stop new Stuxnet-style malware, because it looks like legitimate application software. Also, security patches do not normally get deployed quickly into industrial control systems. Software updates may take years to get installed; this compares to domestic or business computers where security patches may be installed almost immediately they are available.

Firewalls, data diodes, and air gaps should work as protection – but Stuxnet showed how to bypass these by infecting laptops belonging to legitimate contractors. Hence, good security processes are required and must be implemented carefully.

APT1

APT1 is the name given by a Washington-based consultancy, Mandiant, to a Chinese army-sponsored organization that plans and executes cyber-espionage on multinational corporations. It has been doing this from 2007 to the present. Until

2013, ATP1 had attacked some 141 companies in 20 major industries. According to Mandiant [3], the companies were based in the US, UK, Canada, France, Belgium, Norway, Luxembourg, Switzerland, Israel, UAE, India, Taiwan, Japan, Singapore and South Africa, although the vast majority were US-based companies. It has separately been reported that the French multinational company Areva was attacked and that the hackers may have had continuous access to Areva's company files for up to two years.

The cyber-espionage was carried out by the 2nd Bureau of the People's Liberation Army (PLA) General Staff Departments (GSD) 3rd Department, which is most commonly known by its Military Unit Cover Designator (MUCD) as Unit 61398, based in Shanghai. The cyber-espionage consists of large-scale theft of Intellectual Property. Typically, Terabytes of data can be stolen over many months.

According to Mandiant, a favorite technique is "spearphishing": an email purporting to be from a senior colleague's home email address (but actually an artificial address) is sent to a junior colleague at the weekend, requesting that the junior colleague reviews urgently some material in an attachment to the email. Hence it appears to be a legitimate out-of-hours query. The junior colleague is using his work laptop at home over a Virtual Private Network (VPN) link. The attachment appears to be a pdf file, but the filename has many spaces after the "pdf" label before it actually becomes apparent it is an "exe" file – and this is not visible in the small window for the filename on the email. The junior recipient, anxious to please his boss, opens the attachment – and his company laptop becomes infected with malware which enables APT1 to access company files.

In May 2014, the US Government placed formal charges against five named Chinese Army officers associated with the APT1 claims. The charges alleged that the victims of the attacks included Westinghouse Electric, US Steel, Alcoa Inc, Allegheny Technologies, Solar World and the US Steelworkers Union. However, China insists that it is a victim of cyber-attacks and not a perpetrator. Former US intelligence contractor Edward Snowden has indeed published evidence of US attacks into Chinese networks. It therefore seems beyond doubt that the US government is doing the same sort of thing as the Chinese government. The real point of this, therefore, is that no company in the twenty-first century can afford to be complacent about cyber security. Cyber security has become a necessary part of doing business in the globalized commercial environment.

This chapter will not discuss cyber-espionage further. The remainder of the chapter is primarily focused on cyber-attack of industrial control systems.

INDUSTRIAL CONTROL SYSTEM ARCHITECTURE AND CYBER-ATTACK

Most industrial control system architectures should look something like Fig. 3.1, which illustrates the architecture from the perspective of potential cyber-attack of a multisite company with a company Wide Area Network, site-based Local Area

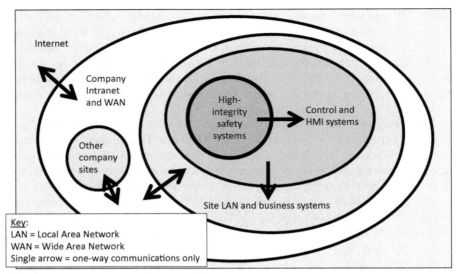

FIGURE 3.1

A typical industrial control system architecture from the viewpoint of potential cyber-attack.

Networks, and control and protection systems on each site with operating industrial plant.

The concern here is that a concerted cyber-attack on a site might pose a threat to the safety of the hazardous plant. The terminology here is important and a summary of basic information security terminology is provided in Table 3.1.

A further layer of terminology relates to the different types of cyber barriers that are employed. These are presented in Table 3.2.

There should be many levels of barrier.

1. There should be firewalls between the Internet and the company Wide Area Network.
2. There should be additional firewalls between the Wide Area Network and the individual sites.
3. There should be unidirectional data flow between the hazardous plant control system and the site offices. This means that non-operational staff can, say, monitor plant conditions from their desks without being able to interfere with plant operation in any way.
4. There should also be unidirectional data flow between the plant high-integrity protection (or safety) systems and the control systems.

This all sounds very good, but we should remind ourselves that the route for the Stuxnet cyber-attack on the Iranian enrichment plant (its "threat vector") was most probably the simple act of leaving memory sticks in the car park. Staff working on the plant found them, put them in their pockets and later connected them to their computers. Hence the barriers listed above may sound impressive, but they can be short-circuited if vigilance is not maintained.

Table 3.1 Some basic information security terminology

Domain	Logical grouping of systems and people within which information can be freely shared.
Environment	Models of the physical world from where people interact with systems via a Portal.
Connections between domains	These define the limits of interaction.
Islands of infrastructure	Single machines or groups of networked machines that operate together to support a business function.
Causeways	Secure connections between *Islands*.
Infosec Architectural Model	This incorporates both the business functions and the technical world in which they operate, and thus enables the different views of system users, security advisers and developers to be discussed.
Risk	This has three elements: threat, vulnerability, and impact. Controls or countermeasures can be applied to each.
Threat actor groups (TAGs)	*Internal*. Operators, engineers with high-level access privileges.
	External. Connected users, maintenance and repair, visitors, those with wider access, regulators and security advisers, natural disasters

Table 3.2 Terminology of cyber barriers

Data diodes and unidirectional networks	These are the methods of ensuring one-way transmission of information from a higher-security ("trusted") system to a lower-security system, but not vice versa.
Network layer or packet filter firewalls	Do not allow packets to pass through the firewall unless they match the established rule set. Packet filters act by inspecting the "packets" which transfer between computers on the Internet. If a packet matches the packet filter's set of rules, the packet filter will drop (silently discard) the packet, or reject it (discard it, and send "error responses" to the source).
Applications layer firewalls	Application-layer firewalls work on all browser traffic, or all telnet or ftp traffic, and may intercept all packets traveling to or from an application. They block other packets (usually dropping them without acknowledgment to the sender). The key benefit of application layer filtering is that it can "understand" certain applications and protocols (such as File Transfer Protocol (FTP), Domain Name System (DNS), or Hypertext Transfer Protocol (HTTP)). This is useful as it is able to detect if an unwanted protocol is attempting to bypass the firewall on an allowed port, or detect if a protocol is being abused in any harmful way.
Proxy firewalls	A proxy server (running either on dedicated hardware or as software on a general-purpose machine) may act as a firewall by responding to input packets (connection requests, e.g.) in the manner of an application, while blocking other packets. A proxy server is a gateway from one network to another for a specific network application, in the sense that it functions as a proxy on behalf of the network user.

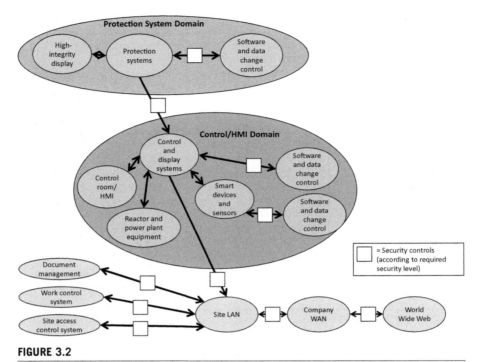

FIGURE 3.2

Typical information security architecture for an industrial plant and its site.

The situation becomes more complicated when full lifecycle and maintenance activities are considered. Cyber security applies to all equipment suppliers and contractors, and also to any software maintenance and modification activities that may be done on the site to digital systems. In addition to the plant control and protection systems, there will be document management systems, work control systems and site access control systems, all of which have some relevance to safety. Hence Fig. 3.2 shows what a typical simplified information security architecture for an industrial plant site might look like. This shows a somewhat higher degree of complexity than Fig. 3.1.

Plant cyber security is assessed according to *plant domains*, such as control systems, protection systems, and business systems. Business systems include document management, work control and site access control systems. In each domain, the full lifecycle needs to be considered (as portrayed in Fig. 2.1). This includes the following.

1. The *development environment*, which comprise contract placement, software specification and writing, system testing at factory, transfer to the site and testing at site.
2. The *operational environment* which covers operational use of software-based systems.

3. The *maintenance environment* which comprises the arrangements for software modification.

For each environment, cyber security needs to be assured. Hence, in this simplified example we have three plant domains, each of which will have three lifecycle environments, so cyber security will have to be considered in nine different cases.

In each environment, the most important issue to consider is the *threat actor groups* (or threat vectors) – in other words, the conceivable routes whereby a virus could be introduced to the domain. This will require consideration of personnel access controls, and physical and management controls of connections to the network and Internet. It will also require security checks on personnel who are involved or have access to secure areas.

Threat actor groups are identified groups of people that could conceivably, either deliberately or accidentally, reduce the security of a domain or environment. In particular:

1. The "threat" may be hidden code which is not formally functioning, in addition to tampering with the main code or its data. (See also the case study "A smart device with a hidden surprise" in chapter 2 on this issue.)
2. The greatest security risks may be from those with detailed inside knowledge.
3. Sabotage may be financially motivated (e.g. to make a project last longer).

The types of threat actor groups that need consideration include the groups listed in Table 3.3.

Transfer of software from factory to site will require special checks to ensure no interference has occurred. Such checks might include (1) the independent transmission of software packages, (2) unambiguous identification information, and (3) assurance that equipment was not tampered with during transportation. In addition site tests should include IT security checks to confirm that the correct configuration has been installed.

Table 3.3 Some important threat actor groups

Threat actor group	Scope of activity
Development and test engineers	Authorized users
Commissioning engineers	Authorized users
System operators	Authorized users
"Connected"	Others who have legitimate connections to other domains and might therefore access or manipulate information
"Location"	Those with legitimate unescorted access
"Handlers"	Maintenance staff and contractors
"Separate"	Visitors, people working on the wider IT network
"Acquisition"	Regulators, security advisors
Disaster	National disasters and major hardware failures

Cyclic redundancy checking (CRC) will normally be carried out as part of installation tests. The CRC can be used to detect accidental alteration of data during transmission or storage. The CRC technique is used to protect blocks of data called Frames. Using this technique, the transmitter appends an extra n-bit sequence to every frame called a frame check sequence (FCS). The FCS holds redundant information about the frame that helps the transmitter detect errors in the frame. The CRC is one of the most used techniques for error detection in data communications.

Finally, the plant and business systems should be graded according to the potential impact of cyber-attacks. Within the nuclear industry, the International Atomic Energy Agency's (IAEA) publication 1527 makes recommendations for minimum countermeasures according to security classification. A simplified version of the IAEA's classification is as shown in Table 3.4.

For high-integrity microprocessor-based systems, one approach to improve cyber security is to "harden" the systems. Hardening microprocessor-based systems involves using reduced version of the firmware to minimize the potential for attack.

The type of approach to cyber security outlined above is termed Domain Based Security, called DBSy, which has been pioneered by Qinetiq Ltd in the UK and is widely used by the Ministry of Defence. It is consistent with the approach of IEC

Table 3.4 Classification of systems according to the potential impact of cyber-attacks

Description	Possible security level and counter measures
Principal role in achieving safety (high-integrity shutdown systems)	*Level 1*. No network data flow of any kind. Strict outward communications only, which excludes handshake protocols such as TCP/IP. No remote maintenance access. Physical access strictly controlled. All data entry is approved and verified.
Other high-integrity control and monitoring systems	*Level 2*. Outward one-way network data flow from level 2 to level 3 systems only. Remote maintenance access may be allowed on a case-by-case basis for limited time only. Physical connections must be strictly controlled.
Other monitoring systems	Level 3: Outward one-way network data flow from level 2 to level 3 systems only. Remote maintenance access may be allowed on a case-by-case basis for limited time only. Physical connections must be strictly controlled.
Safety-related business systems: (i) document management, (ii) work permit and control system, and (iii) access control system	*Level 4*. Only approved users are allowed. Access to the internet from level 4 systems may be given to users if adequate protective measures are employed. Security gateways are implemented. Physical connections are controlled. Remote maintenance is allowed and controlled. System functions available to users are controlled by access control mechanisms. Remote external access is allowed for approved users.
Email	*Level 5*. Only approved users can make modifications to the systems. Internet access is allowed subject to protective measures. Remote external access is allowed subject to protective measures.

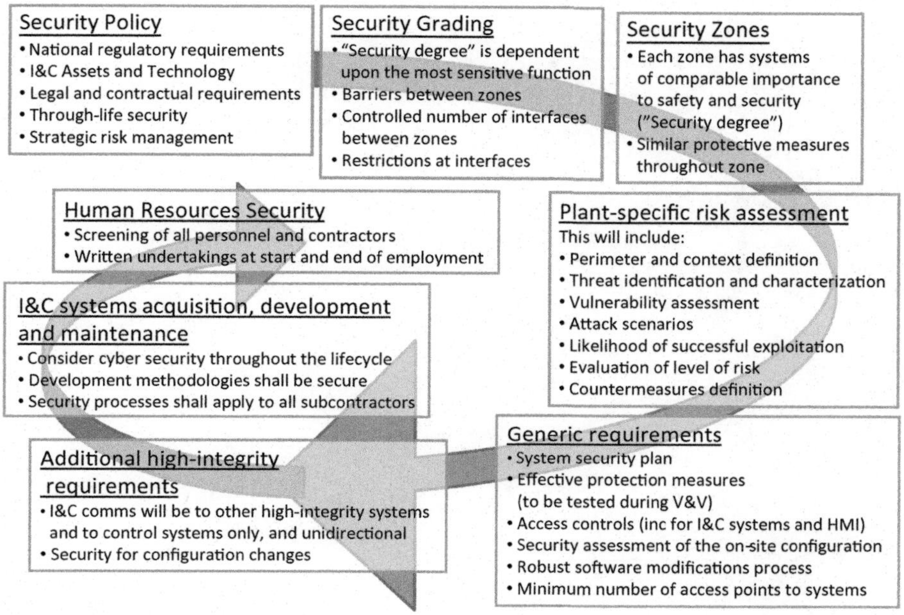

FIGURE 3.3

Some key aspects of high-integrity industrial cyber security.

Adapted from IEC 62645, "Requirement for security programs for computer-based systems", 2012.

62645, which is summarized in Fig. 3.3. The assumption unfortunately has to be that, at every stage of the plant lifecycle, people with malicious intent are interested in attacking the plant IT and control systems. Hence, human resources security has to be an important part of cyber security.

The steps involved in assembling a "security case" using the DBSy methodology involve the following.

1. Develop infosec (**info**rmation **sec**urity) architecture models for each domain
2. Identify the threat actor groups
3. Identify existing security controls
4. Identify any shortfalls
5. Implement additional countermeasures as necessary
6. Residual risk analysis

REFERENCES

[1] To kill a centrifuge, Ralph Langner, November 2013.
[2] Obama order sped up wave of cyberattacks against Iran, NY Times, June 1, 2012.
[3] APT1 – exposing one of China's cyber espionage units, Mandiant, www.mandiant.com, downloaded February 14, 2013.

The Human–Machine Interface

4

"People think computers will keep them from making mistakes. They're wrong. With computers you make mistakes faster."
Adam Osborne

INTRODUCTION

This chapter may appear to be about aviation, but it is actually about how computerized control systems and plant operators (or pilots) communicate with each other, and the design of the communications interface – the *Human Machine Interface* or HMI. The examples are taken from aviation because they are well documented; they also make clear the crucial role of clarity and reliability in ensuring the pilot-operator can control the plant/plane safely.

Good HMI design is about accuracy, clarity and unambiguity in alarms and plant mimic diagrams to ensure that the operators have the best possible *situational awareness* at all times. The worst position for an operator to be in is when something goes wrong and he cannot understand immediately the overall plant situation. "Situation awareness" is an expression used to describe the knowledge that any operator should have of the current circumstances in which he is operating.

Apart from the design of the display screen formats, some other aspects of HMI design deserve a mention.

1. Control rooms should be as peaceful as possible, noting that they are nevertheless the nerve center of any large process plant and must have a lot of through-traffic.
2. Lighting, noise levels and furniture should be designed and managed carefully.
3. At the design stage, the HMI should be carefully specified, and this specification should be done in cooperation with the plant operators.
4. An important area of HMI design is alarm management. Alarm flooding can occur during major plant transients, which may impede situational awareness.

Good HMI design is, to abuse Shakespeare, "more honored in the breach that in the observance", i.e., when good HMI design is achieved we do not notice it; we only notice HMI design when it is done badly. In writing this chapter, a dry presentation of "good practice" in HMI design could have been given. However, by concentrating on a few examples of where bad HMI design has contributed to tragedy, it is hoped that the importance of good HMI design is more strongly illustrated.

High Integrity Systems and Safety Management in Hazardous Industries. 978-0-12-801996-2

What follows is a brief description of three separate air crashes, which took place over 13 years, featuring two completely different designs of airliner (the Boeing 757 and the Airbus 330), and yet the accidents had some similar circumstances. All three accidents were thoroughly investigated and are well documented, and each has even been the subject of its own television documentary.

All three accidents involved aircraft with digital (computerized) cockpits, where the pilots received all their information and alarms about the state of the aircraft from computer displays. All three aircraft suffered blocked Pitot tubes which led to erroneous airspeed indications. In each case the pilots lost situation awareness for a critical short period of time, and aircraft that were otherwise in perfect flying condition crashed with the loss of all passengers and crew.

The aviation and nuclear industries, especially, spend a lot of effort and time worrying about the ergonomics and design of the "human–machine interface", i.e., the layout of the control panels, instruments, warning lights and alarm messages, and the design of the way in which the operator-pilot controls the machine. Older aircraft technology used mechanical-hydraulic control systems with discrete analog-electrical-pneumatic instrumentation. However, these were difficult to maintain, because there were lots of mechanical elements that were prone to failure. The instrumentation required a great deal of wiring and cables. Also, there was no intelligence in the instruments; the pilot had to interpret what he saw to make the right decisions. Finally, the pilot could also fly the aircraft in any way he (or she) chose, which included making mistakes.

The two types of aircraft discussed below (two crashes involving the Boeing 757 in 1996, and an Airbus 330 crash in 2009) belonged to different generations of aircraft. The Boeing 757 was an intermediate mixture of conventional and digital systems; it had conventional flight controls with fully digital Electronic Flight Information Systems. By comparison, the Airbus 330 uses full "fly-by-wire" in addition to digital displays and alarms. In fly-by-wire systems, there is no direct mechanical linkage between the pilot's hands on the control column and the aircraft's control surfaces (ailerons, elevators and rudder). Fly-by-wire technology is gradually entering the automobile industry, with electronic throttles, brakes, and even steering becoming more common. (Mercedes-Benz has even shown a concept car without a steering wheel, where the car is steered with an aircraft-like control stick. Developments also continue with fully automatic cars that steer themselves, although this may still be a long way from commercial viability.)

Modern airliners such as the Airbus 330, and also power stations and other process plant, have microprocessor-based instrumentation and control systems that offer fantastic advantages over old technology. A modern civil airliner will more-or-less fly itself, with the pilot's role reduced to monitoring and oversight under normal conditions. The pilot's job becomes one of ensuring that the flight control systems are doing what they are supposed to be doing, while being ready to assume manual control if necessary. Also, the amount of cables and wires can be greatly reduced because digital signals can be multiplexed with many signals being transmitted on a single cable. Meanwhile, the microprocessor-based systems can include control, indication, alarm, and also protection functions: the software can have the aircraft's "safe flight envelope" within its programing, to ensure that appropriate action is taken if the aircraft is, e.g., flying too slow, or too fast, or at too high an angle of attack.

Older aircraft had controls that were hydraulically linked to the aircraft control surfaces, which meant the pilot got "force feedback", i.e., if the rudder (or elevator or aileron) was being pushed into the airstream, the pilot would feel that he had to push harder. Hence, the pilot could fly by "feel", at least to some extent. In modern aircraft, the mechanical linkage between the pilot's controls and the control surface on the wings or tailplane is completely broken – all signals are electrical – so the pilot will receive no "force feedback" unless the design engineers chose to simulate it in their designs. This aspect – the design of the control column – is important, as we shall see in the case of Air France 447.

Also in older aircraft, each instrument was a "stand-alone" item. It received a signal from a sensor, and it displayed a value. In computer-based instrumentation and control systems, the signals from all the sensors are processed through a few microprocessors, possibly with similar application software, and probably with a common operating system. With modern equipment, the separation between instruments becomes blurred. One disadvantage of this is that, when an instrument fault (or, even worse, a series of faults) occurs, the pilots of modern aircraft may wonder whether the problem is really with the *instruments*, or whether the fault might instead be in the *system*. This confusion may cause brief but important delays in crises, as we shall see. If the pilot is thinking, "Maybe this problem is not just an instrument fault – maybe the whole computer system has gone crazy", it may freeze his decision-making with catastrophic results.

Computer-based instrumentation and control systems are here to stay and offer huge advantages and reduced risks. However, as well as reducing some risks, they also introduce some new ones, and engineers have to be careful when designing such systems. The net effects of computerized instrumentation and control systems are beneficial to safety and costs, but care has to be taken.

A NOTE ON PITOT TUBES

A Pitot tube is a simple device to produce a measure of airspeed. Pitot tubes are mounted on the outside of an aircraft's fuselage, pointing into the airflow. They measure the difference between the dynamic pressure of the air (the pressure measured when pointing into the direction of airflow), and the static pressure (measured perpendicular to the airflow). This pressure difference is proportional to the square of the airspeed (Fig. 4.1).

BIRGENAIR 301, FEBRUARY 6, 1996

This flight was a charter flight taking 176 mostly German tourists home from a holiday in the Dominican Republic to Frankfurt via Gander in Newfoundland. There were 13 crew members on board.

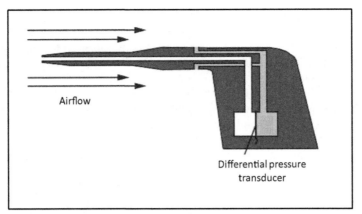

Airflow

Differential pressure
transducer

FIGURE 4.1

Pitot tubes used for airspeed and barometric measurements.

The original aircraft had mechanical problems, so at a late stage a Birgenair Boeing 757 was substituted which had been sitting on the runway at Puerto Plata airport for three weeks. Hence the flight was several hours late and it took off in darkness.

Birgenair was a Turkish-owned airline and the crew were all Turkish; Captain Ahmet Erdem, First Officer Aykut Gergin and relief pilot Muhlis Evrenesoglu.

As the plane was accelerating in darkness for takeoff at 2342:26 h local time, Captain Erdem saw that his airspeed indicator (ASI) was not working.

Five sources of velocity information were available to the crew. They included Captain Erdem's airspeed indicator, the First Officer Gergin's airspeed indicator, a standby airspeed indicator in the center of the instrument panel, a groundspeed read-out on Captain Erdem's Electronic Flight Information System (EFIS) display, and a groundspeed readout on First Officer Gergin's display.

Erdem should have aborted takeoff as soon as he realized his ASI was not working; the purpose of checking the instruments during acceleration is to verify proper operation of the instrumentation. Instead, Erdem asked "Is yours working?" When Gergen said it was, Erdem said "You tell me", meaning that the co-pilot should tell the Captain at which point the aircraft was at the "Vee One" speed of 80 knots. (At Vee One the plane is rotated, i.e., the nose of the plane is raised off the ground.)

The plane then continued to takeoff normally.

If Captain Erdem had aborted takeoff, calculations performed later confirmed that there was enough runway left for safe deceleration.

At 2343:00, Captain Erdem said "It began to operate", meaning that his airspeed indication was working again.

After wheels-up, the autopilot was engaged and the climb continued normally. Unfortunately, the autopilot was selected to use Erdem's (faulty) airspeed indicator.

Two minutes after takeoff, at 2344:25, the captain noted computer alarms *mach speed trim* and *rudder ratio*. The meaning of these alarms was not known to the crew and was not, at that time, included in the flight manual. Erdem said immediately afterward "There is something wrong, there are some problems". Fifteen seconds later he said again "Okay there is something crazy, do you see it?" to which First Officer Gergen replied "There is something crazy there at this moment – two hundred only is mine and decreasing, sir", meaning his airspeed indicator was only showing 200 knots.

The Captain said, "Both of them are wrong. What can we do?" followed by "Alternate is correct", presumably meaning that the standby airspeed indicator in the center of the instrument panel was working properly.

At 2345:04 Erdem said something prescient: "As aircraft was not flying and on ground something happening is (un)usual" (*sic*). Erdem was belatedly showing concern that the aircraft had been sitting on the runway for 3 weeks.

Subsequent investigations of the aircraft wreckage could find no blockages in the Pitot tubes. However, there was a known problem at Puerto Plata airport with a particular species of wasp, the mud dauber wasp, which may have built nests inside the Pitot tubes during the aircraft's 3-week stay on the runway at Puerto Plata.

It appears the aircrew became overwhelmed by the number of conflicting audible warnings and alarms that their flight displays presented to them, some of which seemed almost meaningless. Also, the behavior of Captain Erdem's airspeed indicator was curious; it was not working, then it began to work and indeed started to show excessive speed. It is likely that the blockage in the Pitot tube caused by the mud dauber wasp had completely blocked the Pitot tube, trapping air inside it. As the aircraft continued to climb, the trapped air expanded, causing a false signal indicating *high* speed, and generating more alarms.

Erdem's indicated airspeed reached 350 knots, and this incorrect high-speed signal was used by the autopilot, which therefore raised the nose of the aircraft to almost twenty degrees in order to slow the plane down. At 2345:39, Erdem instructed Gergen to "Pull the airspeed", meaning to silence the overspeed warning alarm.

Faced with confusing alarms and at least one indication that his speed was excessive, Erdem made a bad decision; he decided the aircraft was traveling too fast, and pulled back the throttles.

At 2345:52 the stick-shaker began to operate and continued until the crash. The stick-shaker is a device used to tell the pilots that they are close to stall speed – the control column is made to vibrate as an inescapable warning of low speed. The aircraft was at 7132 feet and Erdem's faulty airspeed indicator was showing 323 knots, when the true speed was less than 200 knots.

As the stick-shaker activated, the autopilot was disengaged automatically because it had reached the end of the range of its "operational authority" – just at the point that Erdem was extremely confused. He had within a few seconds received "high speed" alarms and stick-shaking indicating "low speed". Which should he believe, if any?

The autopilot, before it had disengaged, had raised the nose of the aircraft, and Erdem had pulled back the throttles thinking he was going too fast, when the exact opposite was required; he needed urgently to lower the nose and increase the throttles, but he could not make sense of the conflicting warnings.

The aircraft was almost stalled. At 2346:00, the relief pilot Evrenesoglu said "ADI", referring to the Attitude Director Indicator; he was presumably pointing out the high nose-up attitude of the plane. Erdem continued to struggle with the controls, increasing thrust and trying to lower the nose, but the Angle of Attack was so high that the engines lost thrust. The left engine compressor stalled before the right engine, twisting the plane round and placing it into a full stall.

Erdem's last words were, "Thrust, don't pull back, don't pull back, please don't pull back. What's happening?"

The aircraft hit the sea 20 km from the Puerto Plata at 2347:17, and all on board were killed. The entire flight had lasted less than 5 min, and it had been less than three minutes since the alarms *mach speed trim* and *rudder ratio* had been received.

The official report placed the blame on the crew. The probable cause was "the crew's failure to recognize the activation of the stick-shaker as a warning of imminent entrance to the stall, and the failure of the crew to execute the procedures for recovery from the onset of loss of control." The Boeing 757 Operations Manual did indeed contain procedures for conducting a flight with an untrustworthy airspeed indicator. The procedures included recommended pitch attitudes and throttle settings for climb, cruise and landing.

The accident report said, "While the flight continued to climb, the crew members did not discuss or demonstrate that these procedures were available. They never focussed their attention on the enormous pitch attitude that developed or the alternate sources of velocity information that were present in various indicators in the cockpit…During the final two minutes of the flight, the crew did not take proper actions necessary to prevent the loss of control of the aircraft."

Post-accident tests in a flight simulator showed that a recovery from the stall might have been possible with full power and proper positioning of the flight controls, i.e., normal stall recovery techniques. Doubtless, in a controlled simulator environment, a recovery would have been possible. Also, Erdem should have aborted the flight at takeoff when he saw his airspeed indicator was faulty. However, the alarms generated by the Electronic Flight Information System were so cryptic as to be meaningless: The alarms *mach speed trim* and *rudder ratio* received at 2344:25 were actually intended by the system designers to warn of discrepancy between the airspeed indications, but this was not mentioned in the flight manual, so no pilot could reasonably be expected to know that.

Contradictory alarms led to the aircrew losing situation awareness. In the darkness, they had no other information except their instrumentation, and that instrumentation was not helpful.

The US National Transportation Safety Board (NTSB) issued various Safety Recommendations on May 31, 1996. These included a recommendation that the Boeing 757 flight manual should be revised to notify pilots that "Simultaneous activation of the *mach speed trim* and *rudder ratio* advisories is an indication of an airspeed discrepancy." The NTSB also required Boeing to modify the alarm system to include a "caution" alert when an erroneous airspeed indication is selected. Various other changes to the flight manual were also instructed. Simulator training was changed so that "the student is trained to appropriately respond to the effects of a blocked Pitot tube."

Accident reports, with perfect 20–20 hindsight, often blame the pilot-operator. There were less than 3 minutes between receipt of the incomprehensible *mach speed ratio* and *rudder trim* alarms, and the crash into the sea. This accident was caused by poor design of the human–machine interface, which was then compounded by pilot errors – and not the other way round.

AEROPERU 603, OCTOBER 2, 1996

The Aeroperu 603 accident was a sequel to the Birgenair accident above. It happened a few months later and, crucially, before the NTSB Safety Recommendations arising from the Birgenair accident had achieved wide circulation.

Aeroperu 603 was a scheduled flight of a Boeing 757 from Jorge Chavez International Airport, Lima, Peru to Santiago, Chile, carrying 61 passengers and 9 crew members. On the flight deck were Captain Eric Schreiber and First Officer David Fernandez.

The plane took off at 0042 h local time, i.e., in absolute darkness. The weather was low cloud, with the cloud base at about 270 m, so the pilots will have had no visual reference points.

Immediately after takeoff the crew noticed that the altimeters were not responding. Within a further minute, they realized there was also a problem with airspeed indication also and, at 0043:06, *mach speed ratio* and *rudder trim* alarms were received (as for Birgenair 301). Because the Aeroperu crew had not seen the notifications about the Birgenair crash, they too did not know the meaning of these alarms.

The official report notes "From 00:43:31 the crew start to receive *rudder ratio* and *mach speed trim* warnings, which are repeated throughout the flight, distracting their attention and adding to the problem of multiple alarms and warnings which saturate and bewilder them, creating confusion and chaos which they do not manage to control, neglecting the flight and not paying attention to those alarms which are genuine." The cockpit voice recorder showed the pilots fretting about the significance of these alarms throughout the short flight.

At 0044:32, the crew declared an emergency.

At 0055:07, the flight crew requested, "You're going to have to help us with altitudes and speed if that's possible."

The two flight crew were now over the ocean and trying to fly the aircraft manually to return to Lima in darkness, all the time with abnormal or non-functioning altitude and airspeed indications, and with the Electronic Flight Information System generating lots of alarms.

Aeroperu 603 asked the air traffic control tower at Lima to provide altitude readings from their ground radar, which had recently been returned to service after a major service. The tower responded by providing altitude data from its screens, which the air traffic controller believed were generated from the radar systems, but which were *actually* data provided by the aircraft's own communications data link with the ground; i.e., the tower was simply reading back to Aeroperu 603 its own faulty altitude data.

Some efforts were made to revert to autopilot, but these were unsuccessful and the pilots reverted to manual control. The pilots struggled with knowing which, if any, instruments were credible – at 0052:52 Captain Schreiber said "(Expletive) Basic instruments! Let's go to basic instruments!"

Low-speed stall warnings or overspeed alarms were received several times (0057:12, 0058:25, 0059:08, 0059:27, 0059:35, 0059:41 and 0059:46). The pilots discussed again which airspeed indications they could believe. At 0059:11 Captain Schreiber said "(Expletive) I have speed brakes, everything has gone, all instruments went to (expletive), everything has gone, all of them." Between 0100:19 and 0100:27, there was an exchange between the pilots about whether or not they were stalling. One said "We're not stalling. It's fictitious, it's fictitious."

Lima was meanwhile trying to prepare another plane so that it could fly alongside to guide them back. At 0102:41, Lima advised that the plane would takeoff in about 15 minutes to give help.

Between 0102 and 0104, "Low terrain" alarms, wind-shear alarms, and ground proximity warning alarms all sounded. At 0105:52, they were 50 miles from Lima, heading west. Lima Air Traffic Control said they were at 10,000 feet, but that was based on the faulty data from the aircraft's communication link. They were actually below 4000 feet.

At 0107 h they were at 4000 feet (although they believed they were much higher) and they held this altitude for one minute, before beginning a slow descent. At 0109:36, Lima Air Traffic Control said "Altitude is 9700 and speed is 240 knots, 51 miles from Lima." Again, their altitude was actually much lower. They continued descending but their actual height was now below 1000 feet.

At 0110:17, "low terrain" audible alarms started and sounded twenty-two times for the remainder of the flight, but Lima Air Traffic Control again advised at 0110:18 that their altitude was 9700 feet.

At 0110:57, there was a sound of impact as the plane touched the sea. First Officer Fernandez was able to shout "We are impacting water!" before the fatal second impact at 0111:16. The flight data recorder showed the plane had been descending at a ten degree angle at the time of first impact, when the left wing and engine touched the water. It then climbed to 200 feet before inverting and crashing into the sea.

The flight had lasted 31 min. All 70 people on board were killed immediately and the plane sank into deep water. The crash, in air accident terminology, was categorized as "Controlled Flight into Terrain (CFIT)," since the plane remained more-or-less in the control of the pilots until impact.

Throughout the flight, Schreiber and Fernandez had to cope with multiple, repetitive alarms, many of them spurious, while trying to cope with a full-scale emergency. Both were hopelessly overloaded with information; they were trying to separate genuine information from false. As the official report put it, "The crew were over-saturated with erroneous information."

Their confusion was compounded by Lima Air Traffic Control sending altitude information that the pilots believed was being sourced completely independently using the air traffic control radar, when it was actually just data from the plane's own malfunctioning systems sent to Lima on the aircraft's communications data link. The pilots probably thought the *only* information they could rely on was the height and speed information they were receiving over the radio – yet this was just the plane's own false readings, being recycled to them from Air Traffic Control.

There actually was one reliable instrument – the radio altimeter – but Schreiber and Fernandez were unable to recognize this in the confusion. The radio altimeter had provided the "low terrain" alarms, which sounded repeatedly during the last minute of the flight. (One important issue is not clear from the report: Did the pilots' training mean that they should have known which individual instruments – barometric pressure or radio altimeter – were responsible for each alarm? From the confusion, the answer is probably "no".)

It would not be unreasonable for any pilot in their situation to adopt a working hypothesis that there had been a complete failure of the aircraft's computer-based flight instrumentation systems. Faced with multiple-instrument failure and numerous apparently spurious alarms, no other explanation would have seemed possible on first diagnosis. (Indeed, some initial news reports of the accident stated that "the plane's whole system completely failed.")

What had actually happened was, however, far more prosaic. Debris recovered from the seabed showed that the Pitot tubes (used for airspeed indication) and also the static pressure ports (used for barometric altitude measurement) had been covered by masking tape. This tape was used when the aircraft was polished. Quality control checks should have taken place to confirm the tape was removed – a (unnamed) duty supervisor and line chief were responsible. One of the pilots should also have carried out visual checks as part of pre-flight checks.

The crew were unaware of the meaning of the *mach speed ratio* and *rudder trim* alarms, because they had never seen the National Transport Safety Board (NTSB) Safety Recommendations from the Birgenair 301 accident. The report into the Aeroperu accident noted pithily that the NTSB Safety Recommendations "were not distributed with the necessary urgency."

The recycling of bad altitude information by Lima Air Traffic Control was a further layer of confusion in this accident. Improved training for Lima air traffic controllers was recommended.

Above all, the dreadful story of this accident shows the importance of not overwhelming the pilot-operators with large numbers of alarms, many of them repetitive and/or meaningless, because this distracts them from trying to analyze the problem, and destroys their situation awareness.

A NOTE ON HIGH-ALTITUDE UPSETS AND ANGLE OF ATTACK

Before discussing the last of the three related aircraft crashes, there is a need for a brief aside on "high-altitude upsets." An "upset" is aviation jargon for loss of control, usually through stalling. At high altitude, the "flight envelope" – the scope for the aircraft to change velocity or increase altitude – can be very restricted. This is because

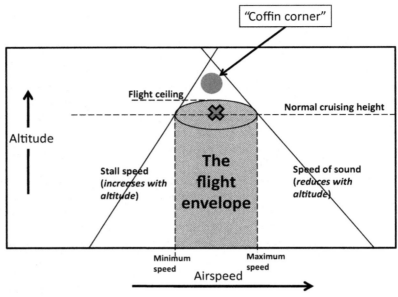

FIGURE 4.2

A schematic diagram illustrating "coffin corner": If the pilot accelerates from typical high-altitude cruise conditions, he will feel sonic buffeting. If he decelerates, he may approach stall conditions. If he tries to climb, the increased angle of attack may cause stall buffeting.

the thin air at altitude has two effects; first, the speed of sound becomes lower at higher altitude and, second, the aircraft's stalling speed is greater in the thin air.

Hence, if an aircraft is flying straight and level at high subsonic speed at high altitude, and the pilot tries to accelerate, he may get close enough to the speed of sound to cause buffeting (the "sound barrier"). Also, if he tries to slow down, the aircraft may approach its stall speed – at which point the pilot will also feel buffeting due to stall effects. "Buffet" feels like vertical vibration, which can reach 0.2 g.

Finally, if the pilot tries to climb upwards from a cruise at high subsonic speed and high altitude, the increased angle of attack in the thin air may also induce buffeting prior to stalling.

The situation described above is known to test pilots as "coffin corner" (Fig. 4.2).

Mishaps due to "coffin corner" have largely been confined to experimental aircraft under test conditions, although some civil aviation accidents did occur in early jet travel, some fatal. Notable incidents with successful recovery include a high-altitude stall of a Pan American Boeing 707 while cruising over the Atlantic at 35,000 feet in February 1959. Happily, the pilot was able to recover control, but by that time the aircraft was at 6000 feet.

Another example where the pilot managed to achieve a successful recovery was in July 1963, when a United Airlines Boeing 720 stalled while encountering

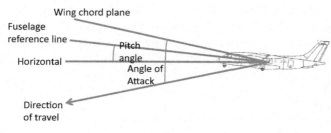

FIGURE 4.3

This diagram illustrates the difference between angle of attack and pitch angle.

turbulence during climbing at 37,000 feet; in this case the pilot recovered control at 14,000 feet. The latter incident was a trigger for change – margins were increased on all jet aircraft so that pilots would have more opportunity to avoid high-altitude stall.

A more recent incident happened in February 1985 when a China Airlines Boeing 747SP lost control after a single-engine "flame-out" when cruising at 41,000 feet over the Pacific Ocean. The pilot recovered control at 11,000 feet, although the plane exceeded its maximum operating speed twice during the dive. It suffered structural damage and two occupants received serious injuries.

Also in the following example, the difference between "angle of attack" and "pitch angle" is significant. This difference is best understood with reference to the Fig. 4.3.

Angle of attack (AOA) is an important parameter for stall avoidance. It is the angle between the wing's chord plane (an imaginary line drawn between the leading edge and the trailing edge of the wing) and the plane's direction of travel. Angle of attack is important for determining stall speed. *Pitch angle*, however, is the angle between the fuselage center line and horizontal. When flying in darkness on instruments, the key difference between AOA and pitch angle is that AOA is not a parameter that a pilot can "feel"; he is dependent on his instruments. However, pitch angle is a parameter that pilots may have some awareness of, since it will affect how they feel in their seats, but it is not directly important for stall avoidance.

AIR FRANCE 447, JUNE 1, 2009

This accident has been the subject of extremely detailed analysis and reporting by the French authorities. It has also been the subject of much news reporting and television documentaries, some of which have been a little hysterical, perhaps with some justification. It is truly one of the most bizarre accidents, where one co-pilot behaved in a very strange manner indeed – apparently unaware what he was doing, almost like he was frozen in complete panic – but the other pilots could not see what he was doing so they could not interpret the instrumentation properly.

Air France 447 (AF447) was a scheduled overnight flight from Rio de Janeiro to Paris on June 1, 2009. (It actually left Rio on the late evening of May 31.) There were 216 passengers and 12 crew members on board. In mid-flight, while over the South Atlantic, the aircraft simply "vanished". No Mayday calls were received and, because the plane was in mid-Atlantic, there were no radar records available. Initially, terrorist action was suspected, especially after some floating debris and bodies were discovered.

The aircraft was an Airbus A330-200, registration number F-GZCP, with a fully digital cockpit and full "fly-by-wire" controls.

The investigation of this accident was an enormous undertaking, and involved the French and Brazilian air forces and navies. One French nuclear submarine was involved as were various remotely operated vehicles (ROVs). Wreckage of the plane was eventually found 3980m underwater in the Atlantic Ocean at about 3 degrees north, 30 degrees west. The debris was spread over an area of seabed 600m by 200m.

Eventually 154 bodies were recovered, either on the surface or in the wreckage deep underwater; the remaining 74 were never found.

The flight data recorders and cockpit voice recorders were not recovered until May 12, 2011, almost two years after the accident.

The basic facts are as set out in the opening paragraphs of the final French report.

"At around 0202 h, the Captain left the cockpit. At around 0208, the crew made a course change of 12 degrees to the left, probably to avoid returns detected by the weather radar. At 0210:05, likely following the obstruction of the Pitot probes by ice crystals, the speed indications were incorrect and some automatic systems disconnected… (The co-pilots) were rejoined 1 min 30 s later by the Captain, while the airplane was in a stall situation that lasted until the impact with the sea at 0214:28."

In a little over 4 min, the plane had fallen from its cruising height of 35,000 feet into the sea. There were no electrical or mechanical malfunctions. A perfectly healthy plane had fallen out of the sky.

In September 2007, Airbus had made recommendations to change the model of Pitot tubes installed in Airbus A320, A330 and A340 aircraft, due to a problem with water ingress. This was not an Airworthiness Directive, so Air France decided to replace the Pitot tubes on A330 planes only when failure occurred. From May 2008, Air France had some incidents involving loss of airspeed data during flights, apparently due to temporary icing of the Pitot tubes. Air France began to accelerate the Pitot tube replacement program on A330 aircraft; this was actually completed by June 17, 2009, but F-GZCP had not been upgraded at the time of the crash on June 1, 2009.

The recovered flight data recorders and cockpit voice recordings enabled the detail of what had happened to be worked out.

Just after midnight, the aircraft was in cruise at 35,000 feet, with autopilot and auto-thrust engaged, with Captain Marc Dubois, aged 58, flying the plane. Dubois had first received a commercial pilot's license in 1977 and had 11,000 flying hours.

His co-pilot in the right-hand seat was 32-year-old Pierre-Cedric Bonin. Bonin had 2936 flying hours and had received his professional pilot's license in 2001.

The relief pilot was 37-year-old David Robert, with 6547 flying hours. He received his professional pilot's license in 1993.

At 0136, the plane was approaching a tropical storm and entered high-level cloud. At 0151, the electrical storm caused the cockpit to be illuminated by St Elmo's fire, where luminous plasma is formed around pointed objects because of the strong electrical field. It is harmless but is often found in thunderstorms near the equator. (This incident with St Elmo's fire should be irrelevant, but there have been suggestions that it "spooked" co-pilot Bonin who had not seen it before.)

Cruise altitude was 35,000 feet, lower than normal, because the plane was heavy with fuel, and the air temperature was relatively high so the air was thinner than normal at that height.

At about 0200, the relief pilot David Robert returned to the cockpit after his break. Captain Dubois stood up and gave Robert the left hand seat. Dubois left Bonin in control, although Robert was the more experienced. At 0202 Dubois went out of the cabin to go for a sleep.

At 0205:55, Robert called one of the cabin crew to warn that the plane would be entering turbulent air shortly. She agreed to forewarn the other flight attendants that there would shortly be an announcement to return to seats and fit safety belts.

Robert began to examine the weather radar for the storm ahead. He realized they were heading straight toward an area of strong storm activity. At 0208:07, Robert said to Bonin "You can possibly pull a bit to the left."

There was a noise interpreted as ice crystals hitting the plane, and shortly after there was an alarm indicating that the autopilot has disconnected. This was caused by the Pitot tubes icing over. Temporarily, the pilots had lost all airspeed indications. This should not have been a problem – other pilots have flown simulations where they have been able to continue quite safely. However, neither Bonin nor Robert had received training in dealing with loss of indicated airspeed at high altitude, or in flying the plane in such conditions.

- Once the autopilot was disconnected, the flight control computer changed from "normal law"; to "alternate law", as programed to do so, in recognition that, because there were some problematic instruments, the pilots should receive more discretion in their actions. "Alternate law" allowed the pilots much greater scope in their actions than would normally be the case.

Until 0210, everything was basically OK. At 0210:06, Bonin said "I have the controls," and Robert replied "OK." At this point, for reasons that are not clear – and never will be – Bonin put the plane into a steep climb. The flight control computer issued a chime warning they were leaving the programed altitude, and the stall warning sounded, "Stall," in English. This alarm thereafter sounded 75 times before the crash.

Throughout the remainder of the flight, neither of the pilots made any reference to the repeated stall alarms.

- A pilot's training is always that, in reaction to an approach to stall, the controls should be pushed forward. Bonin kept pulling his control back. A key feature of

FIGURE 4.4

Airbus A330 cockpit showing the positions of the short control sticks for the two pilots, on the extreme right and extreme left. Small wrist movements by either pilot are enough to make changes to the flight path, and these movements may not be obvious to the other pilot. (Photo copyright Carlos Enamorado. Used by permission.)

the Airbus controls is that the pilots control the plane using small side sticks at their sides, almost like games controllers. The right-hand pilot's control stick is on the right-hand side, and the left hand pilot's control stick is on the left hand side. The two pilots' sticks move independently, so the pilot on the left hand seat cannot feel what the pilot on the right-hand seat is doing. Furthermore, it may not be clear to the non-flying pilot what inputs the flying pilot is making because small wrist movements are enough to cause a control input. Robert will, presumably, have been looking at the instrumentation and not at Bonin's right hand (Fig. 4.4).

- One other crucial point at this juncture was that neither out-ranked the other in seniority. When the Captain was present, it was clear who was calling the shots – but until Captain Dubois returned to the cockpit, Bonin and Robert were effectively equals.

At 0210:07 Robert said "What's that?" Bonin replied, "There's no good speed indication." The plane was now climbing at 7000 feet per minute, and the speed had dropped dramatically to 110 knots. By 0210:25, the altitude had increased to over 36,000 feet.

At 0210:27, Robert said twice "Pay attention to your speed". Bonin said, "OK, OK, I am descending" but he continued to climb. At 0210:31, Robert said, "Descend – it says we are going up – descend." Bonin replied "OK", but Robert said again "Descend". Bonin said "Here we go, we're descending" but the plane continued to climb.

- The official French report is strangely coy about being openly or excessively critical of Bonin. The report refers to "inappropriate pilot inputs." The report does not actually name any of the three pilots.

At 0210:41, Bonin said (bizarrely) "Yeah, we're in a climb".

At 0210:49, Robert was sufficiently worried to use a pushbutton to call Captain Dubois back to the cockpit.

At 0210:56, the engine thrust levers were set to TOGA. "TOGA" means "Take Off, Go Around." Bonin had selected high thrust and raised the nose as if he were climbing away from an aborted landing.

By 0211:03, the ice had melted and the Pitot tubes had unblocked themselves. All the instruments were again functioning normally. From this point onwards there was nothing – *nothing at all* – wrong with the plane, except the behavior of Bonin. Bonin announced, again bizarrely, "I am in TOGA, no?" Robert was clearly extremely anxious: "Damn, where is the captain?"

At this point, shortly after 0211, the aircraft was properly stalled. With the engines at full thrust, the pitch angle reached a maximum of 17.9 degrees. The aircraft reached its maximum altitude at 0211:10 of 37,924 feet. After this time, the plane descended continuously until the crash. All this time, Bonin kept pulling back his control stick. If he had released his stick the plane would have assumed a nose-down attitude and the plane would have recovered from the stall.

At 0211:21, Robert was becoming desperate. Presumably unaware that Bonin was holding his stick back, he shouted "What the hell is happening? I don't understand what is happening."

At 0211:32 Bonin said "Damn, I have lost control of the plane, I have lost control of the plane!" Robert replied "Left seat taking control!" However, Robert too seems to have missed the point that the plane had stalled (despite the "Stall" alarm which has been sounding continuously for the last 90 s). Robert now pulled back on the stick also – but the plane was stalled, the nose was pitched upwards, the plane was falling at about 6000 feet per minute, and the Angle of Attack was approaching 30 degrees. There were continuous alarms going off in the cockpit: stall warning voice alarms, stall warning chime alarms, chimes warning about altitude, a chirp alarm called a "cricket." Shortly after, Bonin resumed control.

At 0211:43, Captain Dubois entered the cockpit. "What the hell are you doing?" he asked, not unreasonably. Both Bonin and Robert said, more-or-less simultaneously, "We've lost control of the plane!" Rate of descent was now 10,000 feet per minute, and the Angle of Attack reached 41 degrees. The plane remained more-or-less in this situation for the whole descent.

Dubois did not try to take one of the pilot's seats – he left Bonin and Robert in control. With the stall alarms still calling out every few seconds, no one discussed the possibility that the aircraft might have stalled. Bonin was still holding his stick back, which Captain Dubois, like Robert, did not notice.

For the next minute and a half, the three pilots were unable to work out what was happening, and whether in fact the plane was stalled, despite all the instrumentation telling a consistent story. They even had some exchanges about whether they were descending or climbing. Meanwhile the stall alarm was repeating every few seconds. The one piece of crucial information that Dubois and Robert failed to notice was that, throughout, Bonin was holding his stick back.

- *A reminder*: it was the middle of the night above the mid-Atlantic, so there were no visual points of reference. Also, the plane was falling at more-or-less constant speed so the pilots will not at this point have been feeling any gross vertical acceleration. Their only input information was as follows:

  1. What they could deduce from their instruments;
  2. They should have been able to feel that the nose was pitched up;
  3. They should also have been aware of buffeting – vertical vibrations – caused by the stalled airflow over the wings;
  4. They should also have felt pressure changes in their ears as the altitude reduced.

At 10,000 feet Robert tried to take control again. He pushed his stick forward but, with Bonin holding his stick back, the control system averaged the two inputs so the nose remained high.

At 0213:40 (when their altitude was about 9000 feet), Bonin suddenly realized what he had been doing. "But I've been at maximum nose-up for a while!" At last, Robert put the nose down and the plane began to regain speed, but it was too late and the plane was now too low to manage a recovery.

At 0214:23, Robert said "Damn we're going to crash, this can't be true!" The aircraft hit the sea at 0214:28. Their vertical speed was about 10,000 feet per minute (about 190 km/h), their horizontal speed was about 100 km/h, the plane was pitched upwards about 15 degrees, and the Angle of Attack was about 40 degrees. The engines were at full throttle.

There had been no Mayday call. There had been no communications with the passengers, most of whom will have been asleep, at least at the onset of the problems. Some passengers will have been woken up by the pitching-up, the buffeting and the changes in air pressure, just in time to wonder what on earth was happening to their plane.

An entirely healthy aircraft, in straight and level high-altitude cruise, had fallen out of the sky and crashed into the sea because one pilot held his control stick back and the other pilots could not work out what was happening.

The Final Report of the Bureau d'Enquetes et d'Analyses concludes, "the airplane went into a sustained stall, signaled by the stall warning and strong buffet. Despite these persistent symptoms, the crew never understood that they were stalling and consequently never applied a recovery maneuvre. The combination of the ergonomics of the warning design, the conditions in which airline pilots are trained and exposed to stalls during their professional training and the process of recurrent training does not generate the expected behavior in any acceptable reliable way."

The immediate causes of this accident were as follows.

1. Temporary freezing of the Pitot tubes caused confusion because of loss of all speed indications.
2. Bonin subsequently (and irrationally) pulled back his control stick and intermittently maintained it in that position for several minutes. This caused the aircraft to climb into a dangerously high, nose-up position and thereby stall. Bonin maintained his stick-back position even after the plane was stalling and losing altitude. By the time he had realized his error, it was too late to avoid the crash.
3. The design of the side-sticks meant that what Bonin was doing was not readily apparent, neither to Robert, nor to Dubois when he returned to the cockpit. Furthermore, the design of the control system meant that Robert could not countermand what Bonin was doing.
4. Neither Bonin nor Robert had received training in high-altitude stall recovery.

A contributory factor appears to have been the sophistication of the computerized flight controls. After the two co-pilots had got into difficulties, they seemed to be blinded by the array of information available, and the sophistication of the different layers of automation. It was as if they were confused whether the loss of control was genuine, or whether the digital instrumentation systems were faulty and were giving them bad information. It was as if they were thinking, "Is this real or have the computer systems gone berserk?" That confusion, combined with Bonin's irrational control stick inputs, caused fatal delays in their reactions.

Captain Solly Sullenberger, the now-retired airline pilot who famously and successfully ditched an Airbus A320 into the Hudson River, New York, on January 15, 2009 after both engines had been wrecked by bird strikes, was interviewed for the magazine *Aviation Week* (December 20, 2011) regarding the AF447 accident. Sullenberger said there was a need for pilots to receive information about Angle of Attack. Pilots have to infer Angle of Attack indirectly by referencing speed, which makes stall recognition and recovery more difficult. The capability exists to display Angle of Attack in the cockpits of most airliners, and it is one of the most critical parameters, yet it is not displayed.

Sullenberger was critical of training, saying that airline pilots practice approaches to stalls, but never actually stall the aircraft. The maneuvres are done at low altitude where they are taught to power out of the maneuvre with minimum altitude loss. In some aircraft, pilots are taught to pull back on the stick, use maximum thrust and let the Angle of Attack protection adjust nose attitude for optimum wing performance. While this may work for *approach* to stall at low altitude (when the stick-shaker is warning that stall is imminent), it will not provide effective recovery after a high-altitude stall.

Sullenberger said that pilots never get the chance to practice recovery from a high-altitude upset, noting that, at high altitude, you cannot power out of a stall without first losing altitude. Hence, Sullenberger was suggesting that Robin was, at least initially, following his training for how to respond to an approach to stall at low altitude – although AF447 was at high altitude.

Sullenberger also was worried about situation awareness in highly automated digital cockpits. "I think the industry should ask questions about situational awareness and non-moving auto-throttles. You lose that peripheral sense of where the thrust [command] is, especially in a big airplane where there is very little engine noise in the cockpit. In some fly-by-wire airplanes, the cockpit flight controls do not move. That's also part of the peripheral perception that pilots have learned to pick up on. But in some airplanes, that is missing and there is no control feel feedback."

SYNTHESIS

The three accidents described above have some aspects in common. The fault sequences of all three were initiated by Pitot tube blockage and loss of indicated airspeed. More importantly, however, there were fundamental flaws in the design of the HMIs, which prevented the pilots from making appropriate responses. In all three cases, the pilots lost situation awareness.

In the first case (Birgenair 301, Boeing 757) the Pitot tube blockage was probably caused by wasps' nests. The alarms generated by the Electronic Flight Information System were unintelligible to the pilots (with simultaneous low- and high-speed alarms) in the less than 3 minutes they had to analyze what was happening. The plane stalled and crashed.

In the second accident (Aeroperu 603, Boeing 757) the Pitot tubes had been blocked by masking tape. The accident happened before the report on Birgenair 301 had been published. The same unintelligible alarms were generated. In addition, the static pressure sensors had been taped over, so the pilots were receiving false barometric altitude indications. The pilots were confused and asked Air Traffic Control to supply information, including readouts of their altitude, from the Lima radar. Unfortunately, the Air Traffic Controller in Lima did not realize that the height data on his radar screen were actually the same faulty data, which the plane was transmitting to Air Traffic Control via a data downlink. The pilots therefore thought they were still several thousand feet in the air when their plane hit the sea. In this case, therefore, the design of the HMI for the Air Traffic Controller was also poorly designed.

In the third accident (Air France 447, Airbus A330), temporary blockage of the Pitot tubes by icing was the starting point but, thereafter, the irrational inputs on his control stick from the right-hand seat co-pilot made the situation fatally worse. These control stick inputs were perhaps a reaction to training in how to recover from low-altitude *approach* to stall – whereas the plane was actually in a high-altitude stall, for which situation training had not been given. Also, on the Airbus A330 control stick, small hand movements can make large control input signals, so it was not evident to the non-flying pilot what the flying pilot was doing.

There are many general lessons for design engineers from these (and other) accidents regarding the layout of the control and instrumentation systems, i.e., the design of HMI:

- In fault conditions, the human–machine interface needs to provide clear, unambiguous information, and the pilot-operators must not be bombarded with too many alarms. The objective is to ensure that the pilot-operators can maintain situation awareness, i.e., they need to be able to retain a good mental model of what state the machine-system is in. (Here "machine-system" means aircraft, nuclear power station, etc.)
- Pilot-operators are usually not engineers; their default position *must* be to believe the data presented to them.
- The Air France 447 accident poses an even more fundamental question for design engineers: Should design engineers have to consider the possibility of pilot-operators making completely irrational control inputs, or is it OK to assume that pilots are always rational? If the design engineer cannot assume a rational operator, what can he assume?

REFERENCES

[1] Flight Safety Foundation, Erroneous airspeed indications cited in Boeing 757 control loss, Accident Prevention, vol. 56 No. 10, October 1999.

[2] Accident of the Boeing 757-200 aircraft operated by Empresa de Transporte Aereo del Peru SA, October 2, 1996, Accident Investigation Board, Ministry of Transport, Communications, Housing and Construction, Directorate General of Air Transport, Lima, December 1996.

[3] Bureau d'Enquetes et d'Analyses, Final report on the accident on June 1, 2009 to the Airbus A330-203 registered F-GZCP operated by Air France flight AF447 Rio de Janeiro-Paris, July 27, 2012. Available from: www.bea.aero/en/enquetes/flight.af.447/rapport.final.en.php.

Some Case Studies of Software and Microprocessor Failures

INTRODUCTION

In Chapter 2, the "software problem" was introduced. To recap, software presents a different set of problems from older analog equipment because (Fig. 2.2) of particular challenges in specifying, verifying and analyzing the failure modes of long and complicated software. There are then further problems with testing software systems exhaustively, determining the failure modes of aging microprocessors, assessing a numerical reliability for software systems, and assuring the quality of bought-in software-based subsystems. Finally, we have the problems of cyber security (Chapter 3) and software diversity.

This chapter will focus on specification, verification, failure mode analysis, and aging microprocessors, by means of a number of case studies.

An issue here is that there is already an extensive body of examples of incidents and accidents related to software failures. We often used to talk about "hypothetical" software faults, but this is no longer really appropriate, since most types of faults have already happened.

QANTAS FLIGHT 72: A SERIOUS INCIDENT INVOLVING SMART DEVICES WITH MIXED CONTROL AND PROTECTION

Fatal accidents or serious incidents in civil aviation or in industry generally lead to some form of independent enquiry and the subsequent report is openly published. Many such accidents and incidents have now occurred where software could be held to blame for the incident.

An interesting example of a non-fatal accident, which nevertheless led to a large number of injuries, and which must also have scared witless everyone else, was the Qantas Airbus A330 "in-flight upset" of October 7, 2008 [1]. The published report is a fascinating read. The aircraft was carrying 303 passengers and 12 crew members on scheduled flight Qantas 72 from Singapore to Perth, Australia in a steady cruise at 37,000 feet, flying on auto-pilot. The events as experienced by the passengers and crew can be summarized quite briefly as follows:

FIGURE 5.1

This photo shows some of the damage to the cabin roof caused by passengers and crew being thrown upwards by the sudden pitch down of Qantas flight 72. (Photo courtesy of Australian Transport Safety Bureau, Report AO-2008-070.)

At 0440:26 UTC (local time 1240:26) one of the Air Data Inertial Reference Units (ADIRUs) started producing multiple intermittent spike signals, and the crew received a number of warning messages many of which were spurious and unhelpful.

At 0441:27 UTC (only one minute later and before any meaningful analysis or interpretation of the alarms could possibly be carried out) the aircraft suddenly and violently pitched down for less than 2 s. The acceleration experienced in the cabin exceeded -1 g, or -10 m per second, and passengers and crew who were not strapped in at the time were physically thrown against the roof of the cabin, causing injuries, in some cases serious ones.

At 0445:08 UTC a second less severe pitch down occurred. The captain declared Mayday and the flight was diverted to Learmonth, Western Australia where it landed successfully at 0532 UTC.

At least 110 of the 303 passengers and nine of the 12 crew members were injured; 12 of the occupants were seriously injured and another 39 received hospital medical treatment. Damage to the cabin roof, from people impacting it due to the high negative "g" force, was significant (see Fig. 5.1).

The circumstances leading to the above events are quite complicated and need some detailed introduction, and there are a lot of acronyms for the different parts of the control systems. Like all modern Boeing and Airbus civil airliners, the Airbus A330 has a digital electronic flight control system, a "fly-by-wire" system which monitors and controls flight parameters continuously and enables the pilot and co-pilot to take a "supervisory" role during normal flight. The electronic flight control system uses three flight control primary computers (FCPCs), one of which is selected to be "master."

Among the many parameters that are monitored is the angle of attack (AOA), the slight nose-up attitude which is maintained in normal flight to generate lift but which can lead to a stall if AOA is too great. AOA is a measurement of the angle of the wing (using a nominal reference line known as the "chord line") relative to the airflow. (This parameter was described in more detail in Fig. 4.2 of Chapter 4.) Angle of attack is a therefore a critical safety parameter for the electronic flight control system, and the flight control primary computers use three independent AOA signals to check their consistency, signals AOA1, AOA2, and AOA3. The AOA signals are created by air data inertial reference units (ADIRUs). The AOA value is then fed into the flight control system and used, in particular, to drive signals to the elevators in the tailplane which control aircraft pitch.

"Inertial reference" means knowing your orientation in space, without reference to external points, by measurement of movements in three dimensions and integrating those movements to calculate orientation. Modern Inertial Reference Units, such as ADIRUs, use *ring laser gyroscopes*, together with accelerometers and GPS to provide raw data. Ring lasers are better than older mechanical gyroscopes because there are no moving parts and they are rugged, lightweight and frictionless. A ring laser gyroscope consists of a ring laser having two counter-propagating modes over the same path in order to detect rotation. An internal standing wave pattern is set up in response to angular rotation, known as the Sagnac effect. Interference between the counter-propagating beams, observed externally, causes shifts in that standing wave pattern, and thus rotation can be inferred.

In so far as the aircraft manufacturer was concerned, the ADIRU is a "smart device", bought in from another supplier. This introduces all the problems of such devices as were discussed in Section "I&C Architecture" of Chapter 2. A simplified version of the A330 control and protection systems architecture is illustrated in Fig. 5.2.

Under normal circumstances in the Airbus A330, the flight control system acted under what is called "normal law", which means the flight control computers act to prevent the aircraft exceeding the pre-defined safe flight envelope. The flight control system had a variety of automatic flight envelope protection functions including high angle of attack protection. Under certain types or combinations of failures within the flight control system or its components, the control law changes to a different configuration law, either "alternate law" or "direct law". Under alternate law, the different types of protection were either not provided or were provided using alternate logic. Under direct law, no protections were provided and control surface deflection was proportional to sidestick and rudder movement.

In this case, however, the aircraft remained under normal law, at least until after the upset had occurred. Under normal law, the master FPCP used the average of AOA 1 and AOA 2 to determine the angle of attack. If either AOA1 or AOA2 deviated (or generated a signal spike) the master FPCP used a memorized value for up to 1.2 seconds; i.e., it remembered the last value before the signal deviation occurred. This was

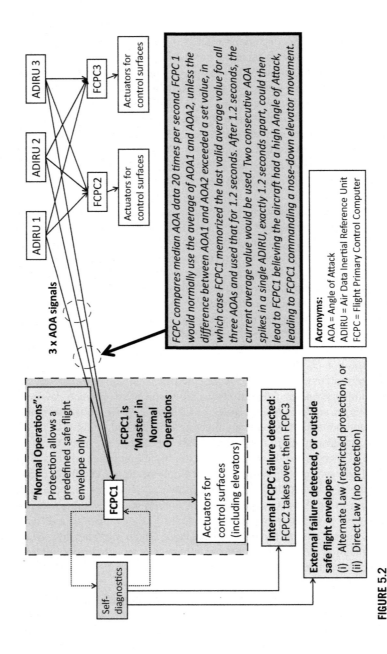

FIGURE 5.2

This diagram presents a much-simplified version of the A330's Electronic Flight Control System, showing its control and protection systems architecture. Because of inadequate design analysis, a particular "spike" failure mode in one single air data inertial reference unit (ADIRU) was able to lead to sudden pitch down of the aircraft when in level high-altitude cruise.

intended to prevent the control system being too sensitive and over-reacting – which of course is exactly what happened in this incident.

The incident investigation showed that the algorithm in the FPCP software had a serious limitation, which meant that in a very specific situation multiple spikes on the AOA signal from only one of the ADIRUs could generate a nose-down elevator command. A timeline for the event is shown in Fig. 5.3.

Quite simply, the FCPC algorithm could not correctly manage a scenario where there were multiple spikes in either AOA1 or AOA2 that were exactly 1.2 seconds apart (i.e., exactly the length of time that the FCPC would memorize old data in the event of a signal spike). The Australian Transport Safety Board report notes that "it

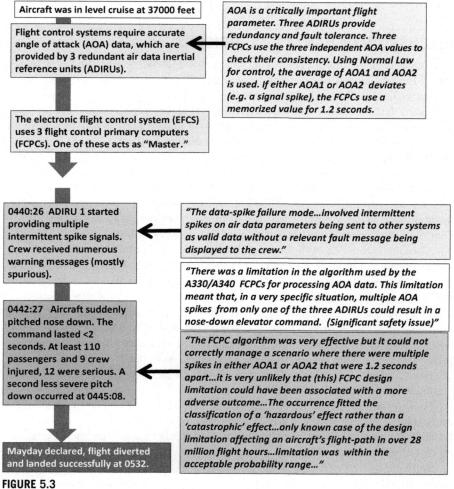

FIGURE 5.3

Timeline of the Qantas 72 incident.

is very unlikely that (this) FCPC design limitation could have been associated with a more adverse outcome," and that "(this was the) only known case of the design limitation affecting an aircraft's flight-path in over 28 million flight hours."

The report also notes that "…the development of the A330/A340 flight control system during 1991 and 1992 had many elements to minimize the risk of design error (and) none of these activities identified the design limitation in the FCPCs AOA algorithm… Overall, the design verification and validation processes used by the aircraft manufacturer did not fully consider the potential effects of frequent spikes in data from the ADIRU."

There had been two other known occurrences of the data-spike failure mode (presumably without inducing auto-pilot anomalies), on September 12, 2006 and December 27, 2008.

DISCUSSION OF QANTAS 72 INCIDENT

The main point of this incident – in which, fortunately, no one died – is that it shows how difficult it has become to analyze and review complex software-based control systems, to identify all potential dangerous failure modes, and to ensure that subsystem failures modes like the ADIRU signal spikes are identified and properly addressed. The identification of the ADIRU-spike failure mode *before* the Flight 72 incident would have required some very clever thinking indeed.

The Qantas 72 in-flight upset is therefore a very good example of (i) the difficulty of analyzing complex software-based systems to determine their possible failure modes, (ii) the difficulty of specifying and verifying complex software-based systems, and (iii) the difficulty of understanding bought-in software systems ("smart devices" such as the ADIRU) where the software and its design are not within the full control of the system integrator – which in this case was Airbus.

Also, the Qantas 72 in-flight upset illustrates the difficulties that are introduced when control and protection systems are mixed. Although the designers had tried to keep a degree of separation within the different systems, nevertheless under "normal law" (average of two ADIRU signals feeding into the high angle of attack protection logic) high angle of attack protection was inhibited by the 1.2 seconds "rule" which ignored signal spikes from a fleeting spike on the master ADIRU signal. However, when two signal spikes on the master ADIRU occurred *exactly* 1.2 seconds apart, the protection then saw an apparently genuine high angle of attack – and this single failure caused spurious operation of the protection. The elevators moved sharply downwards and put the plane into a sudden dive.

Poor or limited separation of control and protection is of course to some extent inevitable in aircraft systems, but in hazardous process plant it should always be possible to separate control from protection to minimize the possibility of common-mode failure between control and protection.

Software-based systems have grown and grown in recent years, and systems with many millions of lines of application software are now common. With software of this size and complexity, it becomes very difficult indeed for any one person to have a complete overview and profound understanding of the system. In that case, it is necessary to break the system down into modules, with precise specifications for each module and the interfaces with other modules. The role of engineering management then becomes, first, to ensure that all the interfaces are properly identified and verified, then to ensure that all the safety and other functional requirements are properly identified and verified, and finally to ensure the interface requirements are identified and verified. Ensuring that "nothing falls between the cracks" still remains a huge task which requires great care and attention.

Finally, it is notable that an earlier case of the data-spike failure mode had occurred in 2006. One is forced to conclude that the events of Qantas 72 indicate a failure of adequate incident investigation in 2006. (We have already seen the consequences of inadequate or tardy accident investigation in Chapter 4. Accident and incident investigation will be discussed further in Chapter 10.)

ULJIN NUCLEAR POWER PLANT, SOUTH KOREA, 1999

In 1999 an incident at Uljin Nuclear Power Station Unit 3 in Korea corrupted data on the performance net of the plant control computer [2]. The incident was caused by the failure of an application-specific integrated circuit (ASIC) chip on a rehostable module, which is part of a network interface module. Several non-operational pumps started without any demand, some closed valves opened and other open valves closed, and some circuit breakers switched on or off. There was also some relay chattering. Due to the response of the operators and because of diverse systems, the incident was mitigated without adverse consequences.

A review of the systems found that a common-cause software error was the likely cause. It was found that there was no provision to protect against foreign writes in the global memories within the communication network. It appears that uncontrolled software modifications had taken place.

Extensive modifications, including hard-wired backups, were subsequently carried out.

This incident illustrates that it is not sufficient to have management controls and procedures in place to prevent uncontrolled software changes. There must also be physical or electronic barriers to prevent operators making uncontrolled changes. In this case, one would have expected that the computer that was configured to help carry out software modifications (the "engineering work station") should have been kept in secure conditions unless authorized.

KASHIWAZAKI-KARIWA NUCLEAR POWER PLANT, JAPAN, 2001

In 2001, a failure of control rod transponder circuit boards at Kashiwazaki-Kariwa Nuclear Power Station Unit 5 (Japan) rendered the control rods inoperable [2]. Following detection of the defective cards, an analysis revealed that the failure mechanism was aluminum wire breakage in the integrated circuits caused by electro migration (the transport of metal atoms induced by high electric current). Aluminum grain size was too small which increased susceptibility to electro migration. The affected integrated circuits had been manufactured between 1985 and 1990.

Failure analysis methods and manufacturing quality control and testing have been improved. This effect is potentially more significant in more modern integrated circuits where the level of miniaturization is much greater.

This is an example of aging phenomena in integrated circuits. This issue will be further discussed in Chapter 6.

NORTH SEA PIPE HANDLING SYSTEM FATAL ACCIDENT, 2008

Pipe-laying vessels come into UK waters for various North Sea oil and gas operations but may be completely outside the regulation of the UK safety regulator, the Health and Safety Executive (HSE). However, when such vessels stop moving around like a real ship and start laying pipes, HSE in principle becomes the regulator. The operators have basic safety management systems in place to ensure, e.g., that people keep out of the way of heavy objects that are being moved around. However, because it is mobile, the lifting equipment's control system cannot really be subject to any single country's safety regulation.

In 2008, a fatal accident was reported when a pipe-laying ship dropped a 20 tonne section of pipe onto operations personnel [3]. The location of the accident was not reported but is presumed to be outside UK territorial waters.

During pipe-laying operations, a system failure in the hydraulic pipe handling system caused two quadruple joints being handled at the same time in two different areas of the tower to drop suddenly. Each piece of pipe was 50 m long with a diameter of 24 inches and weighed approximately 20 tons.

Just prior to the incident the pipe-laying operation had been stopped. Operators reported a system failure and that the hydraulic power had been lost. Such an occurrence was not particularly unusual and, in line with company procedures, this was investigated immediately. A team of technicians led by the chief electrician tried without success to resolve the problems. After these attempts, a more in-depth analysis was made. It was decided, on the basis of input from the system diagnostics, to perform a memory reset. Following this the system appeared to be running correctly. This was the first time that a full memory reset was requested by the internal diagnostics of the control system during a project operational phase.

Only after all indications that everything was in order and all systems were up and running again was the instruction given to the operator to restart the hydraulic packs. As soon as the hydraulic power packs were started, a loud bang was heard along with the noise of the hydraulic systems. Two pipe sections held by the transfer system were released, one of which fell the full height of the crane tower, smashing through an access platform located below.

All the people who were injured had been on the access platform, which was destroyed. The force of impact caused some of the injured persons to fall down on to the lower deck at the base of the tower and some to be thrown overboard. Eight persons were injured, two seriously and two slightly. Four of the injured persons died as a result of their injuries.

The primary cause of the incident was found to be the sudden release of the two quadruple joints was caused by a failure in conceptual design of the control system software. The program relevant to the initializing instruction was pre-loaded in the erasable programmable read-only memory (EPROM) of the programmable logic controller (PLC) with the instruction to open all clamps when re-initialized.

In addition, the unnecessary, uncontrolled presence of working personnel on to the access platform underneath the falling pipes was clearly a bad working practice. People should never be allowed to pass under heavy suspended loads.

This accident illustrates, quite simply, bad design practices. The control system for handling extremely heavy loads, when restarted, defaulted to "jaws open", causing any loads to be dropped. This is an example of poor specification and verification.

AUTO RECALLS FOR SOFTWARE FAULTS

Automobile recalls by the manufacturers to fix software problems have now become commonplace. Three examples amongst many are given below. They have been selected because they are potentially safety-related. It is difficult to be sure how many accidents (if any) occurred before these recalls were made. Unless criminal charges are brought, motor manufacturers will try, wherever possible, to settle out of court in order to minimize publicity.

1. Jaguar recalled some 18,000 X-type cars after it discovered a major software fault, which meant drivers might not be able to turn off cruise control. The problem lay with engine management control software developed in-house by Jaguar. If the fault occurred, cruise control could only be disabled by turning off the ignition while driving, which meant a loss of some control and sometimes loss of power steering. Braking or pressing the cancel button did not work (*Computer World UK, October 24, 2011*).

2. In 2010, the Toyota Prius was recalled for software-related brake problems. Under certain conditions, there could be a slight interruption in the car's brake response, and a software update had to be installed to all of the 133,000 in model 2010 Prius and 14,500 of the similar 2010 Lexus HS250h models on the road. The brake system on hybrid-engine cars is different from conventional cars. The system combines hydraulic brakes with brake regeneration and an anti-lock function. Under light braking, the calipers do not squeeze the rotors. Instead, the resistance of the electric motors provides the deceleration. (This is how the Prius captures the moving car's energy, charges the batteries, and later electrically boosts acceleration.) Most hybrids work this way to keep fuel economy numbers high. Harder braking engages the calipers in a normal fashion, and maximum braking engages the ABS system to prevent skidding. The computer controls the various functions with inputs from several sources

like the wheel-speed sensors, battery charge meter, and brake-pedal stroke. Toyota said that under certain conditions, like on an especially bumpy or slippery road, there could be a brief momentary delay in the brakes response that could increase slightly the stopping distance (*Popular Mechanics, February 9, 2010*).

3. Honda recalled 5626 CR-Z vehicles from the 2011 model year in the United States that were equipped with manual transmissions, to update the software that controls the hybrid electric motor. Under certain circumstances, it was possible, according to the company, "...for the electric motor to rotate in the direction opposite to that selected by the transmission." Honda also recalled 2.5 million other vehicles to update the software that controlled their automatic transmissions. Without the updated software, the automatic transmission could be damaged if the transmission were quickly shifted between positions, e.g., in an attempt to dislodge a vehicle stuck in mud or snow. This could cause the engine to stall or lead to difficulty engaging the parking gear (*IEEE Spectrum September 7, 2011*).

In each of the above cases (and the many, many others that are known about) the software/microprocessor systems will doubtless have been designed and manufactured to high standards, with careful specification, requirements traceability, built-in review processes, and configuration management. Yet failures still occurred. The ability of software systems to manage complex processes is their great strength and the reason designers choose to use them, but that very complexity is also their weakness. Complex software requires great care and attention to achieve true high integrity.

REFERENCES

[1] ATSB Transport Safety Report, Aviation Occurrence Investigation AO-2008-070 Final, 2011.
[2] Descriptions of the Uljin and Kashiwazaki-Kariwa events are contained in Advanced reactor licensing: Experience with digital I&C technology in evolutionary plants, Oak Ridge National Laboratory, NUREG/CR-6842, US NRC 2004.
[3] IMCA Safety Flash 18/08 December 2008. www.imca-int.com/documents/core/sel/safetyflash/.

Managing the Safety of Aging I&C Equipment

INTRODUCTION: THE PROBLEM OF AGING I&C EQUIPMENT

A typical plant design lifetime may be 40–60 years, but I&C equipment is likely to have a much shorter operational life, so it will need replacing during the plant lifetime [3].

A textbook definition of aging I&C equipment would state that it relates to equipment which is approaching the right hand side of the reliability "bathtub curve."

The challenge of managing aging equipment is to know how close the equipment is to "wear-out."

Key issues in determining the achieved life of equipment are the quality of maintenance, and environmental issues. Furthermore, the age of equipment at which it changes from exhibiting random failure rates (i.e., the flat portion of the bathtub) to accelerating (or aging) failure rates will differ for different types of electrical and I&C equipment.

In reality, the presentation in Fig. 6.1 may exaggerate the significance of early failures, whose good pre-service testing should minimize. More importantly, however, the onset of wear-out at end-of-life is generally unpredictable for plant operators before equipment failure rates start to worsen. The original equipment manufacturers (OEMs) may make recommendations for equipment lifetimes, but their recommendations may not be verifiable, nor objective, nor realistic – after all, they want to sell more equipment.

The plant operators may therefore be faced with a dilemma of whether or not to commit to early replacement, thereby perhaps incurring unnecessary expenditure before equipment has fully completed its useful life. However, if the operators choose to defer replacement until failure rates start to rise and replacement becomes unavoidable, it then becomes a race between executing an upgrade project (i.e., going through all the stages of specification, procurement, contract placement, manufacture, testing, installation, and commissioning) versus accelerating failure rates and their impact on plant availability.

Good maintenance is key to getting optimum life from I&C equipment. With care, even quite sophisticated I&C equipment can be kept working to a remarkable age. Other equipment may need replacing after quite short timescales. However, modern digital equipment tends to have significantly shorter lifecycles; for planning purposes, the lifetime of a modern digital control system (DCS) should be taken as typically 15–20 years.

High Integrity Systems and Safety Management in Hazardous Industries. 978-0-12-801996-2

85

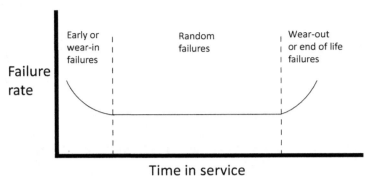

FIGURE 6.1

The classical "bathtub curve" of equipment life and wear-out.

I&C equipment life can be extended if good support arrangements are in place either from the OEM or a specialist contractor. Indeed, as already noted in the discussion on common mode failure in Chapter 2, careful vendor selection is a key factor in plant life management. In general, the following principles apply.

- I&C obsolescence is usually ultimately driven by spares availability.
- A lot of older I&C equipment will be analog.
- Almost all new equipments are digital.
- Transferring ("re-platforming") old software onto new systems can be difficult.
- Demonstrating the safety of new digital equipment requires significant effort.

The operator's capability to manage projects to install new digital equipment and assure its safety is therefore a key factor in managing the obsolescence of older I&C equipment.

By comparison with I&C equipment, cable aging is dependent upon the type of cable (e.g., whether or not it is armored), the type of insulation (e.g., some types of insulation degrade more rapidly when exposed to sunlight), the potential for mechanical or thermal damage, and humidity. It is also a function of cable loading, and environment. Cable joints can be a weakness. However, 60-year-old operational cables are not unknown.

Similarly, high-voltage electrical equipment is typically designed for a life of 40–60 years.

Hence, we can say that, for a typical large plant (such as a refinery or a power station) with a design life of typically 40 or 50 years, most of the cables and high-voltage electrical equipment should be able to remain operational throughout the entire life of the plant. However, the same cannot be said for the I&C equipment. Most large plants will have to plan for some sort of mid-life refit of I&C systems, because either the equipment begins to exhibit wear-out failures (and hence affects plant availability) or else the original equipment manufacturer (OEM) stops supplying spare parts and the equipment becomes obsolete.

The need for I&C upgrade projects can pose major challenges to plant operators. The nature of I&C equipment – control systems, alarm systems, protection systems – is that they are the nerve systems of the plant, connected to everything, and their replacement is costly and time-consuming and demands plant downtime. Also, the skill-set required for such a project may be outside the capabilities of the plant operators, who may have to recruit and out-source to ensure that the end result fits their requirements.

BASIC KNOWLEDGE AND UNDERSTANDING THAT MUST BE RETAINED BY THE PLANT OPERATORS

The list below summarizes some key issues regarding the design and safety justification for I&C equipment which operators need to retain in order to be able readily to set up an I&C upgrade project.

1. What types of I&C equipment are used at the installation?
2. Is there a full understanding of the significance of their I&C systems in:
   a. Immediate process safety such as shutdown systems
   b. Process control
   c. Mitigation of accidents
   d. Monitoring plant for potential safety-related deterioration (process condition monitoring)?
3. Have all existing electrical and electronic systems been assessed for their inherent hazards and safety requirements?
4. Have all existing I&C systems been assessed for their required functional safety integrity levels? Are the site operators aware whether the existing (aging) I&C systems meet these required safety integrity levels?
5. Is there an awareness of potential safety- and business-critical failure modes?
6. Failure to safety – does all I&C equipment fail to a safe state?
7. Have I&C common-mode failures been considered?
8. What are the arrangements for loss of site electrical power?
9. Is there non-operational I&C equipment?

I&C LIFECYCLE ISSUES

The I&C lifecycle (Fig. 2.1) describes the history of equipment from concept and specification through to decommissioning. Some important aspects for I&C obsolescence are as follows.

1. As discussed above, major I&C systems will very likely need refurbishment or replacement during the overall life of the plant. Operators need to plan for this.
2. Operators should ensure that OEM guidance on specific instrument types is built into maintenance policies and procedures as appropriate.
3. Most individual I&C components (e.g. sensors, cards, displays) are designed for replacement at periodic intervals during the overall plant life. In general, therefore, most I&C equipment lifetimes should be determined by obsolescence, spares unavailability, or loss of OEM support, and not by loss of function. Also,

some mitigation of aging is possible in certain circumstances, e.g., aging of digital systems may be temperature-related and mitigation may be possible by means of air-conditioning.

4. Connectivity between modules or components or peripherals can be a difficult or even life-limiting issue, due to lack of spare obsolete plugs and sockets.
5. There should be awareness that major refurbishment projects can have major business impact. This can be mitigated by a project strategy, which replaces the peripherals on a piecemeal basis, although obsolete connectors and software protocols can cause difficulties.
6. Migration of software from old to new equipment is a difficult proposition, which requires care in planning and execution.

MAINTENANCE MANAGEMENT

Good maintenance management processes and practices are central to ensuring that aging safety-critical I&C systems and equipment continue to operate reliably and with good availability.

Some key aspects of maintenance management are as follows.

1. Maintenance planning
2. Procurement – spares and support issues
3. KPIs for management of C&I aging – lagging and leading indicators
4. Specific I&C aging and failure mechanisms

Some of these aspects are discussed further below.

MAINTENANCE PLANNING

Good maintenance planning is absolutely key to life management of safety-critical I&C equipment. Figure 6.2 presents a diagram showing how the safety report for the hazardous plant should relate to maintenance practices.

In particular:

1. Safety-critical I&C equipment should have defined performance standards (in other words, functional and safety specifications).
2. Safety-critical I&C equipment should have maintenance, inspection and test (MIT) plans and policy in place.
3. Testing should be designed to confirm that safety-critical functions are being performed properly.
4. Test frequencies should preferably be risk-based.
5. Non-safety-critical I&C systems, if they fail, may challenge safety-critical equipment.
6. There should be clear evidence of well-planned MIT arrangements and activities, including evidence (key performance indicators, or KPIs) that the maintenance backlog is being properly managed.

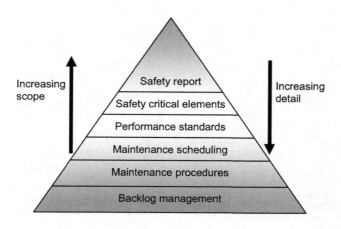

FIGURE 6.2

The relationship between the safety report and maintenance practices.

7. There should be robust managerial controls over the assessment, approval, application and removal of all overrides. Any override will in general reduce levels of safety, at least temporarily, and may constitute a change to the design intent of the plant. This aspect is discussed further in Chapter 10.

PROCUREMENT ASPECTS

So-called I&C "aging" or obsolescence in practice is most often determined by spares availability. OEM contracts may expire after a given number of years, and spares become unavailable. However, I&C equipment can in some instances, and with care and forethought, be operated safely well beyond this point. The key issues (Fig. 6.3) are as follows.

1. Adequate spares holding, in good storage conditions, by the plant operator.
2. Long-term support contracts with the OEM.
3. Specialist contractors for some component repair.
4. Awareness of the risk of lack of robust change control arrangements over time in the supply of spares from the OEM – especially for digital components. It is not unknown for OEMs to modify software specifications on spare components without making this clear to the purchasers.
5. Awareness of the risk of QA issues around buying second-hand components, e.g., over the internet. In particular, the following issues should be considered:
   a. The provenance and history of the equipment.
   b. Its configuration management, i.e., which version of the equipment is being purchased.
   c. The preparation of a test specification.

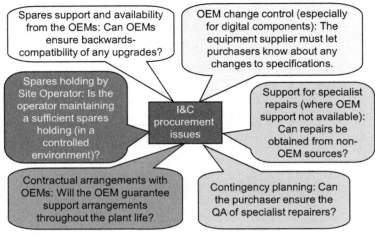

FIGURE 6.3

Procurement aspects of aging I&C equipment.

Counterfeit I&C and electrical components have become a very real threat, especially when purchased from the internet. Such equipment may have credible-looking documentation and test certification. Numerous examples of this have now been reported, and sooner or later an accident may result.

An extreme example was a fake circuit breaker where there was no switch, just a short between input and output. Other examples have included "copper" cables where the copper content is much diluted. The author has even heard of complete fake subsystems being supplied, from apparently reputable suppliers, with apparently kosher QA paperwork.

This problem is unlikely to go away, and indeed it will probably get worse. The lesson here is *caveat emptor*, especially when buying on the Internet. Be sure to use reputable suppliers, even if they are not the cheapest, and try to get as close to the source of the supply chain as possible.

KEY PERFORMANCE INDICATORS (KPIs) FOR AGING I&C EQUIPMENT

The management of aging of safety-related equipment has to, in particular, address carefully the approach to wear-out in the bathtub curve (Fig. 6.1). Management indicators – key performance indicators or KPIs – are essential tools for this [1,2].

Both *leading* and *lagging* indicators should be used as KPIs. Lagging indicators measure *outcome*, whereas leading indicators measure *processes or inputs that are needed to deliver the desired safety outcome*. In general, lagging indicators are easier to measure and more reliable, but they only show what has already happened, whereas good leading indicators should forewarn of future difficulties.

Potential lagging indicators include:

1. Number of defects reports per month for I&C equipment.
2. Rising trends in failure rates for specific components or systems.
3. Overrides: "Spot" values for the number of overrides extant across the plant.
4. Overrides: the total number of newly-applied overrides within (say) the last month.
5. Plant incidents arising from I&C failures.
6. Safety or business-critical system downtime due to I&C failures.

Potential leading indicators include:

1. I&C planned maintenance backlog (i.e., how much planned maintenance has not yet been done).
2. I&C reactive maintenance backlog (i.e., how many reported faults have not yet been fixed).
3. Spares holdings for critical components.
4. I&C plant modifications pending but not yet complete.

Rising trends on leading indicators could provide early warning of wear-out or obsolescence problems. Rising trends on lagging indicators mean the problems have already arrived.

SPECIFIC AGING FAILURE MECHANISMS

Specific information on aging and failure modes of discrete electronic components, or specific instrumentation, is beyond the scope of this book. Aging issues for specific instrumentation should ideally be addressed within each operator's performance standards and maintenance plans, using information acquired from manufacturers, from open literature, from plant experience, or from other sources.

Fire and gas detector heads deserve special mention because they are fundamental to plant safety – and they also show a typical range of potential aging mechanisms that must be considered for field instruments. Their aging mechanisms will include calibration drift, physical damage (wear and tear), detector head movement so it is no longer pointing in the right direction, the installation of other equipment which has impeded the ability of the instrument to monitor the hazardous zone adequately, dirty or damaged detector heads, battery exhaustion, corrosion, seal deterioration, weathering, and (where relevant) catalyst exhaustion.

Solder aging is an issue because of European Union regulations which seek to reduce the use of lead-based solders, and are leading to the widespread introduction of tin-based solders. Tin-based solders can be more brittle than lead-based solders. Tin-based solders can also generate whiskers, which can lead to shorting.

The failure modes of *integrated circuits* (IC) have already been illustrated in Chapter 5. Their significance is special for safety-related equipment since IC failure modes may be undefinable. Some basic terminology for IC failures is presented in Table 6.1.

Table 6.1 Integrated circuit (IC) failure mechanisms – basic terminology

Item	Description of issue	Detail
1	Intrinsic failures vs. extrinsic failure	*Intrinsic failures* are inherent in the design and materials used rather than being caused by an interaction of a stress with some defect. Intrinsic failures are typically wear-out failures, and hence these define the design life.
		Extrinsic failures are due to process defects, e.g., electrical and environmental stresses. Product reliability may be determined by the extrinsic failure rate.
2	Defect classes – by cause	*Production defects*: crystallographic defects, photolithography, surface contamination, oxide defects, packaging defects.
		Lifetime defects – key variables (perhaps acting in combination): RH, temperature, applied voltage across circuit lines, current density through thin-film conductors, thermal gradients, thermal cycles, thermal shock, mechanical stresses, pollutant concentration in the environment, chemically active process residues, oxygen partial pressure.
		Field failures: transport phenomena/electro-migration, surface growth (dendrites and whiskers), dielectric breakdown, conductor breakdown, thermo-mechanical effects (differential thermal expansion), radiation, vibration, hot carrier degradation, electrical noise.
3	Defect classes – by manufacturing process	• Encapsulation failure • Die-attach failure • Wire-bond • Bulk-silicon • Oxide-layer faults • Aluminum-metal faults • Dry and cracked soldered joints are a common cause of PCB failure • Lead-free solder issues – brittleness and whiskers
4	Defect causes – by electrical behavior	• Bridge • Open circuit • Parametric delay
5	Fault models	A defect is the physical deviation of a feature from the required form.
		A fault is the effect of a defect.
		Fault models assign a general behavior as to how a circuit fails and usually do not map to a particular defect:
		Stuck-at fault models
		Stuck-open fault models
		Bridging fault model
		Delay fault model

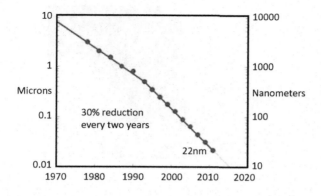

FIGURE 6.4

"Moore's Law" shows how transistor density has more-or-less doubled every two years over the period 1971–2011. This diagram presents the "law" differently, showing how feature size has fallen. Feature sizes of less than about 100 nm may mean the IC is susceptible to age-related electro-diffusion failures with unpredictable failure modes. Feature size of memory chips is expected to reach 10 nm in 2015. (Source: Wikipedia.)

A particular issue for high-integrity systems employing ICs is the rapid reduction of IC feature size as microelectronics has developed. Figure 6.4 illustrates how ICs feature sizes have reduced as the transistor count has increased – the so-called Moore's Law originating in the 1970s. Feature size for memory chips is predicted to reach 10 nanometers (nm) in 2015, according to Intel. ICs with feature sizes less than 100 nm can be expected to suffer life-limiting diffusion failure modes, which will have unpredictable outcomes. Small feature size ICs are also more susceptible to single event upset (SEU) failures caused by cosmic rays.

(For comparison purposes, "nanotechnology" is defined to be less than 100 nm. Feature sizes less than 100 nm are likely to have life-limiting diffusion failure modes. The comparative size of a nanometer to a meter is the same as that of a marble to the size of the earth. The DNA helix has a diameter of 2 nm. The smallest bacteria (mycoplasma) have a length of about 200 nm.)

The significance for high-integrity systems, therefore, is that small feature size ICs must be avoided.

CASE STUDY: BUNCEFIELD 2005

Multiple I&C failures were significant contributory factors to this accident, which occurred following the overfilling of a petrol (gasoline) storage tank. Clouds of petrol vapor were generated which then ignited and exploded (see Fig. 6.5).

At about 0600 hours on Sunday December 11, 2005, several explosions occurred at Buncefield Oil Storage Depot, Hemel Hempstead, in the UK [4]. One of the

FIGURE 6.5

The Buncefield oil storage depot fire, Hertfordshire, UK, 2005. Contributory factors included at least two separate I&C failures. (Photo taken by Chiltern Air Support Unit, as used in the Buncefield Incident Report. Copyright Thames Valley Police. Used by permission.)

explosions was a large detonation with significant shock waves and overpressure. There was a large fire, which engulfed most of the site. There were no fatalities but over 40 people were injured. In addition, significant damage was caused to commercial and residential properties in the immediate area. A large area around the site was evacuated on emergency service advice. The fire burned for several days, destroying most of the site and emitting large clouds of black smoke which were visible over London.

A petrol (gasoline) storage tank was being filled by pipeline when it overflowed. Petrol spilled over the side of the tank and large quantities of petrol vapor were generated.

Normally, tank levels were controlled from a control room using an automatic tank gauging (ATG) system. A level gauge measured the liquid level. The operator in the control room used the ATG system to monitor levels, temperatures and tank valve positions, and to initiate the remote operation of valves.

There was also an independent safety switch on the tank, which provided the operator with a visual and audible alarm in the control room when the level of liquid in the tank reached its specified maximum level (the "ultimate" high level). This alarm also initiated a trip function to close valves on relevant incoming pipelines.

Just after 0300 h on December 11, 2005, the ATG system stopped working, thereafter showing a stationary level at about two-thirds full, which was below the level at which the ATG system would trigger alarms. Nevertheless, flow rate into the tank continued at 550 m³/h, rising to 890 m³/h shortly before the explosion. Automatic shutdown did not take place. The tank would have been completely full at approximately 0520 h, overflowing thereafter. CCTV evidence shows a dense vapor cloud at approximately 0538 hours. The spillage of petrol would therefore have been about 249,000 L by 0600 hours, when the first explosion occurred.

Subsequent investigations suggested that the ultimate high level tank alarms could have worked normally but was probably overridden.

There were many learning points from the Buncefield incident, which could have resulted in large-scale loss of life if it has not happened on a Sunday morning. In particular, there had been a failure of land use planning; offices employing large numbers of people had been built adjacent to pre-existing hazardous plant. However, in the present context of aging I&C equipment, it is evident that there were at least two separate I&C failures of both the Automatic Tank Gauging system (the level control system) and also the "ultimate high level" alarm. The latter failure may have been because the alarm was overridden.

REFERENCES

[1] HSE, Developing process safety indicators, HSG 254, 2006.
[2] Process safety KPI downloads are available from www.aiche.org/ccps/metrics.
[3] P. Horrocks, J. Thomson, et al., Managing Ageing Plant – A Summary Guide, HSE Research Report 823, 2010. (download from http://www.hse.gov.uk/offshore/ageing/ageing-plant-summary-guide.pdf).
[4] The Buncefield Incident, 11 December 2005, The final report of the Major Incident Investigation Team, HMSO, 2008.

Historical Overviews of High-integrity Technologies

Learning from Ignorance: A Brief History of Pressure Vessel Integrity and Failures

"Smart people learn from their mistakes. But the real sharp ones learn from the mistakes of others".
Brandon Mull

"Nothing in the world is more dangerous than sincere ignorance and conscientious stupidity".
Martin Luther King, Jr.

This chapter presents a history of pressure vessel integrity from the First Industrial Revolution to the present day.

Developments in the integrity of pressure vessels have been in progress since the second half of the nineteenth century; however, for many decades, this progress was largely empirical with designers and operators learning from accidents (their own and others'). Today, the frequency of catastrophic pressure vessel explosions is extremely low, but historically the rate was very much higher, because our ability to fabricate high-pressure steam boilers preceded any robust understanding of their failure mechanisms.

This modern day reduction in catastrophic failures has been due to a number of very significant improvements:

- Improvements in the operational control of pressurized boilers
- Improved methods of steel production
- Improved methods of non-destructive examination
- The development of reliable, high quality welding techniques
- Improved understanding of the mechanics of metal fracture.

THE SULTANA DISASTER, TENNESSEE, 1865

Confederate General Robert E Lee surrendered to Union General Ulysses S Grant at Appomattox Court House on April 9, 1865, thus ending what still remains one of the bloodier wars in world history. The American Civil War has been described as one of the first Industrial Wars, a precursor for the even bloodier affairs in the twentieth century. On both sides, mortality rates amongst combatants were comparable to the First World War. Estimates vary, but about three-quarters of a million soldiers died. Tactics and consequences were similar to the First World War – trench warfare with massed assaults, defended by men with repeating rifles, leading to mass slaughter.

With the arguable exception of the Crimean War (1853–1856), the American Civil War was the first major conflict to be carried out in the era of photography, and under the watchful eye of what we would now call "the news media." Although only a few primitive photographs exist from the Crimean War, very many photographs exist of the events of 1861–1865.

During the Civil War, large numbers of prisoners were taken on both sides. Confederate prison camps such as Andersonville and the much smaller Cahaba (both of them in Georgia) gained notoriety. By 1865, Andersonville alone held some 45,000 Union prisoners, of whom about one-third died from malnutrition and disease, including dysentery and scurvy. Conditions were primitive. A small stream flowing through the camp was used for all purposes. There was no shelter apart from a few tents – the men were held within a bare stockade with only a few square feet each, exposed to the elements. Both sides claimed that the others were maltreating their prisoners of war, leading to a spiral of retributions where conditions became even worse. Photographs of survivors taken at Andersonville in 1865 are reminiscent of better-known photographs of survivors of concentration camps in Germany in 1945 – the Union prisoners that were fortunate enough to remain alive until the surrender of the Confederacy were starving and skeletal, barely alive. (The Andersonville Commandant Henry Wirz was subsequently tried for murder and hanged on November 10, 1865. He was the only Confederate official to be executed for war crimes.)

Cahaba prison camp [1] had a much lower mortality rate than Andersonville. It was much smaller (intended for less than 700 prisoners), and it had relatively good sanitation, a prison hospital, and a humane commandant, the Rev Dr Howard Henderson. However in July 1864 a new commandant was appointed, Lieutenant Colonel Samuel Jones, whose intention was to see the "God-damned Yankees" suffer. By late 1864, its population rose to over 2000 and conditions deteriorated, and it became the most overcrowded prison on either side of the war. Rations dropped and rats and lice increased. The camp then suffered serious flooding in March 1865; but shortly after, as the war ended, Jones announced that the prisoners were to be released. Exchanges had been arranged to repatriate soldiers from both sides. These exchanges took place at a neutral site near Vicksburg on the Mississippi River, which had been under Union control since the Battle of Vicksburg in the summer of 1863.

Prisoner repatriation was achieved by means of paddle steamers traveling up and down the Mississippi River between the Union and the Confederacy. One of the paddle steamers contracted by the Union government to carry out prisoner repatriations was the SS *Sultana* [2, 3], a wooden steamship registered at 1719 tonnes with a legal capacity of 376 passengers and a crew of 85. The *Sultana* had therefore stopped at Vicksburg on April 22, 1865 to take on passengers. There were also some quick, temporary, urgent

FIGURE 7.1

This very grainy photograph of the overcrowded paddle steamer *Sultana* was taken at Helena, Arkansas the day before the disaster.

(Source: Wikimedia Commons/Library of Congress)

boiler repairs – its boiler had a section of bulged plate removed, and a thinner piece of plate was riveted on in its place. The repair work was completed in about 30 hours (Fig. 7.1).

Vicksburg on April 24, 1865 was full of released Union prisoners of war from both Andersonville and Cahaba. Many of them were weak from their incarceration, and all of them were desperate to board the *Sultana* to travel north. The exact numbers who boarded the *Sultana* are unclear but it seems likely that more than 2000 passengers were on board. Decks and berths were packed.

The Master of the *Sultana*, Captain JC Mason, was also a part owner. He was paid $5 for each enlisted man and $10 for each officer he took on board. The *Sultana* set sail from Vicksburg on April 24, stopping briefly to take on more coal at Memphis, Tennessee, some 250 miles north. Some 8 miles north of Memphis, Tennessee at 0200 hours on April 27, 1865, grossly overloaded and fighting against a strong spring river current, its boiler exploded and destroyed much of the ship. Burning coals set fire to the remaining superstructure, which burned right down to the waterline. Of those who survived the explosion and fire, many died of hypothermia or drowned in the cold water, and many others later died of burns. Exact numbers have never been reliably established, but it is likely that more than 1500 people died, with fewer than 800 survivors. Captain Mason was one of the dead.

The official inquiry concluded that the boiler had exploded because of poor repair work, combined with low water level and "careening", i.e., the water was sloshing around in the boilers, creating hot spots in the boiler which then were suddenly cooled as the water sloshed back, thereby causing thermal strains. However, as often occurs in accidents, there were conspiracy theories: it has been suggested that a "coal torpedo" – a bomb disguised to look like a piece of coal – had been placed in the coal stock as an act of sabotage against the victorious Union forces.

President Abraham Lincoln had been assassinated by the actor John Wilkes Booth at Ford's Theater in Washington on April 14, 1865, a few days after General Lee's surrender at Appomattox. Booth fled to Maryland where, twelve days later on April 26, he in turn was shot and killed by a Union soldier. The assassination of President Lincoln and the search for John Wilkes Booth put the American press into a feeding frenzy. The final part of the *Sultana* tragedy is that it was almost completely overlooked in the newspapers by the killing of Booth on the day before the explosion. The nation was tired of slaughter, and keen to start its long, slow healing process. The deaths of Lincoln and then Booth were post-scripts. The nation did not want to hear about yet more mass death.

The US loss of life in the *Sultana* accident was greater than in the *Titanic*, *Lusitania*, the USS *Arizona* (at Pearl Harbor) or the USS *Indianapolis* (in 1945), and it remains the greatest-ever US loss of life in a maritime disaster. Despite the scale of the loss, the *Philadelphia Inquirer* dealt with the event in less than 200 words in its April 29 edition (Fig. 7.2).

FRIGHTFUL STEAMBOAT EXPLOSION!

THE "SULTANA" BLOWN UP.

1400 Paroled Union Soldiers Scalded to Death and Drowned!

Sᴛ. Loᴜɪs, April 28.—A telegram has been received by the military authorities from New Madrid, that the steamer *Sultana*, with two thousand paroled prisoners, exploded her boilers, and that fourteen hundred lives were lost.

Second Despatch.

Cᴀɪʀo, April 28.—The steamer *Sultana*, from New Orleans on the evening of the 21st instant, arrived at Vicksburg, with her boilers leaking badly. She remained there thirty hours repairing, and took on one thousand nine hundred and ninety-six Federal soldiers and thirty-five officers, lately released from Columbia and Andersonville prisons. She arrived at Memphis last evening, and after coaling proceeded.

About two o'clock A. M., when seven miles above Memphis, she blew up, and immediately took fire, burning to the water's edge. Of two thousand one hundred and six souls on board, not more than seven hundred have been rescued. Five hundred were rescued, and are now in the hospital. Two hundred or three hundred uninjured men are at the Soldiers' Home. Captain Mason, of the *Sultana*, is supposed to be lost.

At 4 A. M. to-day the river in front of Memphis was covered with soldiers struggling for life, many of them badly scalded. Boats immediately went to their rescue, and are still engaged in picking them up. General Washburne immediately organized a board of officers, and they are now at work investigating the affair.

FIGURE 7.2

The Philadelphia Inquirer, April 29, 1865.

BOILER EXPLOSIONS AND DEVELOPMENTS IN BOILER TECHNOLOGY DURING THE FIRST INDUSTRIAL REVOLUTION

Coal-fired steam power was the motive force of the First Industrial Revolution, driving railways, ships and factories. However, with this new technology came a new risk; exploding steam boilers became regular and inexplicable events. Boiler explosions of the sort that was commonplace in the nineteenth century are now so rare that even the term "boiler explosion" has subtly changed its meaning. In the twenty-first century, if we hear a news item about a "boiler explosion," it generally refers to a gas-fired boiler, where there has been a gas leak that has subsequently ignited.

However, in the nineteenth century, a "boiler explosion" meant a disruptive failure of the steel pressure vessel of a coal-fired boiler, where the explosive force was provided by the sudden release of high-pressure steam. Boiler explosions became, by the end of the century, almost daily random events that were frequently fatal, and were treated as if they were "Acts of God." The mechanisms of failure were simply not understood. This was compounded by a cavalier attitude to operation and maintenance. Operators would increase the pressure above the nominal maximum if they needed more power. Repairs – like those to the *Sultana* in Vicksburg where a metal patch was riveted into place – could be crude and downright dangerous (Fig. 7.3).

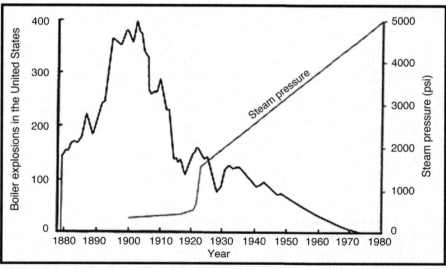

FIGURE 7.3

Rates of boiler explosions in the USA: This graph shows how improved design, a better understanding of the mechanisms causing failure, and better operation, maintenance and inspection regimes, led to a dramatic drop in the rate of boiler explosions. This reduced rate of explosions reached effectively zero by about 1970, despite steam pressures increasing dramatically to achieve improved process efficiency.

(Source: CASTI Guidebook, vol. 3, ASME B31.3 Process Piping, 2nd Ed., 1999. Used by permission.)

The problem was this: the engineering and manufacture of coal-fired steam boilers had moved far faster than the science behind their safe design and operation. Boilers were being designed, manufactured, maintained, and repaired by people who just did not understand the causes of highly dangerous, explosive boiler failures. In the nineteenth century, the design of boilers was completely under the control of the manufacturers – there were no design standards as we know them today.

Nineteenth century boilers could fail catastrophically for various reasons.

- The initial design could be based on a poor understanding of the tensile properties of the grade of steel (or other metal) used.
- There might be stress concentrations around, say, rivet holes, or at sharp angles.
- The quality of steel used in the boiler might be inconsistent, or else compromised in some way – either by virtue of its composition, or by non-metallic inclusions, or by cracks.
- Bad repairs, such as those carried out on the *Sultana*, could introduce weaknesses.
- Corrosion could be caused by impurities in the boiler water.
- Fatigue damage could accumulate through repeated pressure cycles of the boiler, eventually to the point of failure.
- The operators might operate the boiler at pressures greater than the design pressure in order to produce more power.

Any of the above, acting singly or in combination, could lead to boiler explosions. Boiler pressures in the nineteenth century were nevertheless low by today's standards. Trevithick's first "high pressure" locomotive engine in 1804 operated at a pressure of about 3 atmospheres (50 psig or 3 bar). Throughout the nineteenth century boilers would typically only operate at not more than a pressure of 10 atmospheres (150 psig or 10 bar), compared to modern coal-fired and nuclear power stations where steam pressures of 160 atmospheres (2400 psig or 160 bar) are typical. This very large increase in steam pressures during the twentieth century enabled large increases in the thermal efficiency of steam-powered plant.

The control of risks from all of these causes of failure required good engineering (safety valves), good quality control, and good science (metallurgy, chemistry and stress analysis).

Safety valves became an early feature of boiler pressure vessels. However, in the early- and mid-nineteenth century, operators could (and did, often with disastrous consequences) increase the operating pressure of their boilers by tying down the safety valve. Lockable safety valves were developed to prevent this, and also in 1855 John Ramsbottom invented a "tamper-proof" safety valve – one of many developments in safety valve designs in the latter part of the nineteenth century.

Quality control, although not a term that had yet been invented, was addressed indirectly via the exigencies of insurance. Soon after the *Sultana* disaster, in 1866, the Hartford Steam Boiler Inspection and Insurance Company was established, in Hartford, Connecticut – "the first company in America devoted to public safety." This company was the first to merge inspection and insurance, which was undoubtedly a major step forward. In order for the Hartford Steam Boiler Inspection and Insurance Company to insure a boiler, they had first to inspect it – and periodic re-inspection was necessary in order to remain insured with the Company. The Hartford Steam Boiler group still exists, although since 2009 it has been part of Munich Re, one of the world's largest re-insurers.

Companies such as Hartford still employ boiler inspectors to this day, and they perform important work to ensure that the pressure vessels used in industry are in a safe condition. Their inspections are generally to see if there are any signs of mechanical or corrosion damage, to check that pressure gages are fitted and calibrated properly, and to check that overpressure relief valves are fitted, have been tested, and are set to operate at the correct pressure.

Depending on the type of pressure vessel and its application, there may also be a requirement to carry out non-destructive examination (NDE) to look for cracks or inclusions in the metal. NDE takes various forms. In the nineteenth century, the only real inspection technique was simple visual examination – the "mark one eyeball". *Dye-penetration* testing, which offered a major enhancement to simple visual inspection, first came into use in the 1940s. This technique uses a fluid containing strong dye, which is spread over the metal surface and then wiped off. If there are any surface penetrating cracks, the fluid will then re-emerge, exposing the cracks. *Magnetic particle* inspection, another common technique, uses an applied magnetic field and magnetic particles to show if there is any distortion of the magnetic flux as would occur if there were cracks in the metal. Also, beginning in the 1930s, The US Naval Research Laboratory's Robert F. Mehl used *gamma-ray radiography* as a shadowgraphic technique to investigate the extent of suspected flaws in the stern-post castings and welds of new U.S. Navy heavy cruisers. NRL's efforts established gamma-ray radiography's usefulness in NDE of metal castings and welds. Gamma radiography and, later, *ultrasonic* inspection became more common in the second half of the twentieth century.

Steel quality control and consistency improved in the latter part of the nineteenth century. Until about 1860, cast iron was the only economic material available for boilers, which were made in sections. Early boiler pressure vessels would generally be riveted together, which seems incredible to modern engineers. Riveting together the component parts of pressure vessels has very major disadvantages – the rivet holes act as stress concentrators, they might be the source of steam leaks, and they might act as sites for corrosion.

Cheap, good quality steel began to become readily available from about 1860, largely due to the work of Henry Bessemer (1813–1898). Cast iron guns had received bad publicity in the Crimean War – they were extremely heavy and prone to catastrophic failure. The gun barrels were, in effect, pressure vessels and all the same concerns applied, including the real risk of gun barrels exploding when fired. Bessemer's idea was to blow air through liquid pig iron to remove excess carbon. His process was further improved by Robert Mushet (1811–1891) who found that adding manganese helped remove oxides and sulfur. By 1859, Bessemer established his own steelworks in Sheffield, and he subsequently sold licences for his process all over the world. Other improvements followed, notably the open-hearth furnace introduced by the Siemens brothers in the 1860s, which enabled much tighter control of the quality and composition of the steel.

The American Society of Mechanical Engineers (ASME) was founded in 1880 to provide a setting for engineers to discuss the concerns brought by the rise of industrialization and mechanization. ASME established the Boiler Testing Code in 1884.

Meanwhile, the rate of boiler accidents still continued to rise steeply.

Another massive boiler explosion occurred at 0750 hours on March 20, 1905 (Fig. 7.4). It killed 58 people and completely destroyed a four-storey building, the Grover shoe factory ("Makers of the Emerson Shoe"), which was located on the corner of Main and Calmar streets in the largely Swedish neighborhood of the Campello section of Brockton, Massachusetts. The explosion was due to the failure of an old boiler, which was used as a backup during maintenance on a newer model. The boiler rocketed through three floors and the building's roof, and broken beams and heavy machinery trapped many workers who survived the initial explosion and collapse. Burning coals thrown from the boiler landed throughout the crumbling superstructure, starting fires that were fed by broken gas lines.

A report in the New York Times on March 30, 1905 (Fig. 7.5) is interesting in what it reveals about the lack of understanding of boiler failures at the time. No evidence of criminal liability was found. The explosion was attributed to a "hidden defect" in the boiler, which had been inspected in December 1904. "There was no knowledge that it was not in good condition." Note the double negative: they did not know that it was not safe.

A local history website, www.brocktonma.com, reports how the exploding boiler tore through the four-storey building, causing the factory roof to collapse and the four floors to crash down on each other. Those workers who survived the explosion and collapse became trapped beneath heavy timbers, flooring and heavy shoe manufacturing equipment. Unable to move, the workers were helpless as the ruins caught fire, fueled by the broken gas pipes that fed the factory. Also, the Grover building had more than three hundred glass windows, which had allowed the factory floor to be bathed in sunlight. These were now shattered and helped cause a chimney effect. Air was pulled in, and the fire burned "hotter and faster than any fire the city's fire department had ever encountered." Of the 300 plus workers who were in the building, only some 100 made it out unharmed. Fifty-eight people were killed, including some from surrounding buildings that also caught fire, and a further 150 people were injured.

Brockton at that time was a major center for shoe manufacture, which employed some 35,000 people, and steam was used for leather processing in various local factories. A little over 18 months after the Grover shoe factory explosion and fire, another boiler exploded on December 6, 1906, this time in the factory of the P.J. Harney Shoe Company in Lynn, Massachusetts. As a result of these two major accidents in Massachusetts in quick succession, a five-man Board of Boiler Rules was convened by ASME, whose charge was to write a boiler law for the state; this board published its boiler laws in 1908 and the state of Massachusetts enacted these laws, which were the most rigid boiler inspection laws in the United States to that date.

AFTER THE EXPLOSION AND FIRE. Looking west. Smokestack and ruins of Grover Factory at right; Dahlborg Block at left; ruined dwellings in background. Curved white line shows path of boiler from base of smokestack to house of Miss Pratt, where it lodged.
View from Churchill & Alden Shoe Factory on east side of Main Street.

FIGURE 7.4

The boiler explosion at the Grover Shoe Factory, Brockton Massachusetts, on March 20, 1905. The only constant in these "before and after" photographs is the chimney at the rear. This explosion in particular led to the development of the ASME Boiler Code (public domain).

BROCKTON, Mass., March 29.—A statement that no evidence had been adduced to show that any person was criminally liable for the explosion in the R. B. Grover Company's shoe factory here last week, which caused the death of fifty-eight employes, made this afternoon by District Attorney Asa P. French, concluded the inquest.

Mr. French said he believed it proved that the explosion was due to a hidden defect in the factory boiler. The witnesses examined included a representative of the Grover Company, who testified that the boiler was inspected in December, and that there was no knowledge that it was not in good condition. Several Boiler Inspectors said that they had found a crack in the boiler, which was of obsolete type. Several factory employes declared that David W. Rockwell, engineer in charge of the boiler, who was one of the killed, seemed capable of attending to his duties on the morning of the accident.

FIGURE 7.5

New York Times, March 30, 1905. "There was no knowledge that it was not in good condition."

ASME then convened its ASME Boiler Code Committee, formed in 1911. This committee led to the creation of the first edition of the ASME "Boiler Code – Rules for the Construction of Stationary Boilers and for the Allowable Working Pressures," which was issued in 1914 and published in 1915. This first 1914 edition was later incorporated into laws throughout most of the United States and Canada. The ASME Boiler Code has continued to be developed over time into the ASME Boiler and Pressure Vessel code, and its use worldwide has spread. Today, there are over 92,000 copies in use, in over 100 countries around the world.

The first, 1914 edition of the ASME Code did basic things like defining the term "maximum allowable working pressure," and defining the minimum allowable distance between rows of rivets. The 1914 Code stipulated that the maximum allowable pressure on cast iron boiler headers should not exceed 160 psig (about 10 atmospheres or 10 bar). Minimum capacities for boiler pressure relief valves were defined.

Welding technology development began in the latter part of the nineteenth century. (See e.g., www.weldinghistory.org.) Welding is a process that we take for granted today but it was developed over many decades by many people and many companies. In 1881–1882 a Russian inventor Nikolai Benardos created the first electric arc welding method known as carbon arc welding, using carbon electrodes. Arc welding uses a power supply to create an electric arc between an electrode and the base material to melt the metals at the welding point. Advances in arc welding continued with the invention of metal electrodes in the late 1800s by a Russian, Nikolai Slavyanov (1888), and an American, C. L. Coffin (1890). However, the development of arc welding also relied on other technologies; e.g., the electric power required was so great that it required a boiler (with its pressure vessel), driving a steam engine and an electric generator. In other words, arc welding, which enabled better pressure vessel construction, itself required pressure vessels. (This type of interaction between different technologies was a key aspect of the First Industrial Revolution – each technical development fed directly into other technical developments.)

Acetylene had been discovered in 1836 by Edmund Davy, but its use was not practical in welding until about 1900, when a suitable blowtorch was developed. At first, oxyacetylene welding was one of the more popular welding methods due to its portability and relatively low cost. As the twentieth century progressed, however, it was largely replaced with arc welding, as metal coverings (known as flux) for the electrode that stabilize the arc and shield the base material from impurities continued to be developed.

The first heavy-walled pressure vessel built entirely by welding was fabricated in 1925.

A common problem with both arc welding and blowtorches was that the metal surface would become burnt and brittle on either side of the weld, due to oxidation. Gas-shielded arc welding, where the arc was shielded by an inert gas to stop metal oxidation, was first developed in about 1940. During the Second World War, Gas-Tungsten Arc Welding (sometimes known as Tungsten-Inert Gas or TIG welding) was developed. At this point welding really entered the modern era, as three separate developments came together; a *tungsten electrode* (since tungsten has a high melting point), *inert gas shielding* (normally argon) to prevent oxidation of the metal being welded, and *automatic feeding* of filler wire. These three developments together led to an industrial process that could produce consistent high quality welds in thick steel pressure vessels. TIG welding could also be applied to other metals including aluminum alloys used for aircraft.

Occasional serious boiler accidents continued to happen through the twentieth century. Another workplace tragedy occurred at the New York Telephone Company, New York, on October 3, 1962, when a boiler explosion left 23 dead and 94 injured. The failed boiler was fairly low-pressure, and weighed seven tons empty. The boiler operator had reportedly started the boiler then left the building to have lunch. In addition, however, three different safety systems were not operational. First, the operator started the boiler with the steam shutoff valve on the steam supply header pipe in the closed position. Second, a pressure switch had been inadvertently unplugged by a maintenance worker. Third, a pair of safety valves on top of the boiler apparently did not function when they exceeded pressure limits.

The force of the explosion propelled the boiler forward about 40 metres, through two concrete walls, killing and disfiguring employees who were dining in the nearby cafeteria. Arising from this, one of the most deadly boiler accidents of the time, New York enacted an improved law for low-pressure boilers.

Advances in quality control, a move from riveting to welding, consistency of design standards, and improved operational controls regarding maximum pressures, enabled dramatic reductions in the rates of boiler pressure vessel explosions.

However, during the nineteenth century and the early part of the twentieth century, all of these parallel developments (together with improvements in steel manufacture, pressure vessel inspection and welding) were essentially happening empirically. People were learning from mistakes, not from improved understanding. The scientific understanding of the actual causes of metal failure was still poor, and the rate of boiler explosions, although much improved, continued to be unacceptably high. The problem with the science was that, whenever a "first principles" effort was made to understand why materials failed, the results were just plain wrong. Calculations that could be made of inter-atomic forces – which might have been thought to relate in a fairly direct way to material tensile strength – seemed to bear no relation whatsoever to the actual, observed tensile strength; such calculations generally showed that metals *should* be very much stronger than they were measured to be.

The problem was that science was still asking the wrong questions. Materials' strengths are not determined only by inter-atomic forces, they are largely determined by the existence of micro- and macro-defects in the materials.

THE COCKENZIE STEAM DRUM FAILURE, 1966

In the second half of the twentieth century, boiler explosions did become very rare, mostly because of empirically derived improvements in design, fabrication, inspection, and operation. However, occasional boiler explosions still happened, and this low residual rate was unacceptable for new industries such as aviation and nuclear power. Developments in these new industries demanded extremely high standards of reliability. They demanded proof of integrity; there had to be a scientific basis for saying that a pressure vessel was safe.

Pressurized fuselages of high-flying jet aircraft became, in effect, thin-walled aluminum pressure vessels. The two Comet 1 crashes in 1954 [4] were caused by stress concentrations around a square-shaped window on the top of the fuselage used for navigation. The stress concentration was enough to cause fatigue with each cycle of climbing to cruise height and returning to ground level. Two Comets disintegrated in mid-air, both after about 1000 pressure cycles.

In process plant, it had become normal practice to subject new, large pressure vessels to hydraulic proof pressure tests. The distinction between *hydraulic* and *pneumatic* pressure tests is an important one. Water is essentially incompressible – its volume barely changes when it is pressurized. Gas (or steam) is, however, highly compressible. Hence, in a hydraulic pressure test, the pressure vessel is filled with water, and if any failure should occur, as soon as the enclosed volume begins to increase because of the developing failure, the pressure immediately collapses. This means there is no explosion – just a "bang" as the metal fails, followed by a flood of water. By contrast, if a pressure vessel fails when pressurized with gas or steam, as in the boiler explosions described above, the energy of the compressed gas/steam drives the failure to create a violent explosion, and pieces of the failed pressure vessel become missiles.

The advantages of hydraulic proof pressure testing are best demonstrated by reference to the failure of a large steam separator "drum" at Cockenzie Power Station in East Lothian, Scotland, on May 6, 1966. (Separator drums are large, horizontal, cylindrical, thick-walled steel pressure vessels used to separate steam and water in coal-fired or nuclear power stations.)

Cockenzie was a coal-fired power station in construction at the time. (Cockenzie power station went into service in 1967. It operated until March 2013.) To satisfy the designers and the boiler insurance company, each of the boiler drums was put through a number of hydraulic pressure cycles. Each drum weighed 164 tonnes, had an inside diameter of 5 feet 6 inches (1.67 m) and an overall length of 74 feet 9 inches (22.78 m). The drum was an all-welded construction, fabricated from forged steel plates 5.5625 inches thick (14.13 cm). The material was a mild manganese steel alloy called Ducol.

The Cockenzie steam drum failure of May 6 1966 [5] happened during the final, seventh cycle of hydraulic pressure testing before the steam drum would have been declared ready for service. It failed adjacent to one of the "economizer nozzles," the transition pieces to the high-pressure pipes in operational service, would have led water from the underside of the drum to the boiler tubes (See Fig. 7.6.)

The drum had been manufactured by Babcock and Wilcox Ltd (now Doosan Babcock) at Renfrew, near Glasgow, to a high quality using the standards of the day, in particular BS 1113 (1958) which had much in common with the then-current version of the ASME Boiler Code. The drum shell plates were hot-formed at about 1000°C. Welding was carried using the Electro-Slag welding technique, a variation of arc welding with an inert cover gas and continuous feed of filler wire. Afterward, the welded plates were heat-treated at about 920°C.

The economizer nozzles had been welded into the main pressure vessel using manual metal-arc welding. However, one of the economizer nozzles was found to be defective during inspection at works, and it was removed (by "hand chipping and grinding") and replaced. A replacement nozzle was welded into place using the same procedure. The welds were inspected using magnetic particle inspection and reported clear of defects.

It was this replacement nozzle where the drum failure originated.

On completion of all welding, the drum was given a final heat-treatment at 650°C. The drum was subjected to independent inspections at various stages of manufacture, in accordance with BS 1113 (1958). The inspection techniques included both ultrasonic and magnetic particle inspections.

At the time of the failure, the drum had already been through six successful pressurization tests, which had taken place at both the manufacturing works and Cockenzie. The pressure tests had included four cycles to about 4100 psig, which was about 1.5 times its maximum normal operating pressure. The first such test had taken place at Babcock and Wilcox's works on June 26, 1964.

The failure occurred during the drum's seventh cycle of pressure testing, at about 3980 psig, on May 6, 1966 at Cockenzie. The drum was filled with approximately 48 tonnes of water at about 8°C. A large portion, about 5 m long and at its maximum

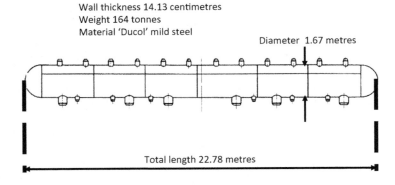

Wall thickness 14.13 centimetres
Weight 164 tonnes
Material 'Ducol' mild steel

Diameter 1.67 metres

Total length 22.78 metres

FIGURE 7.6

The Cockenzie steam drum after failure, with the failure site (inset). The dark tarnished area shows that the crack already existed during final heat-treatment, because the crack surface had discolored (public domain).

almost the full diameter of the drum, suddenly broke away, releasing the water onto people working underneath. The report notes laconically that "the deluge of water caused fright and shock to a number of men...but there were no serious injuries." There will also have been a very loud bang when the drum failed.

The failure was caused by a crack next to the replaced economizer nozzle, created during the manufacturing process, which had penetrated part-way through the thick wall of the pressure vessel. The report does not reach definitive conclusions about the exact cause of the crack, but it suggests that it could have been caused by differential expansion created during the heat-treatment process. The crack, which was 33 cm long and 9 cm deep, must have been there during at least part of the heat-treatment, because its surfaces had been tarnished by the heat-treatment (as can be seen in the photo).

There are two key lessons to be drawn from this very expensive event, which fortunately harmed no one.

- If this failure had happened when the drum was operational (pressurized with hot steam) or if the testing had been carried out using pressurized gas or steam pressure, there would have been a large, disruptive, and probably fatal explosion, with large missiles traveling significant distances. The use of hydraulic testing was therefore justified.
- However, when it comes to demonstrating the required high integrity of pressure vessels for other applications, notably nuclear reactor pressure vessels, difficult questions are posed regarding the philosophical basis for their safety.

If the Cockenzie drum had survived this hydraulic pressure test (the seventh and last) it would have entered service. Sometime later, when pressurized with steam, it would have failed, and the explosion would have caused massive damage, and probably injuries and deaths. The question is this: How can we be really sure that there are no significant cracks in thick-walled pressure vessels?

MODELING THE FRACTURE OF PRESSURE VESSELS

To model the failure of pressure vessels, we must first differentiate between *brittle fracture* and *ductile failure*. The easiest way to do this is to think of the child's toy called "Potty Putty" or "Silly Putty." If this material is pulled slowly, it will stretch to tens of times its original length before it breaks (*ductile failure*). However, if it is pulled sharply, it snaps with hardly any stretching (*brittle fracture*). In the right circumstances, metal components can also fail in either a plastic or brittle manner.

AA (Alan Arnold) Griffith was a young PhD mechanical engineer working in the Royal Aircraft Establishment during the First World War who realized that the important issue was not how strong a metal component should "theoretically" be but, instead, the nature of the imperfections that actually caused metal components to fail. Metal fails most often because of cracks, and Griffith was able [6] to show that, for *brittle fracture* to occur, the key issue was the energy required for crack propagation within the metal. He was able to demonstrate how to calculate the critical crack size before the crack could propagate to cause sudden, catastrophic failure – the "Griffith crack equation." His ideas were published in 1921 although it was actually many decades before Griffith's work became widely used.

The principles of *linear elastic fracture mechanics* were developed in the 1950s by George Irwin at the US Naval Research Laboratories [7]. Irwin demonstrated that a crack shape in a particular location with respect to the loading geometry had a stress intensity associated with it. He demonstrated the equivalence between this stress intensity concept and Griffith's criterion of failure. More importantly, he also described the systematic and controlled evaluation of the toughness of a material. *Fracture toughness* is defined as the resistance of a material to rapid crack propagation and Irwin showed it can be characterized by one parameter. The fracture toughness of a material is generally independent of the size of the initiating crack.

Next, in 1963, a means of reconciling the differences between theoretical and measured metal strength was published, which modelled *ductile failure* as arrays of atomic-level dislocations. This model, known as the "BCS model" after its authors Bilby, Cottrell and Swinden [8], would later be combined with the Griffith crack equation to enable a mathematical model for both ductile failure and brittle fracture, the R6 method, which we will come back to shortly.

In the late 1970s, the UK Central Electricity Generating Board was considering the construction of pressurized water nuclear reactors. The first requirement for licensing of this reactor design – in which the nuclear reactor is contained in a large steel pressure vessel – was to be able to put the safety and integrity justification of the steel pressure vessels onto a firm scientific footing.

The Cockenzie incident was a near-thing: any such failure had to be ruled out with extremely high confidence in the manufacture of pressure vessels for nuclear power stations. Confidence had to be so high that an in-service failure of a nuclear reactor pressure vessel would, in effect, be an incredible event. This meant that there had to be clear answers to some basic questions, including the following:

- For a given set of conditions, what size of defect can cause failure?
- How are the mechanical properties of the metal affected by temperature or by aging processes?
- What size of defects in thick-walled metal pressure vessels can be detected reliably by non-destructive examination (NDE)?
- How fast can cracks and defects grow in-service?

For a credible safety case to be made, the operator has to be able to say as a minimum the following.

1. the pressure vessel has been inspected using a technique which will have detected (with high confidence) all cracks above a given size;
2. a mathematical fracture analysis has shown that the minimum critical crack size, to cause a catastrophic pressure vessel failure, is significantly bigger than the detectable crack size; and
3. an unseen crack below the detectable size, if it exists, cannot feasibly grow to the minimum critical crack size during the operating life of the pressure vessel.

The UK Atomic Energy Authority carried out a detailed review under the chairmanship of Walter Marshall [9], which made recommendations about design, manufacture, materials, pre-service inspection, and in-service inspection and analysis. By and large, these recommendations supported the ASME approach. The report also made recommendations for improving the confidence in pressure vessel integrity, by enhanced analysis, inspection, and quality control.

The report included detailed consideration and review of various fracture mechanics analysis and assessment techniques. The ultimate mechanical failure of a loaded metallic structure can occur in several ways. At one extreme, *ductile failure*, "deformation can occur throughout the structural section resulting in large dimensional and possibly shape changes. Alternatively, the deformation may begin homogeneously but become localized as the section thickness is reduced thereby increasing the effective stress under a fixed load." Ductile failure thus corresponds to the stress in the metal locally exceeding the ultimate strength of the metal. At the other extreme, *brittle fracture*, "the failure process may involve only a very small volume fracture of the structure, with deformation being extremely limited, resulting in the propagation of a crack across the material section without gross dimensional changes."

In metal structures, both of these failure modes are strongly influenced by the presence of pre-existing cracks or flaws in the structure.

About the same time as the Marshall Report, a method for the analysis of a pre-existing defect, to determine whether it would cause either ductile failure or brittle fracture, was developed by the UK Central Electricity Generating Board, called the R6 method [10]. This technique was a development of earlier methods but can still be thought of as a breakthrough. It has now been further developed and is in wide use. The R6 method uses a *Failure Assessment Diagram* developed from a combination of theoretical and empirical work. The original R6 Failure Assessment Diagram was a combination of Griffith crack equation and the BCS model, which were discussed above. (Later versions have refined this somewhat.)

The R6 methodology enables the integrity of pressure vessels to be validated, using a technique that is underpinned by science. R6s pre-requisites are that reliable, verifiable data about material properties and stress distributions have to be available, and that all significant defects in the metal have been identified and measured. The R6 methodology generates a line on the Failure Assessment Diagram (Fig. 7.7), which gives a lower bound to all ductile or brittle fracture, regardless of detailed crack geometry, which thereby enables the avoidance of failures.

The R6 method is not the only method for assessing the integrity of pressure vessels with known defects, but it is widely known and well validated. The Marshall Report compared the R6 methodology with other methods (such as the "COD" method and the "J-design curve" method) and concluded that the R6 was the preferred route for the assessment of nuclear pressure vessels.

By the time that R6 was in use, the First Industrial Revolution was about two centuries old; it had taken that long for the engineering of thick-walled pressure vessels to be properly underpinned by science.

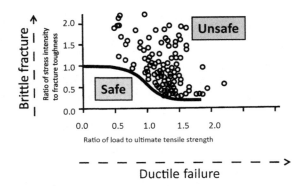

FIGURE 7.7

A failure assessment diagram for the R6 method. A point on the graph can be plotted corresponding to a known defect and loading conditions. The points are experimental results from validation tests where failure occurred – the key requirement to validate the R6 method is that no failure should occur on the side of the line labeled "safe".

(Source TWI, "BS 7910: History and future developments", Isabel Hadley, ASME pressure vessels and piping conference, ASME, Prague, July 26–30, 2009. Used by permission.)

THE ACCURACY AND RELIABILITY OF NON-DESTRUCTIVE EXAMINATION

One of the other "basic questions" given above was: "What size of defects in thick-walled metal pressure vessels can be detected reliably by non-destructive examination (NDE)?" The answer to this question will depend on the NDE method used (dye-penetration, magnetic particle, X-ray or ultrasonic), the skill and experience of the NDE technician, and the geometry of the crack within the steel (e.g., surface or non-surface penetrating, flat steel plate or near other features, and shape of defect).

The first real efforts to answer this question were made in the 1980s in the program for the inspection of steel components (PISC), which was coordinated through the Council for the European Commission and the Organization for Economic Cooperation and Development (OECD) [11]. Steel plates with known different types of defects (but of unknown size) were passed around NDE inspection teams in various countries, who carried out inspections in accordance with the requirements of the American Society of Mechanical Engineers (ASME) Code. The teams also used some advanced ultrasonic NDE techniques that were still undergoing research. The countries involved in the set of "round-robin" trials included various European countries, the USA and Canada. At the end of the tests, the steel plates were destructively examined so that the exact sizes of the defects could be measured. The measured defect sizes were then compared with what the NDE teams thought they had found.

The results of this lengthy and expensive program can be summarized very briefly: The PISC program proved it was possible to detect a 3 cm deep crack with very high confidence, but it is only possible to detect a 1.5 cm deep defect with about 50% confidence.

Of course, these conclusions related to non-destructive examinations being carried out under laboratory conditions, so they may be unrepresentative of "real-life" conditions. However, they did give confidence that critical cracks in thick-walled steel pressure vessels, which would typically be significantly bigger than 3 cm, would be detectable. The work of the PISC trials has since been supplemented by more trials of a similar nature.

DOUNREAY, 1981 – SAFETY VALVE TESTING ON A STEAM DRUM

(The following is a personal account of pressure vessel safety valve testing at a nuclear power station, which I observed early in my career. I hope this may be of interest to design engineers who perhaps do not get to see this type of activity).

My first encounter with the operational challenges of ensuring pressure vessel safety came at Dounreay in northern Scotland, at the prototype fast reactor (PFR), in 1981. PFR was a highly unusual power station since the reactor was cooled by liquid sodium. PFR had three identical sodium-heated boiler circuits, each containing a boiler, a steam drum, a superheater and a reheater (Fig. 7.8).

The three steam drums were similar to the Cockenzie steam drum, although somewhat shorter.

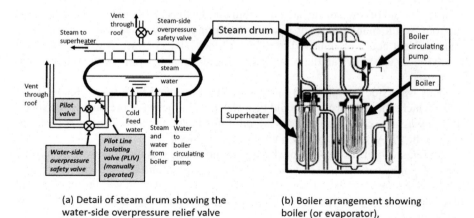

(a) Detail of steam drum showing the water-side overpressure relief valve and its manually-operated pilot valve

(b) Boiler arrangement showing boiler (or evaporator), superheater, and steam drum

FIGURE 7.8

This schematic diagram of the prototype fast reactor (PFR) illustrates the steam drum and its unique waterside overpressure safety valve.

Sodium-heated steam boilers are, to say the least, difficult to design and tricky to operate, because sodium and water react violently on contact to produce hydrogen, heat, and sodium hydroxide. Hence the power station's designers had to think very hard about the integrity of the boundary between the sodium and the steam (the boiler tubes), and they also had to plan for contingencies if that sodium-steam boundary were to fail for any reason. By and large, the designers did a great job, incorporating some very clever features into the design:

- There were non-radioactive sodium circuits which acted as buffers between the reactor and the boilers.
- Steam pressure was much higher than the sodium pressure, so any leakage would be from the steam into the sodium. The designers ensured there were instruments to detect tiny amounts of hydrogen in the sodium, giving warning if small leaks occurred.
- If a large steam leak occurred, the affected sodium circuit was designed to quickly empty into storage tanks. Simultaneously, the reactor would shutdown.
- In a worst-case scenario, the designers imagined that there could be a need to get rid of water from the system as quickly as possible, so pressure relief valves on the steam drums could be electrically actuated to open and let the steam and water blow out through the roof.

The designers then went one step further – in addition to the "normal" spring-loaded overpressure safety valves on the top (steam-side) of the steam drums, they also installed a mechanical overpressure safety valve on the underside (the water-side) of the steam drum. This was highly unusual – the overpressure safety valves on steam drums are normally only on the steam-side (the top). The waterside overpressure safety valve meant that, in extreme overpressure fault conditions, high-density water – instead of low-density steam – from the steam drum would be blown through the roof, so the many tonnes of water in the drum available to feed a steam–sodium leak accident could be removed in a few seconds. (By contrast, it would take very much longer to remove all the water if only the steam-side overpressure relief valves were used.)

This waterside overpressure safety valve was operated by a spring-loaded pilot valve which, for test purposes, could be isolated from the water pressure in the steam drum by a valve called the PLIV – the pilot line isolating valve. During normal operation the PLIV was locked open with a large padlock.

So far, so good. The operational problem, however, was this: *the waterside overpressure safety valve had to be tested every three years to satisfy the boiler inspectors who insured the plant.* Watching this operation being carried out was fascinating, and also one of the hairiest operations I have ever witnessed in my career, as I shall now describe.

The waterside overpressure safety valve was set to operate at a pressure such that it was the last one on its steam drum to operate in a high-pressure fault – the steam-side Safety Valves would operate first. The normal steam operating pressure at PFR was about 160 atmospheres. The superheater overpressure safety valves operated at 170 atmospheres, the steamside overpressure safety valves on the steam drums operated at 180 atmospheres, and the waterside overpressure safety valves operated at 190 atmospheres. These conditions were saturated steam conditions, so the steam/water temperatures were as follows.

Conditions	Pressure (atmospheres)	Steam/water temperature (°C)
Normal operation	160 atmospheres	347
Superheater safety valve operates	170 atmospheres	352
Steam drum steam-side safety valve operates	180 atmospheres	357
Steam drum waterside safety valve operates	190 atmospheres	361

The procedure was that, with the reactor shutdown, heat was gradually added using the large sodium coolant pumps. These pumps were driven by massive electric motors and in normal operation they circulated the sodium, thereby transferring the heat from the reactor core to the boilers. For this test, however, the reactor was shutdown, but the action of the pumps churning the sodium in a closed circuit was enough to heat up the sodium. Under the action of the pumps, the sodium temperature gradually crept up, in a slow and controlled way. Meanwhile, the water level in the steam drums was being maintained constant by the electrically driven Boiler Feed Pump driving pressurized, cold water through feed-water control valves. The temperature of the sodium, and hence the temperature of the water in the steam drum, increased until each of the overpressure safety valves was tested in turn. Adjustments were made if necessary so that each valve operated at exactly the right pressure. The tested valve was then isolated so that the pressure could be increased further for the next Safety Valve.

First was the superheater Safety Valve, since it required the lowest test temperature and pressure. After it had been tested, it was isolated, and we moved onto the steam drum Steam-side Safety Valve. Eventually conditions were right for its test; the steam-side Safety Valve blew off some steam, the pressure dropped a little, and the valve re-seated. All was OK.

We then waited for the temperature and pressure to rise again until conditions were right for the waterside safety valve test.

All of these tests took many hours. I guess about a dozen people were present; shift staff, day staff and two people from the independent boiler inspectors to witness the events. We had started on day shift, gone right through back-shift, and it was now in the small hours – may be 2 o'clock in the morning. We mostly just hung around between tests, standing next to the steam drum, listening to radio communications from the control room giving temperature and pressure read-outs.

Eventually we had got to the point of being ready to test the waterside overpressure safety valves. The other safety valves had been tested and had been isolated, and the steam drums were at the highest pressure and temperature they had ever experienced.

The Shift Manager was in the Control Room looking after the plant and monitoring the temperatures and pressures. Meanwhile, at the steam drum, the Operations Manager was in control. An old-fashioned-looking manual pressure gage had been installed on the steam drum; it was a specially calibrated gage for the boiler inspector.

I was a trainee, watching all that was going on from a walkway 2 m or 3 m from the steam drum. My feelings were a strange mixture of sensory over-stimulation, cheerfulness, and suppressed panic:

- It was hot. The steam drum, although heavily insulated, radiated heat and we were near the top of the building where the heat collected anyway.
- The noise when a power station safety valve opens is thunderous, even when wearing ear-defenders.
- There was a chummy camaraderie amongst all present. We were all doing something vaguely macho together: Noisy, sweaty, and uncomfortable.
- We all knew about the Cockenzie event – the whole electricity generating industry had learned a lot from it. Right now, we were raising the temperature and pressure in the steam drum to the highest it had ever experienced since commissioning and, furthermore, it was pressurized with steam. If it were ever going to fail, now was the time; and if it failed, it was not hydraulic pressure, it was steam pressure, so it would blow us all away.
- However, I knew the steam drum had been carefully inspected at the shutdown which had just been completed. I told myself it would be alright. I suppressed my imagination and tried to maintain an attitude of looking interested but not overwhelmed.

The waterside overpressure safety valve test was more complicated than the others. The problem about testing the waterside overpressure safety valve was that, because it let liquid water out very quickly (instead of just steam), the steam drum would empty all its contents in a few seconds. It might all happen so quickly that the drum would be empty before the safety valve had a chance to re-seat itself. Why was that a problem? Well, the feed-water entering the drum was much colder than the bulk water in the drum, so if the drum were emptied it would re-fill with cold water. This would "cold-shock" the drum, as the bottom half cooled rapidly while the top half was still very hot. Huge thermal strains would be created – basically, the drum would try to take on an inverted-U shape. This was called "hogging the drum," which would cause significant fatigue damage and make any existing cracks grow. It would mean we would have to shutdown again and review the safety case for the drum. At worst, we might have to do more inspection work.

The only way to avoid the risk of hogging the drum was to have someone next to the Pilot Line Isolating Valve, the PLIV, to shut it immediately after the main waterside overpressure safety valve had opened. By closing the PLIV, the Pilot Valve would close, making the Safety Valve close also. The PLIV was a small handwheel-operated valve directly under the main steam drum, which was not normally readily accessible. For the purposes of the Safety Valve tests, a small scaffold platform had been installed so that someone could lie on his back, directly under and only a few inches away from the steam drum, from where he would close the PLIV as soon as the waterside safety valve opened.

Whoever got the job of closing the PLIV was going to be extremely close to the drum, in a very hot and uncomfortable position; he would have to wait until he heard the very loud noise of the waterside overpressure safety valve opening, and then immediately close the PLIV. Tonnes of scalding pressurized water were going to flow through the safety valve a few inches away from him.

I suppose everyone, including me, thought it would be a job for a young apprentice fitter from the workshop. Or may be I would be asked to do it, since I was the junior engineer present. However, to my surprise the Operations Manager said he was going to do it. Even now, this seems a strange decision: the senior guy on site was going to twiddle a valve. I assume his decision-making was based on the significance of the test and the importance of doing it right. I also suspect, however, that he felt he had something to prove. The Operations Manager, to whom all the shift managers reported, had never himself actually worked on shift operations; his background was on the engineering and maintenance of the station control and computer systems. I think this was his way of saying to the shift guys present, Look, I can do this macho stuff too.

The test went fine – it was quite anticlimactic, really. The Operations Manager climbed onto the scaffold platform under the steam drum and carefully opened the PLIV. We then waited as the control room gradually raised the temperature and pressure. After a few minutes, quite suddenly, the safety valve opened, there was a huge noise and the pipes shook. The Operations Manager closed the PLIV despite the shaking pipes, and the safety valve re-closed. It was all over in a few seconds. He climbed out, sweat running down his face, and we congratulated ourselves on doing the job. After a bit of chat, we all left the power station to go back to our homes. The night shift team would bring the pressure and the temperature back down, and de-isolate all the steam-side pressure valves, ready to return the power station to generating electricity after its planned shutdown.

It was about 3 o'clock in the morning. In the high-latitude Caithness summer it was already daylight as I left the building. The wind had been blowing from the power station toward the car park. When the safety valves had opened, the insides of their vent pipes had been untouched since the last test three years previously, so the pipes had a layer of rust inside them. The discharge of the vent pipes also had seagulls' nests on them – the warmth rising from the power station must have been comfortable for them. Brown rusty water and the remnants of the seagulls' nests were scattered across the car park, streaking our cars with bits of straw and feathers and rust marks.

I got back to work at about 11 o'clock the following morning after a reasonable sleep. Someone told me the Operations Manager had been at his desk at 0745 h, as usual.

THE SS *NORWAY* BOILER EXPLOSION, MIAMI, 2003

There may now be a good understanding of the mechanisms of pressure vessel failure, and the technology for monitoring the condition of thick steel pressure vessels may be available, but there cannot be any complacency. Pressure vessels, in all process industries, remain a significant industrial hazard, and care in their design, operation and maintenance is required. A reminder of this occurred May 25, 2003 when the Bahamas-registered passenger ship SS *Norway*, carrying 2135 passengers and 911 crew, suffered a boiler explosion shortly after arrival in the Port of Miami, killing 8 crew and injuring 17 others. The accident on the *Norway* provides an example of the complexity of managing pressure vessel integrity, and what can go wrong if the management processes are poor – or even, as in this case, criminally negligent (Fig. 7.9).

The SS *Norway* was one of the last steam-powered cruise liners. It had been launched in Saint-Nazaire, France, in 1960 as the SS *France*, but was sold in 1979 to Klosters, one of Norway's oldest shipping companies. The ship was refitted for Caribbean cruising operations and renamed SS *Norway*. In 1987 she was reflagged to the Bahamas.

At 0637 h on May 25, 2003 one of the *Norway*'s four oil-fired boilers exploded with a loud bang. The rupture occurred along 2 m–3 m of the longitudinal welds of a 29 inch (0.736 m) header – a large thick-walled steel tube that connected banks of smaller boiler tubes. The boiler contained 20 tonnes of water at 310°C and 60 bar pressure, which flashed to steam and expanded into the boiler room. The fatalities and injuries were due to scalding by hot steam – the deceased were killed by second and third degree burns to 50%–100% of their bodies.

FIGURE 7.9

A poor quality video still of the SS Norway just after the boiler explosion which killed eight crew members.

(Source: http://ships-no-mar.blogspot.co.uk/2012/07/blue-lady.html)

Subsequently, the US National Transportation Safety Board investigated the accident. Their investigation included a detailed review of the boiler's operational and maintenance history. The boiler header had failed because of extensive fatigue cracking. The NTSB report [12] is interesting because of the number of causal factors, the depressing chain of missed opportunities when actions could have been taken to achieve a better outcome, and, notably, the deliberate effort to deceive boiler inspectors by hiding boiler cracking problems:

- Cracks had been detected in the original longitudinal header welds as far back as December 1970 and several weld repairs had been carried out. The cracks had started at corrosion pits on the original welds. The corrosion pits had occurred because of poor water chemistry during periods of shutdown. Specifically, higher-than-specified oxygen levels had occurred which should have been prevented by dosing the water with small quantities of hydrazine, which is common practice for boiler pressure vessels, but the NTSB investigators found there had been long periods (10–40 days) when hydrazine dosing had been inadequate.
- The cracks grew from the corrosion pits because of stress cycling. Boiler startup and shutdown procedures apparently did not specify the time for raising or lowering pressure, but typically the operators did this in about 3 hours in either direction. Boiler cycle time was known to be a problem but nothing was apparently done to control this. This rapid power raising and shutdown will have imposed significant stresses on the boiler, thereby causing fatigue crack growth. Furthermore, the feet of the boiler, which were supposed to slide to allow for thermal expansion, had not been lubricated for sometime and were stuck, thereby imposing further stresses on the boiler.
- The boilers were periodically inspected visually and by non-destructive tests (such as dye penetration, magnetic particle or ultrasonics), and any cracks found were ground out. Eventually, however, the minimum wall thickness was reached. At that point, weld repairs were done to build the wall thickness backup again. The repairs were done by Lloyd Werft in Bremerhaven, Germany, between 1987 and 1990. The weld repair procedure was written by Deutsche Babcock and reviewed and approved by Bureau Veritas.
- The NTSB report criticizes the weld repair procedure, saying that it was not specific, and the welders were not necessarily qualified. Also, the *Norway* operators did nothing to eliminate the causes of the cracking – namely oxygen corrosion during shutdowns, and high fatigue stresses during startups and shutdowns. The weld repair left enhanced residual stresses, which allowed subsequent cracks to initiate sooner and propagate faster than the original cracks.
- Lloyd Werft wrote to Klosters (owners of the Norway) in January 1991 to point out that "preservation of the boilers in the shutoff condition should be given the utmost attention."
- An engineer had written in 1997 that "the boilers on the SS Norway have reached a state where a decision must be made," and the options offered

included boiler replacement. Boiler re-tubing was carried out, but the boiler drums and headers were left unchanged. Several other reports pointed out that oxygen corrosion was taking place, and expressing concerns. The boilers were apparently inspected thoroughly for the last time in 1996. Subsequent inspections were carried out by removing the manhole at the end of the header and looking inside with a flashlight, instead of crawling inside and inspecting thoroughly.

- Finally, and most damningly, the NTSB examination found large, isolated copper nuggets on the fracture surfaces. NTSB concluded that the copper had been deliberately introduced to the cracks when they were already about 0.1 inch (2.5 mm) wide. "From a metallurgical point of view, copper would offer no structural benefits and would not be considered a repair. The only explanation for the presence of copper is that it was introduced to mask the crack, impede inspection, and avoid necessary repairs… The presence of copper…indicates that…engineers were aware of the cracking condition but did not take appropriate action to fix the problem." The implication of this bizarre conclusion in the NTSB report seems to be this: At some unknown point, probably many months before the accident, someone had used brazing (a process that uses copper–zinc alloy as a filler material) to cover over cracks in the boiler, in order to prevent the independent boiler inspector from noticing the cracks. This was a stupid, deliberate and criminal act, perpetrated by individuals who remain unnamed – they have literally got away with murder.

Five years later, in May 2008, the Norwegian Cruise Line (a subsidiary of Klosters), pleaded guilty to criminal negligence in a court in Miami. The company was fined $500,000 in criminal penalties. The accident was described as a "preventable tragedy." As part of a plea bargain, the families of dead and injured crew members (mostly Filipinos) were not able to press further charges.

The *Norway* never saw any more commercial service. She was ultimately taken to India for scrapping, amid legal challenges because of concerns over asbestos on board.

Engineers and scientists may now understand the mechanisms of pressure vessel failures, and have the tools and techniques for monitoring for cracks and preventing corrosion, but it still requires a great deal of vigilance to ensure that their integrity is maintained.

REFERENCES

[1] For a description of conditions in Cahaba prison, see http://www.historynet.com/surviving-a-confederate-pow-camp.htm.

[2] There have been several books written about the *Sultana* tragedy. The Sultana Tragedy: America's Greatest Maritime Disaster, Jerry O Potter, Pelican, 1992, provides a fascinating account.

[3] Nine boiler accidents that changed the way we live, Bulletin of the National Board of Boiler and Pressure Vessel Inspectors, vol. 58, no. 2, 2003.

[4] Report of the Court of Inquiry into the Accidents to Comet G-ALYP on January 10, 1954 and Comet G-ALYY on April 8, 1954, HMSO, 1955.

[5] Report on the Brittle Fracture of a High-pressure Boiler Drum at Cockenzie Power Station, South of Scotland Electricity Board, January 1967.

[6] A.A. Griffith The phenomenon of rupture and flow in solids, Phil Trans Roy Soc A221 (1921) 163–198.

[7] GR Irwin, Fracture dynamics, in Fracturing of Metals (ASM, Cleveland, 1948), pp. 147-166.

[8] B.A. Bilby, A.H. Cottrell, K.H. Swinden, The spread of plastic yield from a notch, Proc Roy Soc A272 (1963) 304.

[9] An Assessment of the Integrity of PWR Pressure Vessels, 2nd report, (The Marshall Report) UKAEA, March 1982.

[10] R.P. Harrison, I. Milne, Assessment of defects: the CEGB approach, Phil Trans Roy Soc A299 (1981) 145–153.

[11] Evaluation of the PISC-II trials results, PISC-II Report No 5, Final Issue, Sept 1986, CEC/OECD.

[12] National Transportation Safety Board, Marine Accident Brief NTSB/MAB-07/03, October 29, 2007.

The Second Industrial Revolution – A Brief History of Computing

"The digital revolution is far more significant than the invention of writing or even of printing".
Douglas Engelbart

In the twenty-first century, it is difficult to imagine a world without computers. Almost every aspect of our lives is now tied to computers in some way or other. They are used throughout industry for the control of industrial plant and processes, manufacturing, aircraft, automobiles, washing machines, telephones and telecommunications, etc. Most domestic electrical hardware now contains some sort of microprocessor control.

Microprocessors are now so completely ubiquitous, and so much of our lives is completely dependent on them, that it is surprising that the people and the laboratories that led their invention and development are not household names in the same way as, say, Newton's discovery of gravity or Darwin's theory of evolution. Microprocessors have caused an ongoing revolution in all our lives. The first industrial revolution in the nineteenth century led to steam power for factories, ships, and trains, and also to the ready availability of substances like steel, which transformed the lives of our forebears. This new, second, industrial revolution may even be more overwhelming than the first.

The starting point for the history of computers is unclear, and it can be selected almost arbitrarily, because so many separate technical developments were necessary before they could all come together to create the modern computer. Also, unlike many other technical developments, one cannot point to a single "inventor" to give the credit – it is really much more complicated than that. The following chapter is intended as a very brief summary of key developments and the names of those principally involved. However, any condensed summary such as this will almost inevitably be invidious, and doubtless some key names have been omitted, for which I apologize.

The history of computing really begins with ideas for mechanical calculating devices and machine programing, long before electronics existed. These ideas originate at the time of the first industrial revolution, or even earlier. A very brief and selective historical overview follows. Many will no doubt think this is too selective, or too short.

Blaise Pascal, the French philosopher, mathematician, and inventor who died in 1662 aged only 39, is nowadays remembered mostly for his contributions to mathematics. Pascal achieved some truly amazing things when he was still a teenager, such as a mathematical analysis of conic sections. He also produced, at the age of 19, a mechanical calculator capable of addition and subtraction. It became known as the Pascaline, the only functional mechanical calculator of the seventeenth century, and some 20 machines were built.

In 1801, Joseph Marie Jacquard demonstrated a means of controlling looms using punched cards, thereby creating the world's first programable machine. If a different set of punched cards was used, the pattern of the loom's weave was changed.

In 1812, Charles Babbage conceived the idea of what he called a "difference engine," a mechanical device for computing and printing tables of mathematical functions. He demonstrated a small version in 1822 that enabled him to secure funding from the Royal Society and the Government of the day to develop a much more sophisticated version. He worked on this for 20 years before government support was withdrawn. Part of the problem was that Babbage had lost interest in the Difference Engine and had become interested in a more ambitious scheme to produce a device he called the "Analytical Engine." (He had other distractions, too. In 1828 he became the Lucasian Professor of Mathematics at Cambridge University, although it is said he never gave any lectures.) The Analytical Engine was never built, but he conceived it as a "universal machine," a general-purpose device that could perform any sort of calculation whatsoever. His Analytical Engine would have comprised several different parts.

- There would have been a memory, for holding numbers to be used as data in the calculations, and also temporary storage of numbers involved in intermediate stages of the calculations.
- There would have been an arithmetical unit, which performed the arithmetical operations on the numbers.
- There would have been a control unit for ensuring the desired operations occurred in the correct sequence.
- There would have been input devices for supplying numbers and operating instructions to the machine.
- There would have been output devices for displaying the outputs from the calculations.

The analytical engine would have comprised a huge number of shafts, gears, and linkages. He expected the memory would hold 1000 numbers of 40 decimal digits each, which equates to about 16 KB. He also expected to be able to supply data to the machine in punched-card format. (A portrait of Joseph Marie Jacquard hung in Babbage's drawing room.)

The whole device was impossibly complex for its time, although Babbage continued to work on it up till his death in 1871. His ideas were too ambitious, and the required precision in manufacture was too great for the nineteenth century. In later life, he apparently became a quarrelsome old man, a misunderstood genius, who was intolerant of criticism and embittered by his own failure.

The real problem, of course, was that Babbage was trying to achieve things with nineteenth century mechanical technology that needed twentieth century electronics to be realizable. However, his ideas were visionary, and the overall architecture of the Analytical Engine sounds very familiar to those who work with electronic computers.

In 1991, the Science Museum in London successfully built a working model of Babbage's earlier Difference Engine. In 2010, an organization called Plan 28 announced plans to raise funds to build an Analytical Engine.

Lady Ada Lovelace was born in 1815, the only legitimate child of the poet Lord Byron. He abandoned her and her mother Anne Isabella Byron shortly after her birth, and her mother later deliberately encouraged her interest in mathematics to counteract what she thought of as her father's madness. As a young adult, she formed a friendship with Charles Babbage and an interest in the Analytical Engine. In 1843 she published an article on the Engine, translated from an article in French by the Italian Luigi Menabrea which he had written after attending a seminar given by Babbage in Turin. Her own notes were appended to the translation, and these notes included a description of an algorithm which could have been used on the Analytical Engine (if it had ever been built) to calculate Bernoulli numbers. In effect (and, OK, this is maybe stretching the point) this algorithm was the world's first computer program. She also recognized that the Analytical Engine might be used for things other than just calculations. As an example, she suggested that perhaps it could be used to compose music.

In 1936, the British mathematician Alan Turing proposed a machine, now known as the "Universal Turing Machine," which he described thus: "It is possible to invent a single machine which can be used to compute any computable sequence." This was a piece of mathematical philosophy at that time, but it formed the basis for many developments in World War Two and later. Turing himself went on to wartime work on cryptography at Bletchley Park, Buckinghamshire, in England – he specified the electromechanical (relay) devices, called "Bombes", used in decoding intercepted German radio messages which had been encrypted using the Enigma devices. In 1946, he proposed one of the first designs for a stored-program computer, the Automatic Computing Engine (ACE), which could carry out subroutine operations, with a memory of some 25 KB and a speed of 1MHz. A prototype was built using 1450 thermionic valves (or "vacuum tubes" as they were known) and mercury delay lines for memory. It first ran in 1950. Production versions were produced by English Electric Ltd and sold from 1955.

(The tragedy of Turing was that he was homosexual at a time when it was illegal in the UK. In 1952, he was prosecuted for "gross indecency" and found guilty, and he was given chemical castration as an alternative to prison. He died from cyanide poisoning in 1954, which the inquest determined was suicide. In 2009, Prime Minister Gordon Brown gave an official government apology for the way he had been treated.)

Also at Bletchley Park during wartime, a semiprogramable electronic digital computer called "Colossus" was developed using 1600 thermionic valves (with later versions rising to 2400 valves), to help with decryption of German coded messages that used the Lorenz cipher (known in the UK as "Tunny"), which was more complicated and sophisticated than Enigma. The first Colossus became operational in January 1944 and immediately speeded up the decryption process which had previously been done manually. Ten Colossus machines were built before the end of the war. Colossus was developed by Tommy Flowers, Max Newman and Allen Coombs, although little credit was given to Flowers and his team until several decades later because of wartime and Cold War secrecy [1,2].

Among their many other achievements, the Colossus machines were used to decrypt German signal traffic around the time of the D-Day invasion of Normandy on June 6, 1944. On June 5, Colossus decrypted a message from Hitler to Field Marshal Rommel that ordered Rommel not to move any troops away from the Pas de Calais area. Hitler was expecting the real invasion to be near Calais, starting five days after the Normandy landings. This confirmed that the Nazis had been misled by Allied efforts to have them believe that the main invasion would be near Calais and that the Normandy landings were just a feint. General Eisenhower, who was in overall command of the Allied invasion forces, saw this decrypted message on June 5 and said, "We go tomorrow." Much later, Eisenhower said that, without the information Bletchley Park had supplied, the war could have gone on for at least two years longer than it did.

Flowers' key contribution was probably his realization that high-speed switching processes (as are used in computer calculations) could be carried out much more quickly by circuits using thermionic valves than by circuits using relays. Colossus was the world's first electronic computer but it was not designed as such – it was strictly for decryption. However, the basic technologies used in modern computers – data storage and retrieval, fast processing, variable programing, data output by printer – were all anticipated by Colossus.

Flowers' invention remained secret for decades after the war possibly because the advancing Soviet army had captured Lorenz cipher machines and had begun to use them for their own purposes post-war. Hence, Bletchley Park was able to continue to monitor Russian Lorenz-enciphered messages well after the end of World War Two, and for that reason Colossus remained "Top Secret." Flowers' very success ensured his lifelong near-anonymity.

Flowers was the son of a bricklayer who had studied electrical engineering at the University of London in the 1920s, and had subsequently worked on the design of electronic telephone exchanges. He is almost forgotten but he should be an extremely well known name indeed, as well known as other famous British wartime engineers such as Frank Whittle and Barnes Wallis. He developed the world's first semiprogramable electronic computer, under extremely difficult wartime conditions, which enabled some very secret Nazi communications to be decrypted. He was unable to patent his developments. Of the imposed secrecy which lasted until the 1980s, he wrote, "The one thing I lacked was prestige." Flowers died in 1998.

After the war, work proceeded in parallel in the UK and the United States, with multiple developments, and a simple chronological order becomes almost impossible.

In 1946, Max Newman (who had worked with Flowers on Colossus) went on to establish the Royal Society Computing Machine Laboratory at Manchester University, where with colleagues he built the world's first electronic stored-program digital computer, called the "Manchester Baby," which was first operational in June 1948. Their development led to the world's first commercially available general-purpose electronic computer, called the Ferranti Mark 1. The first was delivered in February 1951.

Whereas much of the UK-based wartime work was kept secret, work progressed in the USA in parallel with the UK work, and in a more open fashion. Electronic numerical integrator and calculator (ENIAC) was conceived and designed by John Mauchly (1907–1980) and John Adam Presper "Pres" Eckert (1919–1995), following a 1942 memo from Mauchly proposing a general-purpose electronic computer. A contract was received from the US Army in 1943, and the completed machine was announced to the public in February 1946. It used some 17,000 thermionic valves and weighed about 27 tonnes. Vacuum tube failure rate was such that its availability was only about fifty percent. Input was via an IBM card reader.

The successor to ENIAC was called electronic discrete variable automatic computer (EDVAC), which was proposed by Mauchly and Eckert in 1944. The Hungarian-American mathematician, polymath and all-round-genius John von Neumann acted as a consultant on the EDVAC project, and in 1945 he wrote a report summarizing the project and proposing some improvements [3]. In particular, von Neumann described a design architecture for electronic digital computers, separating the computer into a central processing unit (CPU), a memory, and input and output mechanisms. EDVAC included a magnetic tape reader (or wire recorder), a control unit with an oscilloscope, a dispatcher unit for receiving instructions from the control with memory to enable redirection to other units, a computational unit, a timer, a memory unit comprising two sets of 64 mercury delay lines, and temporary memory. It had some 6000 thermionic valves and consumed 56 kW of power. EDVAC was delivered in 1949 and operated until 1960. Its delivery was delayed because of disputes between Eckert and Mauchly and their then-employers, the University of Pennsylvania; such disputes have been a recurring theme in computer developments right up to the present day.

Back in the UK, Maurice Wilkes (1913–2010) at Cambridge University was able to read a copy of von Neumann's 1945 report which inspired him to develop a working stored-program computer called EDSAC, the Electronic Delay Storage Automatic Calculator, which was operational in May 1949 and actually pre-dated EDVAC.

The first US-produced commercially available electronic stored-program digital computer was UNIVAC 1, which was developed by Eckert and Mauchly and was delivered to the US Census Bureau in March 1951.

The name of John Von Neumann (1903–1957) appears above in relation to EDVAC and EDSAC; he made contributions to many aspects of science and technology in the twentieth century, including mathematics, game theory, quantum mechanics, linear programing, the Manhattan project that led to the atomic bomb, and the postwar development of the hydrogen bomb. He applied game theory to the Cold War nuclear stand-off, and helped develop the strategy of Mutually Assured Destruction

(MAD) which (arguably) helped avoid global destruction during the period from the 1950s until 1990.

Von Neumann was also a founding figure in the development of computer science. With Stanislaw Ulam he developed the Monte Carlo method for statistical modeling of complex problems such as nuclear criticality; these problems can really only be solved by computers, since they require many thousands of repeat calculations. He also proposed improvements to ENIAC that enabled it to run stored programs. In 1949, he designed the first self-replicating computer program, thus anticipating computer viruses.

The above paragraphs trivialize John von Neumann's achievements. He did so much, in such a wide variety of topics, that it is difficult to describe his achievements without sounding hyperbolic. His name should be extremely well known indeed, as much as Einstein or Newton; perhaps he will receive wider recognition at some point in the future [4].

A final name-check in this first, pre-transistor stage of computer development is for Harry Huskey. Harry Huskey was an American who had worked on ENIAC, and then worked with Alan Turing at the UK's National Physical Laboratory on development of the ACE computer in 1947. He then moved back to the USA and worked on the EDVAC project and, later, he designed the G15 computer for the Bendix Corporation in Los Angeles. The Bendix G15 used 450 thermionic valves, weighed 950 pounds (430 kg), and was the size of a small cupboard, about 1.5 m^3. It cost $49500 to $60,000 and could also be rented for $1500 per month. It went on sale in 1954, and over 400 were manufactured with some remaining in use until 1970. With the Bendix G15, the computer truly entered the marketplace.

Hence, by 1950, the basics of computer science and engineering were in place – logic elements for computation, memory, programing, means of control, and input and output devices. The ability to scale up the early computer designs was limited, however. The logic elements used in computation and processing normally employed thousands of thermionic valves, which were large, consumed a lot of power, and were unreliable. Memory size was limited because of the space requirements and general clumsiness of mercury delay lines. Fundamental improvements in technology had to be made for the computer to increase in capability and reduce in size.

Mercury delay lines and thermionic valves have now gone the way of the dodo, the steam engine, the gramophone and the video cassette recorder. Their replacements came ultimately from the invention of the transistor in the Bell Laboratories (formerly the Bell Telephone Laboratories, originally named after and created by the inventor of the telephone, Alexander Graham Bell), in New Jersey, in the period 1946–1951. This was the product of a team including William Shockley, John Bardeen and Walter Brattain, for which they were jointly awarded the Nobel Prize for Physics in 1956. John Bardeen is the only person to have been awarded the Nobel Prize for Physics twice (he won it again in 1972 for work on the theory of superconductivity). John Shive was another important team member, although he was not honored by the Nobel committee. However, it is Shockley who is most often associated with the invention of the transistor.

Shockley had worked during the war on radar and anti-submarine warfare. In July 1945, he was asked by the US War Department to prepare a report on the likely casualties if Japan were invaded. His report predicted millions of both Japanese and Allied casualties, and influenced the decision to use the atomic bomb.

After the war, Shockley and his team at Bell Labs first developed the point-contact transistor, and then the junction transistor, using germanium, and for which they received the Nobel Prize. However, Shockley was a difficult person to work with – he had a domineering, abrasive style that caused him to fall out with his co-inventors and subsequently to leave Bell Labs. He then set up his own Shockley Semiconductor Laboratory, which became part of what would later be known as Silicon Valley in California. Shockley Semiconductor Laboratory was sold to ITT in 1968.

In later life Shockley was Professor of Engineering and Applied Science at Stanford University. He became interested in eugenics, and made some fairly bizarre statements about intelligence, race and reproduction, including advocating that people of low intelligence should be sterilized. He donated his sperm to a sperm bank that only supplied sperm from high IQ individuals. In 1981 he sued a newspaper for calling him a "Hitlerite".

Following the invention of the transistor, some visionary people began almost immediately to conceive of an integrated circuit, where transistors, resistors, diodes and connecting links could be constructed on a single piece of semiconducting silicon. Geoffrey Dummer (1909–2002) was a graduate of Manchester College of Technology who worked at the Royal Radar Establishment during the Second World War. In 1952 he presented a paper at a US conference in which he suggested that "it now seems possible to envisage electronic equipment in a solid block with no interconnecting wires." He is recognized as the "Prophet of the Integrated Circuit" and he went on to propose a simple design for an integrated circuit in 1957, although he never produced a functioning circuit.

A group of Shockley's co-workers left Shockley Semiconductor Laboratory in 1958 to set up Fairchild Semiconductors, because Shockley would not allow work to continue on the silicon transistor. The group included Robert Noyce (1927–1990) and Gordon Moore (b.1929).

Noyce filed a patent for the integrated circuit in July 1959, a few months after Jack Kilby (1923–2005) had produced the world's first integrated circuit while working for Texas Instruments. Kilby received the Nobel Prize for Physics in 2000.

Moore and Noyce together founded the Intel Corporation in 1968, which today is the world's largest and highest valued semiconductor chip manufacturer. Moore's name is probably best known for his proposal of what has become known as "Moore's Law" which is based on a presentation he gave in 1965 showing that the number of components in integrated circuits had doubled every year and he predicted at that time the trend would continue for at least a further 10 years. In fact, the trend has continued to the present day, and Intel expects to be offering memory chips with a "feature size" of 10 nm by 2015. This astonishing rate of technical development has enabled integrated circuits, microprocessors and memory chips to have the power and the ubiquity they currently enjoy. However, it is becoming clear that the limits of

Table 8.1 The development of the computer – a highly selective roll of honor

Blaise Pascal	1623–1662	France	Mechanical calculator
Joseph Marie Jacquard	1752–1834	France	Programable loom
Charles Babbage	1791–1871	UK	"Difference Engine" and "Analytical Engine"
Lady Ada Lovelace	1815–1852	UK	First "program"
Alan Turing	1912–1954	UK	"Universal Turing Machine" concept, ACE
Tommy Flowers	1905–1998	UK	Colossus semiprogramable electronic computer
Max Newman	1897–1984	UK	"Manchester Baby" Ferranti Mark 1
John Mauchly	1907–1980	USA	ENIAC/EDVAC/UNIVAC
Pres Eckett	1919–1985	USA	ENIAC/EDVAC/UNIVAC
John von Neumann	1903–1957	Hungary	"Von Neumann architecture"
Maurice Wilkes	1913–2010	UK	EDSAC
Harry Huskey	1916	USA	Bendix G15
William Shockley	1910–1989	USA	Transistor (Bell Labs)
John Bardeen	1908–1991	USA	Transistor (Bell Labs)
Walter Brattain	1902–1987	USA	Transistor (Bell Labs)
John Shive	1913–1984	USA	Transistor (Bell Labs)
Geoffrey Dummer	1909–2002	UK	"Prophet of the integrated circuit"
Robert Noyce	1927–1990	USA	Integrated circuit (Intel)
Gordon Moore	1929	USA	Integrated circuit (Intel)
Jack Kilby	1923–2005	USA	Integrated circuit (Texas Instruments)
Ted Hoff	1937	USA	Microprocessor (Intel)
Federico Faggin	1941	Italy	Microprocessor (Intel)
Chuck Thacker	1943	USA	Personal Computer (Xerox PARC, project leader)
Steve Jobs	1955–2011	USA	Apple
Bill Gates	1955	USA	Microsoft

Unlike many other technical developments, the computer cannot be said to have any single inventor. Here is a personal selection of some of the key innovators whose work led to computers as we know them today. In some instances the work will have been the efforts of large teams; assignment of credit to individuals maybe invidious.

Moore's Law are being reached and any further reduction in feature size will require another radical change in technology, if that is possible.

Ted Hoff (b. 1937) joined Intel in 1967 and is credited with the insight that led to development of microprocessors during the 1970s. Federico Faggin (b.1941) also worked for Intel and, in 1970–1971 developed the methodology and chip design which led to the first microprocessor. Faggin had previously worked for Fairchild Semiconductor where he led the project that developed silicon gate MOS (Metal Oxide Silicon) technology that enabled the production of semiconductor memories and microprocessors.

In the early 1970s, the Xerox Corporation's Palo Alto Research Centre (PARC) developed what could probably be called the world's first personal computer as a research project. Chuck Thacker was the project leader. It was never a commercial product although some two thousand were built. Called the Xerox Alto, it featured a Graphical User Interface (GUI) on a television screen and a mouse with point-click functionality. It used a Texas Instruments 74181 chip and had 128 KB of main memory, with 2.5 MB hard disk. In short, it had many of the features we recognize in today's personal computers. It was undoubtedly highly influential in the later developments of Apple and IBM.

(Like many others, I worked on mainframe computers in the 1970s, cursing their inflexibility on a frequent basis. The first personal computer I used (in 1979) was a Commodore PET, which used audio cassettes for data storage. The IBM PC was introduced in 1981 but required the user to be fluent in the Disk Operating System (DOS) computer language. DOS was an early Microsoft development, but it was very clumsy. I remember the first time I used an Apple Macintosh in 1984. You switched it on and it just worked. I remember thinking, "This is what computers should be like!")

One early use of computers in industrial control systems was the Apollo Guidance Computer, used in the Apollo spacecraft for the moon landings between 1969 and 1972. The astronauts communicated with it via a keypad, and the computer gave messages to the astronauts using a numerical display. The computer carried out computation for guidance, navigation and control. It was designed by MIT Instrumentation Laboratory and manufactured by Raytheon, using Fairchild Semiconductor integrated circuits. During the first moon landing, the computer famously issued "1202" alarm messages during the latter stages; Mission Control advised Neil Armstrong to ignore them.

In the UK, the Prototype Fast Reactor at Dounreay in northern Scotland was an early adopter of computer control systems in industrial safety-related applications. The PFR was designed in the early 1970s and used dual-redundant Ferranti Argus 500 computers for plant indications, alarm messages and some control functions.

In the 1970s and 80s, computers such as PDP-8, PDP-11, and VAX from the Digital Equipment Corporation (DEC), or other manufacturers such as Honeywell, became widespread in industry. These were generally known as minicomputers (to differentiate them from the older, larger mainframe computers) although they were still each the size of a large cupboard. (DEC was acquired in 1998 by Compaq. Compaq merged in 2002 with Hewlett-Packard.)

However, by the late 1970s, some concerns were appearing about software safety and reliability. Academic work to try to analyze or predict software reliability was not very productive. Software was not amenable to the approach used in hardware systems for reliability calculations, where failure rate data for each type of component could be used to produce an overall estimate of failure rate for a system or subsystem. The problem with software is that it is generally unique to each application.

During the 1980s and 90s, many companies began to produce microprocessor-based industrial control equipment for use in factories, oil platforms, refineries, and power plants. Through the normal process of mergers and acquisitions, this sector of industry has "shaken down" until today most of the main players in microprocessor-based industrial digital control systems can be identified in a single table (see Table 2.3).

The point of the above short history is that we are still really just at the beginning of the Second Industrial Revolution, i.e., the revolution brought about by computers. The pace of change is still incredibly rapid, and we are all still learning fast. It is just this sort of environment where mistakes are made, and when computers are used in safety-related applications we need to proceed with caution.

In the 1980s, concerns about the application of software-based systems for the control of hazardous industrial processes began to be expressed. A first attempt to produce guidelines on the design and verification of software for high-integrity applications, the programable electronic systems (PES) guidelines, was produced by a cross-industry committee. The PES guidelines subsequently formed a contribution to the development of international standards such as IEC 61508.

The Second Industrial Revolution is still in its infancy. Computers have entered everyday lives but they will become even more pervasive and ubiquitous, and they will assume more and greater roles. During the First Industrial Revolution, many painful experiences had to be learned – people were killed and injured and environmental damage was caused, and these lessons have had to be learned anew as developing countries industrialize. Our hope for the Second Industrial Revolution is that we can learn more quickly and we can communicate our learning points to others so that mistakes are not repeated.

REFERENCES

[1] J. Copeland, et al., Colossus: The Secrets of Bletchley Park's Code-breaking Computers, OUP 2006, which includes some recently-declassified information.
[2] M. Ward, The road to uncovering a wartime Colossus February 11, 2013, http://www.bbc.co.uk/news/technology-21384672.
[3] J. von Neumann, First Draft of a Report on the EDVAC, Moore School of Electrical Engineering, University of Pennsylvania, June 30, 1945.
[4] W. Poundstone, Prisoner's Dilemma, Anchor Books, 1992.(an account of the life and work of John von Neumann).

Safety Management

Introduction: Organization and Safety Culture

"Good judgment comes from experience, and experience comes from bad judgment".
Rita Mae Brown

"What it takes to do a job will not be learned from management courses. It is principally a matter of experience, the proper attitude, and common sense – none of which can be taught in a classroom... Human experience shows that people, not organizations or management systems, get things done".
Hyman G. Rickover

As discussed in Chapter 1, accidents happen when people under pressure make mistakes. The mistakes may have either immediate or delayed consequences. They may be mistakes of commission (doing something in error) or omission (not doing something they should have done). They may be cognitive errors ("I didn't know that"), or confirmation bias ("I misunderstood the situation"), which can lead to overconfidence.

The pressure can be time, budget, novel situations, too many responsibilities, or something else, or some combination of these – or there may be no pressure at all, in which case we call it "carelessness".

One of the roles of the design engineer is to try to minimize the likelihood that plant operators will make mistakes, by means of sound design that makes mistakes less likely.

One of the roles of engineering and operations managers is to try to set up processes that minimize the likelihood that design engineers or plant operators will make mistakes.

One of the roles of senior managers is to ensure that adequate resources are made available to ensure safety. (This includes training, maintenance, and staffing levels.)

One of the roles of regulators is to ensure that best practices are being followed.

All of the roles have to be underpinned by training, experience, and safety culture. Fig. 1.2 tried to portray this graphically.

"Safety management" is a very wide-ranging activity within an organization. It should not be a stand-alone activity, because (to use another clichéd-sounding safety slogan) safety really is everyone's business. Hence a key role of safety management is actually to check that everyone is really looking after safety in the way they should. This means that a great deal of safety management consists of audit and review of safety-related business processes, with a very clear route for "exception reporting", i.e., the escalation of any deficiencies to senior management so that actions can be placed for problems to be fixed.

It is common in high-hazard industries for the Director (or Vice-President) of Safety to report directly to the Chief Executive Officer; this means that, if the Director (or V-P) of Safety has a valid concern, he can take it straight to the CEO.

There are many business processes that require routine auditing and review, but the key safety-critical processes include the following:

Safe working arrangements
Engineering change arrangements
Personnel recruitment, competence assurance and training, and control of contractors
Accident and incident investigation and corrective actions
Emergency arrangements
Periodic reviews and updating of operating and maintenance procedures
Plant and IT security arrangements

Audits and reviews do not guarantee safety, but they may help if they are done thoughtfully and intelligently. However, there are other prerequisites:

The organization has to have a good *safety culture*. What does this mean? One definition is that safety culture is "*a part of the overall culture of the organization and is seen as affecting the attitudes and beliefs of members in terms of health and safety performance.*" So, a good safety culture corresponds to an organization where everyone genuinely believes that safety is important, and that it requires care and nurturing. There must be clear accountability for safety, strong safety leadership, safety must be part of everyday business, and the company must be a "learning organization" that continually seeks to improve its way of doing its business, and not to be complacent.

Recruitment, training and development of technical personnel have to meet business needs to ensure that operations and maintenance departments are staffed with suitably qualified and experienced people, and to ensure that good technical advisors are available when required.

There must be adequate financial planning and financial resources to ensure that the plant and equipment is maintained in a safe operating condition.

A structured arrangement has to exist for employees and contractors to "whistle blow" if they feel that safety is being compromised in anyway. This process should include an option for complaints about safety-related issues to be made anonymously. (This is now done as a matter of routine in many high-hazard industries.) All workers should have a right to stop all work – to call a time-out – on a particular task if they have safety concerns.

One must be careful about the conflation of two different types of safety, "personal safety" and "major accident safety" (or "process safety"). Some people used to take the view that the number of lost time accidents – i.e., people taking time off work because of, say, hurting their backs or twisting their ankles or falling down stairs – was in some way a good indicator of major accident safety. Notably, some Chief Executive Officers took this view – it made the CEO's job easier if he convinced himself that he had a single metric for "safety." This viewpoint was thoroughly de-bunked in the Baker report on the Texas City oil refinery fire in 2005, which will be discussed in more detail in Chapter 14. Industries where the processes involved are inherently hazardous must retain a very firm understanding of the fundamental technical safety issues of those processes, and that understanding must be passed on in training to all relevant staff. It is quite wrong to assume that, because the personal accident and injury rate at a hazardous plant is low (or even zero), this means the process safety genie is firmly in its bottle.

THE SWISS CHEESE MODEL

James Reason devised a good metaphor for how accidents happen: the "Swiss cheese model [1]." In general, any hazards arising from an industrial process will have a number of controls or barriers in place, each of which should suffice to stop the haz-ard becoming an accident. (Here, "hazard" is used in its technical sense. A hazard is a conceivable occurrence or process failure (or series of failures) which could lead to an accident.) It is the job of engineers and operators to ensure that a hazard does not become an accident. These can be physical barriers such as pressure vessels or control systems, or they can be procedural barriers such as routine test arrangements (to, say, confirm that a fire alarm is working), or they can be personnel training ar-rangements to ensure people do their jobs properly. Reason's model is to picture each barrier as a slice of Swiss cheese, with holes in it. The holes represent the almost in-evitable imperfections in each barrier. Between the hazard and an accident, there are various barriers, but each barrier has deficiencies or holes. Accidents happen when the holes are aligned (Fig. 9.1).

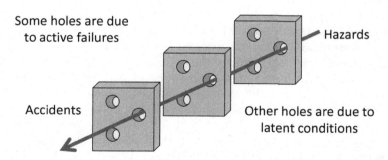

FIGURE 9.1

The "Swiss cheese" model of accidents.

The holes can be caused by *active* failures, such as a control system suffering a dangerous failure or pressure vessel leaking, or an operator making a mistake. The holes can also be caused by *latent* failures, meaning something that happened, perhaps a long time ago, and has gone unrecognized or undetected ever since; it is only when other things go wrong also that the latent failure is discovered.

The role of safety management should be to ensure that all the hazards are identified, that appropriate barriers are in place, and that the barriers remain in place.

Most of the accidents described in this book are about failures of safety management, in some way or other. One example of particularly bad safety management over many years is described next – the sad story of the Royal Air Force Nimrod fire and crash in Afghanistan in 2006, where multiple organizational and safety culture failings were to blame.

ROYAL AIR FORCE NIMROD CRASH, AFGHANISTAN, SEPTEMBER 2, 2006

On September 2, 2006, RAF Nimrod MR2 XV230 caught fire in mid-air over Afghanistan shortly after completing an air-to-air refueling operation. The crew had time to declare Mayday and begin an emergency descent, but the aircraft exploded at 3000 feet altitude and all fourteen crew died.

This accident led to a detailed review by Charles Haddon-Cave into "the broader issues surrounding the loss of RAF Nimrod MR2 aircraft XV230 in Afghanistan in 2006." His report, published in 2009, is one of the most outspoken and savagely critical of its kind [2]. He described a long saga of bad original design, poorly conceived and implemented design changes, and manufacturing defects. This was combined with a series of incidents which should have generated strong warning messages. Lastly, when the Ministry of Defence (MOD) decided belatedly to produce a safety case for the Nimrod between 2001 and 2005, the result was described by Haddon-Cave as a "lamentable job" which missed the key dangers. The production of the safety case was described as a story of "incompetence, complacency, and cynicism."

Haddon-Cave did not hold back from naming individuals and organizations that were culpable – he specifically named, and criticized, key organizations and individuals who bore a share of responsibility for the loss of XV230. In particular, he named those individuals whose conduct he considered to fall well below the expected standards, where their ranks, roles and responsibilities indicated they could be held personally to account (Fig. 9.2).

The RAF's Nimrod MR2's role was anti-submarine warfare and anti-surface unit warfare. Its airframe was derived from the De Havilland Comet, so the Nimrod's lineage dated back to the 1950s. The Hawker Siddeley Nimrod MR1 entered service in 1969, and the upgraded MR2 variant entered service in 1979. Hawker Siddeley Aviation became part of BAE Systems in 1977.

By 2006, the Nimrod was nearing the end of its operational life.

FIGURE 9.2

RAF Nimrod MR2 XV230.

(c)Crown copyright 2013. Reproduced under the terms of the Open Government Licence.

Weaknesses in the original Nimrod MR1 design dated back to 1969. A fundamental problem with the Nimrod's layout was that the fuselage space (the "No. 7 Tank Dry Bays") between the two pairs of Rolls-Royce Spey engines was extremely congested, and that the congestion included both hot gas ducts and fuel pipes.

Because the Nimrod was required to cruise around at low speed for many hours while airborne, it was designed so that two of its four engines could be shutdown, and then re-started quickly when needed. The Nimrod also had to cope with loss of a single engine when operating on only two engines. Hence, to enable mid-air engine re-start, very hot air could be bled from any one engine to start any of the other engines. This was achieved by means of a three-inch diameter steel duct known as the "Cross-Feed Duct." This was insulated with a glass fiber blanket and covered with a stainless steel sheath, where possible; elsewhere it was covered by metallic shrouds. The cross-feed duct was fitted into the space between the engines. In the same space were fuel pipes supplying fuel for the engines. Despite this proximity of hot air ducts and fuel pipes, the No. 7 Tank Dry Bays were not fitted with fire detection or fire suppression equipment.

Design changes were implemented over the decades which made both the congestion and the fire hazard worse in this space between the engines (Fig. 9.3).

The first notable design change was in 1979, when the MR2 variant entered service, it was fitted with a "Supplementary Conditioning Pack" (SCP) to provide cooling air to the Nimrod MR2's improved electronic systems. This Supplementary Conditioning Pack was powered by hot air bled off the Cross-Feed Duct – now called the "Cross-Feed/SCP Duct." The normal condition of this Cross-Feed/SCP duct when in flight was therefore that it would be hot and pressurized.

In practice, the insulation of the hot Cross-Feed/SCP Duct was imperfect; in service the glass fiber insulation got squashed causing the stainless steel sheath to get hot. Bellows units at bends in the duct were, in any case, less well insulated, or even

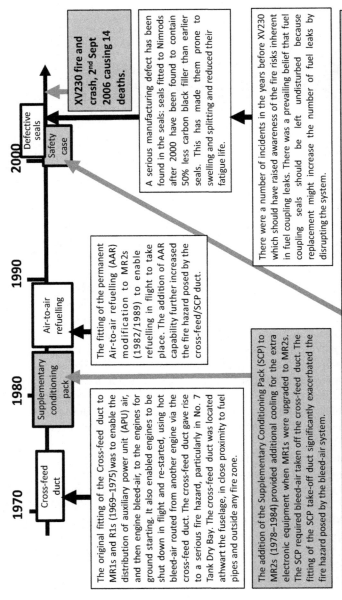

1970

Cross-feed duct

The original fitting of the Cross-feed duct to MR1s and R1s (1969–1975) was to enable the distribution of auxiliary power unit (APU) air, and then engine bleed-air, to the engines for ground starting. It also enabled engines to be shut down in flight and re-started, using hot bleed-air routed from another engine via the cross-feed duct. The cross-feed duct gave rise to a serious fire hazard, particularly in No. 7 Tank Dry Bay. The cross-feed duct was located athwart the fuselage, in close proximity to fuel pipes and outside any fire zone.

1980

Supplementary conditioning pack

The addition of the Supplementary Conditioning Pack (SCP) to MR2s (1978–1984) provided additional cooling for the extra electronic equipment when MR1s were upgraded to MR2s. The SCP required bleed-air taken off the cross-feed duct. The fitting of the SCP significantly exacerbated the fire hazard posed by the bleed-air system.

1990

Air-to-air refuelling

The fitting of the permanent Air-to-air refuelling (AAR) modification to MR2s (1982/1989) to enable refuelling in flight to take place. The addition of AAR capability further increased the fire hazard posed by the cross-feed/SCP duct.

2000

Defective seals

Safety case

XV230 fire and crash, 2nd Sept 2006 causing 14 deaths.

A serious manufacturing defect has been found in the seals: seals fitted to Nimrods after 2000 have been found to contain 50% less carbon black filler than earlier seals. This has made them prone to swelling and splitting and reduced their fatigue life.

There were a number of incidents in the years before XV230 which should have raised awareness of the fire risks inherent in fuel coupling leaks. There was a prevailing belief that fuel coupling seals should be left undisturbed because replacement might increase the number of fuel leaks by disrupting the system.

The question is: why then did nobody spot these design flaws during the intervening years? The answer lies in an understandable assumption by operators that aircraft are designed properly and delivered in an airworthy condition. **The best opportunity to capture these flaws, during the Nimrod Safety Case produced between 2001 and 2005, was lost.** A careful safety case would, and should, have highlighted the catastrophic risks to the Nimrod fleet presented by the Cross-feed/Supplementary Conditioning Pack duct and the Air-to-Air Refuelling modification. If the Nimrod safety case had been properly carried out, the loss of XV230 would have been avoided. **Unfortunately, the Nimrod Safety Case was a lamentablejob from start to finish.** It was riddled with errors. It missed the key dangers. Its production is a story of incompetence, complacency and cynicism. There was a widespread flawed assumption that the Nimrod was "safeanyway" and this fatally undermined the safety case process.

FIGURE 9.3

Timeline of key events leading to the Nimrod XV230 accident, with extracts from the Hadden-Cave report.

completely un-insulated. Nevertheless, the existence of regions of insulation produced a false sense of security – it was perceived that fire risk had been minimized. Thus there were now normally hot surfaces adjacent to fuel pipes, in a region without fire detection or fire suppression systems. Haddon-Cave was critical of BAE Systems for not identifying the fire risk and not installing fire detection and fire suppression equipment. Even worse, fuel leaks within the No. 7 Tank Dry Bays could accumulate in-flight without any facility for drainage.

BAE Systems sought to share blame for these design mistakes by stating that the design modifications had been accepted by the Ministry of Defence in 1977–1978. Haddon-Cave did not consider that this absolved BAE Systems of its responsibility in respect of the poor design of the aircraft. The cross-feed/SCP duct represented a fundamental flaw in the design of the Nimrod aircraft and was the primary physical cause of the accident.

The next relevant design change was the installation of air-to-air refueling (AAR) capability. This was implemented as an urgent design change at the time of the Falklands War in 1982; it was conceived, designed and installed in 18 days. The temporary (1982) modification involved a refueling hose passing through the aircraft cabin, which the crew had to step over. A permanent modification was implemented in 1989; both the temporary and permanent modifications were carried out by BAE Systems.

The fuel tanks were fitted with pressure relief valves ("blow-off valves") to prevent the tanks over-pressurizing during AAR. Blow-off during AAR was seen as a potential hazard, because of the risk of fuel spillage. Also, there was significant concern about the design of the fuel pipe seals, which might cause leakage especially within the No. 7 Tank Dry Bays. Either of these sources of fuel – blow-off during AAR or leaking seals – could have led to the fatal fire, although Haddon-Cave favored blow-off during AAR as the cause.

So, in short, Haddon-Cave concluded that either AAR blow-off or fuel pipe seal failure led to fuel leakage, and the hot Cross-Feed/SCP ductwork provided a source of ignition that led to the fire and subsequent explosion.

There had been several other incidents of fuel leaks, damaged fuel pipe seals, and failure of the SCP ductwork, from which clear lessons should have been learned. However, it appeared no one was taking a wide view of these incidents, which were each treated as independent "one-off" events.

So far, so bad: a less-than-perfect design with some less-than-perfect design changes over a 37 year operating history.

Haddon-Cave then turned his attention to the safety case for the Nimrod MR2, which was intended to identify, assess and mitigate catastrophic hazards, which might occur to the aircraft. In plain language:

What could go wrong?
How bad could it be?
What controls are in place to stop it happening?
What if it happens? (i.e., contingency planning)
Is the level of risk acceptable?

The definition of safety case given by Haddon-Cave is more legalistic: A safety case is "a structured argument, supported by a body of evidence, that provides a compelling, comprehensible and valid case that a system is safe for a given application in a given environment."

The Nimrod safety case was produced by a team consisting of people from BAE Systems and the Ministry of Defence's Nimrod Integrated Project Team (IPT), with QinetiQ Ltd as an independent advisor. Haddon-Cave concluded that the Nimrod safety case was "a lamentable job from start to finish. It was riddled with errors. It missed the key dangers. Its production is a story of incompetence, complacency and cynicism." Few other inquiry reports have used such excoriating words as these.

It should be noted that the drive to produce safety cases within Ministry of Defence projects came as a result of a political directive that military projects should be designed with civil safety standards in mind. Military projects should be "as civil as possible, as military as necessary." However, much of this change of approach was aimed at new projects; in the case of Nimrod, however, it was necessary to produce a safety case for an aircraft that had been operating for over 30 years with a generally good safety record.

A key part of the safety case for Nimrod was its Safety Management System, the objectives of which were described as follows.

1. To establish and maintain an effective management structure and organization for implementing and promulgating airworthiness policy;
2. To assess the safety performance of the equipment and the safety management system itself by measurement and audit;
3. To provide for the documentation of the evidence for airworthiness in a safety case; and
4. To establish mechanisms for learning from the Ministry of Defences and others' experience in safety and airworthiness.

The safety case itself was to be written in accordance with a Ministry of Defence Standard, Def Stan 00-56, which included a requirement for system hazard analysis and risk assessment. The safety case was to have a pyramid structure to be "credible, complete, consistent, and comprehensible," comprising four levels:

- Safety evidence – the foundation of the safety case, including all analysis, records and data.
- Safety argument – justification that the safety evidence is sufficient to demonstrate that the equipment is tolerably safe.
- Safety case report – a summary of the safety argument, including all salient issues and any recommendations for future work.
- Safety statement – a certificate of acceptance from the Integrated Project Team Leader, stating that the equipment is tolerably safe.

For old aircraft such as Nimrod, there was an "implicit safety case" by virtue of its previous operation, which Haddon-Cave describes as "something of an oxymoron," i.e., a safety case implied detailed analysis, and an "implicit safety case" did not.

A review of the "implicit safety case" was to be completed by April 2004, and a Safety Report was to be issued.

A problem here is that the Ministry of Defence was both the "operator" and the "safety regulator", so the Ministry of Defence could re-interpret these requirements as it saw fit.

A Nimrod Project Safety Working Group (PSWG) was established, reporting to the Integrated Project Team Leader, to produce the Nimrod safety case in the period 2001–2005 (i.e., before the accident). The Nimrod Project Safety Working Group therefore had a marvelous opportunity to identify and perhaps rectify the fire hazards in the No. 7 Tank Dry Bays, but it failed to do so.

The main steps in production of the Nimrod safety case were as follows.

1. First, the Nimrod PSWG issued, in February 2002, a Safety Management Plan in which the conclusion of the Nimrod safety case was given – before the work had been done! – as follows. "By virtue of a range of traditional methods, there is a high level of confidence in the safety of the Nimrod aircraft." This made it clear that those involved in producing the Nimrod safety case thought they knew the answer at the outset; i.e., the Nimrod had been in service since 1969, so it had to be safe.

2. Next, The Nimrod PSWG issued a "Hazard Identification Report" in April 2003. Six further reports on hazard analysis and hazard mitigation were issued in September 2004, completing BAE Systems' input to the exercise. Regarding the fire risk in No. 7 Tank Dry Bays, where the fatal fire arose, BAE Systems provisionally assessed the risk of fire as "improbable", although this was one of the "open" items. At no stage was the catastrophic risk of fire created by the hot Cross-Feed/SCP Duct and the Air-to-air refueling modifications ever properly identified, assessed or addressed. Nevertheless, at a two-day meeting with the Ministry of Defence Nimrod Integrated Project Team and QinetiQ Ltd to present the results of the Nimrod Hazard Identification Report at a "Customer Acceptance Conference" on August 31 and September 1, 2004, BAE Systems said the Nimrod was "acceptably safe to operate." The Ministry of Defence and QinetiQ Ltd simply accepted that BAE Systems had completed their task, and so the reports were signed off as complete.

3. Finally, although over 40% of the hazards remained "open" in the BAE Systems reports (meaning there were unresolved issues), the Nimrod Integrated Project Team sentenced these issues, and the Nimrod safety case was declared complete in March 2005.

So, on what basis did the Nimrod Integrated Project Team accept the level of fire risk in the No. 7 Tank Dry Bays? This was the crux of the issue – how did this fire risk get missed? The answer to this is that the approach adopted was highly superficial indeed. No detailed assessment was ever carried out:

a. During inspection visits to operational Nimrod aircraft, the BAE Systems team had indeed noted that in the No. 7 Tank Dry Bays were congested areas with "a potential for hot air, fuel and hydraulic leaks and possible fire." However, and without any further analysis, this risk was assigned an "initial probability" of "improbable." This risk category was important: an "improbable" likelihood of occurrence was equated to a "tolerable" risk, so no further action was required.

b. This initial review was later used by BAE Systems' Nimrod Safety Manager, Frank Walsh, as the basis for sentencing the "open" hazards on behalf of the Nimrod Integrated Project Team (step 3 above of the production of the Nimrod safety case). Hence, on the basis of an initial assessment during an inspection visit, and without any further analysis, the risk of fire in No. 7 Tank Dry Bays was deemed to be "tolerable."

As Haddon-Cave noted, "All three phases of the Nimrod safety case were fatally undermined by an assumption by all the organizations and individuals involved that the Nimrod was safe anyway." In particular, the BAE Systems work was incomplete and contained "numerous systemic errors."

He also noted the budget constraints that the Nimrod safety case project was working within. In October 2001, at the outset of the project, the Ministry of Defence only wanted to spend £100,000 to £200,000 over six months, whereas the Harrier safety case cost £3 million and took three years. Although the final costs and timescales for the Nimrod safety case did exceed these initial Ministry of Defence aspirations, it is nevertheless an indication that the Nimrod safety case was considered from the outset to be a "rubber-stamping" exercise; the Nimrod had been operational since 1969 – so it must be safe. And, in any case, Nimrod MR2 was near the end of its operational life.

Haddon-Cave named three individuals in BAE Systems, three individuals in the Ministry of Defence Nimrod Integrated Project Team, and two individuals in QinetiQ Ltd as the key people in the debacle of the Nimrod safety case. "The best opportunity to capture these serious design flaws in the Nimrod fleet, that had lain dormant for the decades before the accident to the XV230, was squandered."

The role of QinetiQ Ltd as Independent Safety Auditor throughout this exercise is of interest. When QinetiQ's representative Martin Mahy raised concerns about the Nimrod risk assessment methodology in November 2003, the Nimrod Integrated Project Team Leader, Group Captain (later Air Commodore) George Baber, said, "I don't need to get independent advice from QinetiQ, I can go elsewhere," and referred to "bloody QinetiQ" and said "QinetiQ is just touting for business."

QinetiQ's response to this warning from Group Captain George Baber seems to have been conciliatory. Subsequently, QinetiQ went to "extraordinary lengths" to keep the Nimrod Integrated Project Team Leader happy and were "eager to please."

Nevertheless, in 2004, QinetiQ's Martin Mahy again challenged aspects of the Nimrod safety case, including the need to provide sufficient identification and risk mitigation of hazards. BAE Systems' Nimrod Safety Manager, Frank Walsh, replied in an email, "Your full guidance would produce a 'gold-plated' solution that...would not represent value for money."

Later, however, QinetiQ clearly softened their tone, and in June 2004 Martin Mahy wrote to Frank Walsh to say that, "Provided all the risk mitigation evidence is included in the final safety case report, I don't foresee any difficulties...in the sign-off of the baseline safety case."

QinetiQ's Martin Mahy could not be present at the Customer Acceptance Conference on August 31 and September 1, 2004. Another QinetiQ person stepped in at the last moment, and the only briefing he received was that Mahy told him, "66% of the Nimrod risk mitigation work is outstanding." In this conference, BAE Systems gave very optimistic presentations about the work done, while remaining light on details. Haddon-Cave notes that at no stage in the Customer Acceptance Conference did BAE Systems give information about how many hazards had been "closed", how many were still "open" or "unclassified," and how much work remained to be done even in broad percentage terms. In particular, he concluded with regret that it was, in fact, a deliberate and conscious decision by the senior BAE Systems representatives present not to mention or otherwise draw the Customer's attention to the large percentage of hazards which had been left "Open" and "Unclassified." Their motive was to avoid an argument with the Nimrod IPT representatives as to whether the task had been properly completed by BAE systems and whether final payment could be made.

This was a most damning conclusion. Basically, BAE Systems connived to present as positive a picture as possible so that a payment milestone could be deemed to have been passed. In the meeting, both the Nimrod Integrated Project Team and QinetiQ agreed that "the aims and objectives of the project had successfully been achieved." So why did neither QinetiQ nor members of the Nimrod Integrated Project Team challenge the BAE Systems' presentation?

The answer seems, amazingly, to be that, at the time of the presentation, the final BAE Systems Nimrod safety analysis reports had not yet been issued, and they were not actually issued until three weeks later. So, the minutes of the Customer Acceptance Conference recorded the agreements of both the Nimrod Integrated Project team and QinetiQ to a set of safety analysis reports, although neither party had yet seen the final version of those reports, and those final reports still left a number of hazards as "open" or "unclassified."

There were some more steps in the completion of the Nimrod safety case (and yet more opportunities missed) but basically the deal was done, and in March 2005 the Nimrod Integrated Project Team declared that the Nimrod safety case was complete. A strong impression of completeness was given: "....all potential safety hazards have been identified, assessed and addressed...the aircraft type is deemed acceptably safe to operate..."

Eighteen months later, Nimrod XV230 caught fire and exploded over Afghanistan.

None of the individuals and organizations (MOD, BAE Systems and QinetiQ) involved in this sorry saga deserves any credit. All were, to varying degrees, complicit in a story of "incompetence, complacency, and cynicism." And, of course, everyone had felt the Nimrod was "safe anyway" because it had been operating since 1969.

Final sign-off had been carried out by Group Captain (later Air Commodore) George Baber, who therefore carried the responsibility, although he later claimed he had been "hoodwinked" by BAE Systems' Frank Walsh. Haddon-Cave said that Baber "bears the lion's share of the blame", but that this must be put into context. The project was complex and over-stretching, and he also had to deal with increasing and extremely challenging operational demands supporting aircraft in conflicts in both Afghanistan and Iraq, including unscheduled modifications and other operational requirements. He was responsible for an annual budget of £200 million and an acquisition program of £500 million. He was traveling a great deal. There is, therefore, probably some implied failure of delegation.

Haddon-Cave also pointed his finger at organizational causes of the accident. There had been some significant organizational changes in the RAF – a move from a function-based organization to a multidiscipline project-based organization, and larger structures created by Joint Service, (i.e., navy/army/air force) organizations, and "whole-life" equipment management, and finally a lot of outsourcing of work to industry. Thus there had been a period of intense, major organizational changes, which left many people unclear about where responsibilities really lay. On top of this there had been significant budget cuts following the 1998 Strategic Defence Review.

As a result of these changes, "airworthiness" of aircraft within the RAF fleet became a casualty, and it was no longer always uppermost in the minds of personnel in the Ministry of Defence and RAF. In particular, the changes led to expectations that the Nimrod Integrated Project Team would deliver the Nimrod safety case at minimum cost and with minimum necessary upgrades to the aircraft. George Baber told Haddon-Cave that, following the re-organizations, there was a "lack of supervision" from his superiors and he felt "abandoned." "Airworthiness" just became another part of the tri-Service SHEF (Safety, Health, Environment and Fire) function. Haddon-Cave blamed the Chief of Defence Logistics for this loss of focus. The post of Chief of Defence Logistics was held by two people during the relevant period: General Sir Sam Cowan, and Air Chief Marshal Sir Malcolm Pledger.

The RAF's initial Board of Inquiry (which preceded Haddon-Cave's report) made recommendations, amongst which were:

- Nimrod air-to-air refueling was stopped.
- In-flight use of the cross-feed/SCP duct was also stopped.
- A complete review of the Nimrod safety case was begun.

Haddon-Cave made further, very wide-ranging recommendations to improve safety and airworthiness. These recommendations included aspects of safety principles, military airworthiness, safety cases, aging aircraft, personnel, engagement with industry, procurement, and safety culture.

Haddon-Cave noted parallels between Nimrod and other major accidents such as the capsize of the *Herald of Free Enterprise* ferry, the King's Cross fire, BP Texas

City and, in particular, the Columbia Shuttle accident in 2003. He espoused the adoption of "Four Key Principles":

1. Leadership – strong clear leadership from the very top.
2. Independence throughout the regulatory regime
3. People (not just Process and Paper)
4. Simplicity – regulation, processes and rules must as simple and straightforward as possible.

He recommended a new Military Airworthiness Authority (MAA), with clearly identified Airworthiness "Duty Holders", to bring coherence, governance, and responsibility. The MAA was established in April 2010.

The Nimrod was retired from RAF service in 2010. Since then, for the first time since the Second World War, the UK has no long-range maritime reconnaissance capability.

THE MEANING OF SAFETY IN A MILITARY ENVIRONMENT

Accidents such as Nimrod lead to very thorough investigations and great soul-searching amongst all the people involved, directly and indirectly.

Prior to the accident, concerns were raised by people who knew what they were talking about (QinetiQ) to those in positions of authority – and those concerns were rebuffed. The apparent confidence in the safety of the Nimrod was based on prior successful operation, but the apparent confidence was at least partially a bluff; project pressures about costs and programs were almost certainly the real drivers.

"Normal" civilian safety certification (or licensing) practice does not apply – nobody had ever suggested that military aircraft should be licensed by the UK Civil Aviation Authority, or by equivalent bodies in other countries. Combat aircraft have their own safety licensing regimes. And yet there is also, somehow, a political and public expectation that the standards of safety should in some way be comparable to civil standards. In the United Kingdom, the Ministry of Defence had gone so far as to say that safety standards should be "as military as necessary, as civil as possible." How realistic is this? Everyone knows that combat aircraft have to face (potentially) much greater risks than commercial airliners. Is it realistic to think about applying a similar approach to their licensing and operation? (Compare the UK Ministry of Defence approach with the US Air Force approach. The USAF "Air Force System Safety Handbook" has a mission statement "Designing the safest possible systems consistent with mission requirements and cost effectiveness.")

In a military environment there is an expectation of risk because, inescapably, that is what combat involves. Does this make those engineers who are involved in military safety assessment and analysis less thorough in their approach? Perhaps BAE Systems engineers working on the Nimrod safety case had another "reason" not to regard the Nimrod safety case as a hugely important piece of work; namely, they may have thought that any identified risks in their safety analyses will have been much less significant than combat risks, e.g., the risk of a Nimrod being shot down by a ground-to-air missile. (This false "reason" would be in addition to two other false "reasons": First, the Nimrod was "safe anyway", and second, it was near the end of its operational life. The latter implies some sort of judgment about "time at risk.")

Also, it is implicit in safety analysis that identified weaknesses will be rectified; this will cost money, yet ultimately there are overall budget constraints. After all, Group Captain George Baber had potentially much bigger things to worry about than the Nimrod safety case – he had responsibilities for Nimrods flying in hostile environments over Iraq and Afghanistan. For George Baber, could it have come down to a choice between either spending money on urgent operational (combat) requirements, or else spending money on the urgent mitigation of "theoretical" risks?

The principal requirement in any safety justification is that the design is *fit-for-purpose*. The Nimrod No. 7 Tank Dry Bay designs failed a basic "fitness for purpose" test.

Haddon-Cave wrote that the Nimrod's "implicit safety case" was an oxymoron. May be this comment can be extended more generally: What does it mean to have a safety case for combat aircraft or, say, for manned spaceflight – i.e., for activities that are inherently dangerous? Is "combat aircraft safety case" also an oxymoron? Do we really understand what "safety" means in a combat aircraft setting?

A difficulty here can be the conflation of two issues: *fitness for purpose* and *risk-based safety justification*. Sometimes (though not necessarily in these examples) it can appear to senior management that a particular safety concern has been derived solely from risk-based safety analysis, and not from a review of fitness for purpose. Senior management may think the concern is about a hypothetical fault sequence expressed in terms of probability or fault frequency, unless it is made very clear where the engineering issues lie.

A further difficulty for senior management is that they may be asked to make decisions about expenditure on design improvements based on detailed technical analyses which they will almost certainly not have sufficient time to read and fully understand.

Safety analysts need to be very clear and concise about their safety concerns; they need to be able to express them in basic engineering terms such as "Nimrod No. 7 Dry Tank Bay is a major fire risk yet there is no fire detection or fire suppression." The real problem for Nimrod was that the design was not fit-for-purpose. (Just for clarity: It is not necessarily true that a failure to explain the deficiencies in clear engineering terms was a problem in this case – but I am suggesting it can be.)

Regarding risk-based safety criteria, the philosophical basis of safety and risk in hazardous civilian industries is well understood and numerically defined. Perhaps, for combat aircraft, there should also be a first-principles-based set of risk criteria, particular to combat aircraft and equipment, using a cost–risk–benefit approach, which addresses both initial design and subsequent modification work in the context of combat risk. Existing defence safety standards, such as the UK's Def Stan 00-56, rely heavily on civilian safety standards. This seems unrealistic. (At time of writing, Def Stan 00-56 is under review.)

REFERENCES

[1] J. Reason, Managing the risks of organisational accidents, Ashgate, (1997).

[2] C. Haddon-Cave QC, The Nimrod Review, HMSO, London, (2009). The philosophical basis of risk and safety management in hazardous industries is set down, e.g., in the following two documents.The Tolerability of Risk from Nuclear Power Stations, HMSO;1; 1992. Reducing Risks, Protecting People, HMSO;1; 2001.

Management Systems to Prevent or Mitigate Accidents

10

This chapter provides a high-level overview of some of the key safety management processes, techniques and tools that should be in place for high-hazard industries. It is intended that the processes presented here are more-or-less generic to any high-hazard industry, including nuclear, oil and gas, and petrochemical process plant. The processes are mostly presented as diagrams and flowcharts, which are to some degree self-explanatory (although of course the devil is in the details).

Safety management processes include:

Personnel recruitment, competence assurance and training
Safe working arrangements
Design engineering and safety functional requirements
Technical safety and technical risk assessments
Engineering design change (including temporary modifications)
Accident and incident investigation
Emergency planning

Each of these processes is addressed in more detail below. Other important safety management processes that are not included here include operating and maintenance procedures, project quality assurance arrangements, the control of subcontractors, and security arrangements (including information technology security, as discussed in Chapter 3).

Senior management's role is not just to ensure that appropriate management arrangements are in place – there is also a crucial role in high-hazard industries for senior managers to send clear and unambiguous messages to personnel at all levels about the importance of safety. This must mean more than just repeating the tired old cliché "safety is our top priority"; senior managers have to live the values, and be seen to send consistent messages at all times. This is especially important at times when budgets are under review: safety has to maintain its importance, and senior management have to be quite clear about this in all they say and do.

Safety management arrangements differ from country to country, and between industries. What follows in this chapter is generic. In particular, there are differences in regulatory approach. For example, in offshore oil and gas, the arrangements in the United States are prescriptive, i.e., the government regulator imposes detailed mandatory requirements on operators, but countries such as Norway and the United Kingdom have performance-based regimes which set goals but leave the responsibility to industry to formulate the details. Other countries such as Canada have hybrid regimes containing both prescriptive and performance elements.

THE HEALTH, SAFETY AND ENVIRONMENTAL MANAGEMENT SYSTEM

All of the management systems listed above are important; there are no "optional" items, although extra items may be appropriate. In an effective organization, these items are captured within a health, safety and environmental management system (HSE-MS) or safety and environmental management system (SEMS) and are kept under regular review (Fig. 10.1). The HSE-MS or SEMS consists of enabling arrangements such as leadership and organization, competence assurance and training, work planning, audits, corrective action tracking, and security, as well as the details of health, safety and environment management arrangements.

The SEMS addresses worker safety and major accident risk and controls. Hence the SEMS includes, e.g., safe working arrangements (including the permit to work arrangements), the technical safety justification of the plant, the quality assurance arrangements for design engineering and site construction work, and the arrangements for ensuring that there are sufficient suitably qualified and experienced people. The

FIGURE 10.1

Key elements of a company's health, safety and environmental management system (HSE-MS).

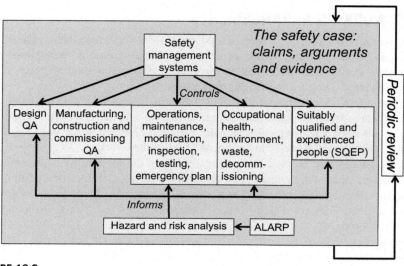

FIGURE 10.2

Safety management systems and the safety case.

SEMS will also include arrangements for carrying out periodic reviews of the technical safety justification – the "safety case" (Fig. 10.2).

PERSONNEL RECRUITMENT AND TRAINING

Ultimately, all safety comes down to people doing the right things, which means that recruitment, training, and succession planning arrangements need to be reviewed and maintained from a safety perspective. Companies need to recruit, develop and retain suitably qualified and experienced people (sometimes known as SQEPs).

In recruitment and training, risk-takers do not make good employees in front-line roles in the operations and maintenance for high-hazard industries. Situations where individuals are placed in the position of having to make risk-related decisions under time pressure can lead to accidents. Staff selection processes and training need to ensure that any risk-taking tendencies are firmly discouraged in front-line roles. All personnel need to be reminded regularly that delays because of safety concerns are alright; it must always be perfectly acceptable for employees to check "up the line" whether a situation is safe, and to double check elsewhere if they are unhappy with the answer received. (The paradox here, of course, is that operators in front-line roles also need to "get things done" – and a shift operations supervisor who acquires a reputation for, say, repeatedly delaying plant start-ups because of needless safety concerns will not remain in post for very long. The challenge for managers is to get this balance right.)

SAFE WORKING ARRANGEMENTS

Wherever there is risk to personnel or a potential for a major accident on hazardous plant, safe working arrangements are necessary. These require conscientious, diligent people working through procedures, which are designed to minimize risk, either to the individual doing a job or to the plant itself.

These safe working arrangements must begin by having clear statements of how certain specific types of activities will be carried out on site. The range of specific activities addressed will typically include: welding; confined space working; scaffolding; radiography; working at height; and underwater work.

There should also be clear definitions regarding the availability of essential or safety-related plant – e.g., emergency shutdown systems, fire detection equipment, fire suppression equipment, or backup electricity supplies. These types of procedures and definitions are usually called "operating rules" or "site safety policies" (Fig. 10.3, top right).

Next, there has to be controlled means of introducing new work into the work planning system. Plant defects must be logged and tagged, to ensure they have not forgotten about. Urgent work must be given priority. Non-urgent work must be scheduled for a suitable later time. Work to be done in scheduled shutdowns (both routine maintenance and inspection activities and non-urgent defect repair work) must be carefully planned, often long in advance, to ensure that the planned shutdown takes place in the most efficient way possible (Fig. 10.3, top left).

For each job, there should be a clear definition and scope of work, a method statement, an isolation and de-isolation procedure where necessary, and a job risk assessment. The job risk assessment addresses the hazards of the specific job at the workplace, and should be done by people who are familiar with the plant and the type of work (Fig. 10.3, center). The output from this process is the permit to work for doing the specific job.

Next, the necessary isolations are completed and an isolation certificate is issued. The isolations are all locked and keys put in a locked cabinet, with the key held by the person in charge of doing the work. At that point, a pre-job brief can be given to the workers doing the job, and they complete the work. Afterwards, the plant is de-isolated, and the permit to work is withdrawn (Fig. 10.3, bottom).

The isolation process deserves special consideration (Fig. 10.4). Here, "isolation" refers to the secure containment of any hazards so that technicians can work on the isolated plant in safety. However, the process of isolation has a dual role; not only does the isolation ensure the safety of the technicians, it also acts to ensure that the plant cannot be returned to operational status until the technicians have completed their work, and the work has been inspected and deemed ready for return to service. The isolation process ensures that all hazards are securely separated from the work to be done – this includes electrical systems, high-pressure fluid systems, and any sources of toxic or inflammable material. Any residual high pressure or high voltage must be vented or earthed. Valves are locked shut,

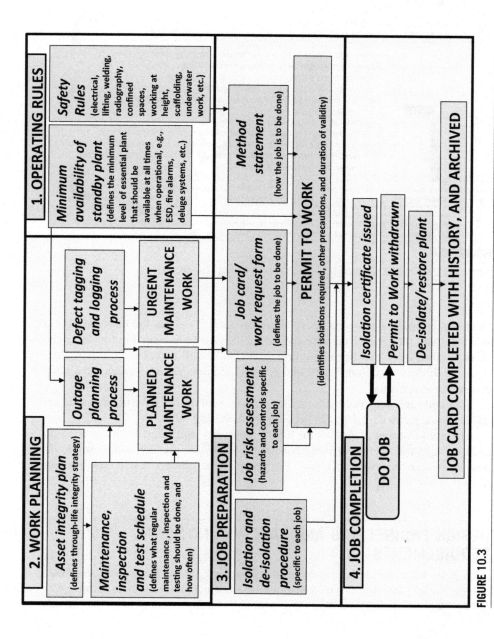

FIGURE 10.3

Safe working arrangements and permits to work.

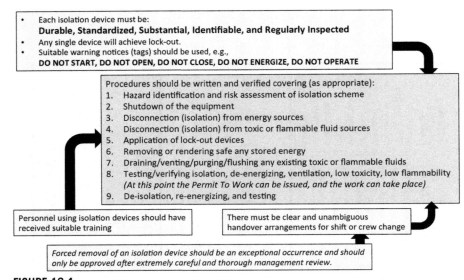

- Each isolation device must be:
  Durable, Standardized, Substantial, Identifiable, and Regularly Inspected
- Any single device will achieve lock-out.
- Suitable warning notices (tags) should be used, e.g.,
  DO NOT START, DO NOT OPEN, DO NOT CLOSE, DO NOT ENERGIZE, DO NOT OPERATE

Procedures should be written and verified covering (as appropriate):
1. Hazard identification and risk assessment of isolation scheme
2. Shutdown of the equipment
3. Disconnection (isolation) from energy sources
4. Disconnection (isolation) from toxic or flammable fluid sources
5. Application of lock-out devices
6. Removing or rendering safe any stored energy
7. Draining/venting/purging/flushing any existing toxic or flammable fluids
8. Testing/verifying isolation, de-energizing, ventilation, low toxicity, low flammability
   (At this point the Permit To Work can be issued, and the work can take place)
9. De-isolation, re-energizing, and testing

Personnel using isolation devices should have received suitable training

There must be clear and unambiguous handover arrangements for shift or crew change

Forced removal of an isolation device should be an exceptional occurrence and should only be approved after extremely careful and thorough management review.

FIGURE 10.4

Some key elements of isolation (lock-out/tag-out) for maintenance.

electrical breakers are locked open, earths are locked closed, the keys are placed in a safe, and the key for the safe is held by the person responsible for doing the work. Before work can begin, confirmation checks will be carried out that electrical equipment is de-energized and toxic or flammable substances have been cleared from the area.

Standards and guidance for isolation (or "lock-out/tag-out") for maintenance activities are, e.g., 29 CFR 1910.247 (USA) and HSG 253 (UK).

The above seems like an awful lot of paperwork and effort, sometimes to do even a simple task. However, bitter experience has shown that this sort of thoroughness is essential to avoid accidents. Each step in the process is potentially safety critical.

DESIGN ENGINEERING AND SAFETY FUNCTIONAL REQUIREMENTS

Parts of this have already been introduced in Chapter 2. Design engineering as applied to high-hazard plant begins with a design concept, which develops in an iterative way until a "frozen" design specification is available; at that point, detailed design can begin. This process is sometimes called *front end engineering design or FEED*.

In broad terms, the process for producing a specification for a safe design involves *hazard identification (HAZID)*, which asks "what sort of accidents do we need to worry about?", followed by detailed analysis to identify the magnitude of potential accidents. From a safety perspective, a most important step is the clear and robust definition of the *safety functional requirements*, i.e., the requirements for the control and protection systems on the completed plant. The history of accidents involving design failures shows a frequent root cause to be inaccurate or inadequate definition of the safety functional requirements (see Fig. 2.3).

Thereafter the designers have to identify the necessary barriers and controls for the identified hazards and safety functional requirements. (A barrier or a control, which prevents a hazard is called a *safety critical element (SCE)*.) The barriers and controls will typically consist of a mixture of mechanical barriers, instrumentation and control (I&C) systems, and fire-fighting systems, combined with administrative controls. *Functional requirement specifications* for all the necessary SCEs are then included in the overall design specification.

Operability and maintainability studies may be necessary before a final design specification can be issued. Input from experienced operations and maintenance engineers will be required.

The flowchart in Fig. 10.5 presents an idealized view of the conceptual design process.

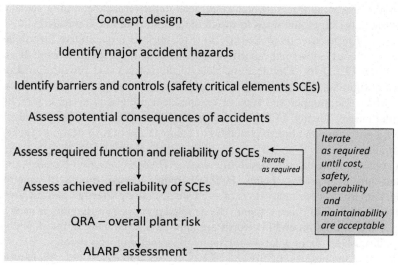

FIGURE 10.5

Safety aspects of front end engineering design (FEED)
QRA = quantified risk assessment
ALARP = as low as reasonably practicable.

TECHNICAL SAFETY AND TECHNICAL RISK ASSESSMENTS

The principal objective of technical safety assessments is to demonstrate that the risk associated with plant operations is "As Low as Reasonably Practicable" (ALARP). First, the assessed frequency of major accidents has to be better than the applied "tolerable risk" threshold. Thereafter, there must be a balance between the cost of further risk reduction, and the assessed frequency of major accidents causing fatalities.

In addition to the requirements of front end engineering design, technical safety and risk assessments are needed at other stages of a plant's lifecycle, e.g., to justify plant modification work or as part of periodic plant safety reviews. The range of jargon that has developed around technical safety and risk analysis can be quite bewildering for the ingénue: a wide range of acronyms is used, and it is quite common to attend meetings with experts where it seems like entire sentences can be constructed from acronyms. This section gives an extremely brief overview of technical safety and risk assessment methodologies – a quick reference guide, if you like, for non-experts who have found themselves caught up in a blizzard of acronyms.

The techniques can be divided into two types: *qualitative methods* are shown in Fig. 10.6, and *quantitative or semi-quantitative methods* are shown in Fig. 10.8.

HAZID was introduced in the discussion above about front end engineering design (FEED). A group of suitably experienced people will meet to identify the range and likely magnitude of hazards associated with a hazardous plant. This can then lead to a *hazard and effects register*, which identifies: the hazards and their likely magnitudes; potential threats which could lead to the hazard becoming an accident; and (ultimately) suggested ways in which the hazard could be controlled.

Layers of protection analysis (LOPA) (Fig. 10.7) is a qualitative technique that is used to identify how many barriers are available for a given hazard. It can be used during FEED to examine basic design alternatives and provide guidance to select a design that has lower initiating event frequencies, or a lower consequence, or for which the number and type of independent protection systems are "better" than alternatives. Ideally, LOPA can be used to design a process that is "inherently safer" by providing an objective method to compare alternative designs quickly and quantifiably.

In contrast, *Hazard and operability studies (HAZOP)* are used for analyzing the detailed design of process plants. HAZOP studies involve groups of experienced people, sometimes over many days or even weeks, sitting together and reviewing in detail the operation of process plant. The detailed process flow diagram is used as the basis for the discussion and keywords are applied at each node in the flow chart to prompt discussion about what would happen if something went wrong.

Two other techniques, ESSA and EERA, are used to assess the effects of major accidents. *Essential systems survivability analysis (ESSA)* is as its name implies; analysis is carried out to see whether essential systems such as emergency shutdown (ESD) or fire-fighting systems can survive, say, flood or localized fire. *Escape, evacuation and rescue analysis (EERA)* is used in particular in offshore platforms

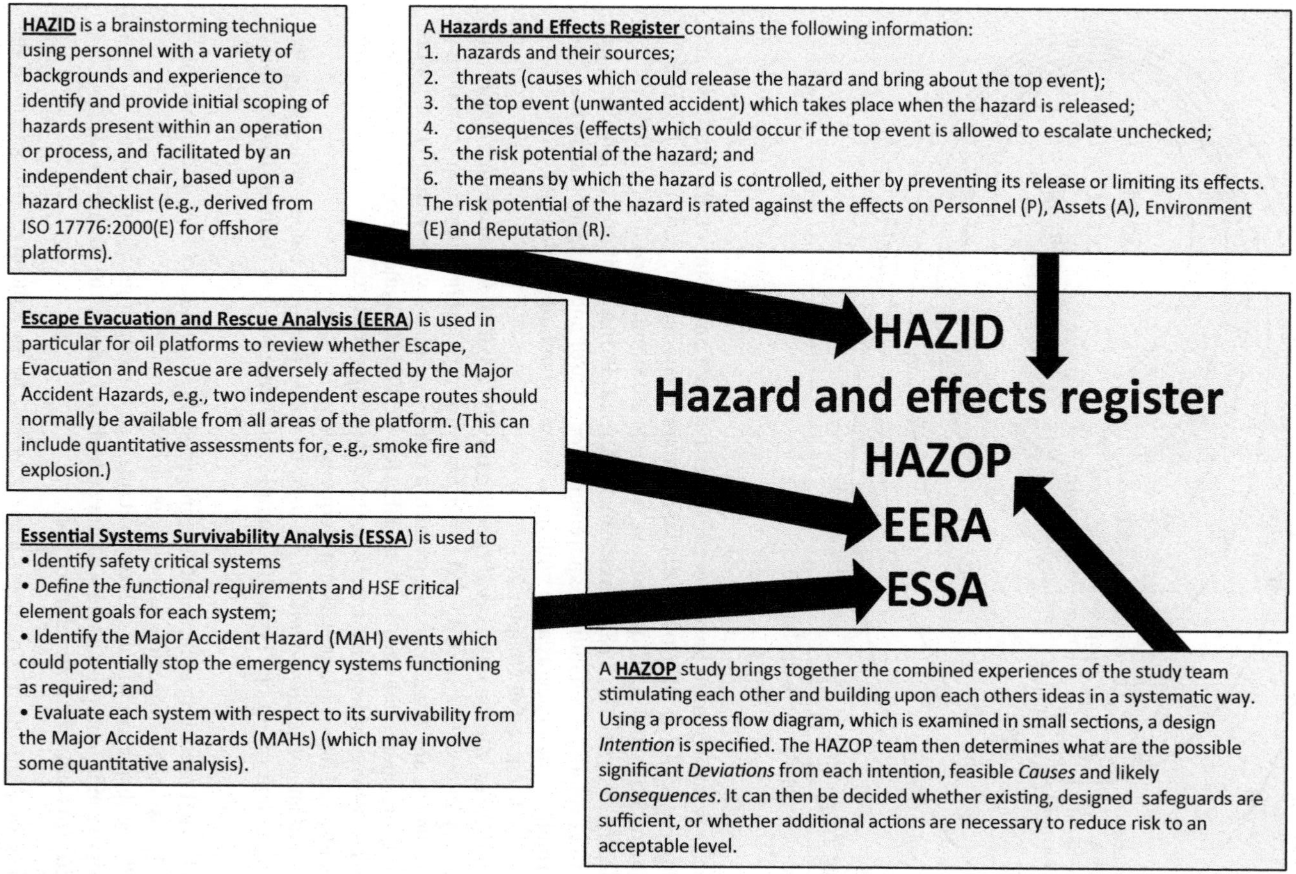

HAZID is a brainstorming technique using personnel with a variety of backgrounds and experience to identify and provide initial scoping of hazards present within an operation or process, and facilitated by an independent chair, based upon a hazard checklist (e.g., derived from ISO 17776:2000(E) for offshore platforms).

A **Hazards and Effects Register** contains the following information:

1. hazards and their sources;
2. threats (causes which could release the hazard and bring about the top event);
3. the top event (unwanted accident) which takes place when the hazard is released;
4. consequences (effects) which could occur if the top event is allowed to escalate unchecked;
5. the risk potential of the hazard; and
6. the means by which the hazard is controlled, either by preventing its release or limiting its effects.

The risk potential of the hazard is rated against the effects on Personnel (P), Assets (A), Environment (E) and Reputation (R).

Escape Evacuation and Rescue Analysis (EERA) is used in particular for oil platforms to review whether Escape, Evacuation and Rescue are adversely affected by the Major Accident Hazards, e.g., two independent escape routes should normally be available from all areas of the platform. (This can include quantitative assessments for, e.g., smoke fire and explosion.)

Essential Systems Survivability Analysis (ESSA) is used to
- Identify safety critical systems
- Define the functional requirements and HSE critical element goals for each system;
- Identify the Major Accident Hazard (MAH) events which could potentially stop the emergency systems functioning as required; and
- Evaluate each system with respect to its survivability from the Major Accident Hazards (MAHs) (which may involve some quantitative analysis).

Hazard and effects register

HAZID

HAZOP

EERA

ESSA

A **HAZOP** study brings together the combined experiences of the study team stimulating each other and building upon each others ideas in a systematic way. Using a process flow diagram, which is examined in small sections, a design *Intention* is specified. The HAZOP team then determines what are the possible significant *Deviations* from each intention, feasible *Causes* and likely *Consequences*. It can then be decided whether existing, designed safeguards are sufficient, or whether additional actions are necessary to reduce risk to an acceptable level.

FIGURE 10.6

Some qualitative techniques used in technical safety and risk assessments.

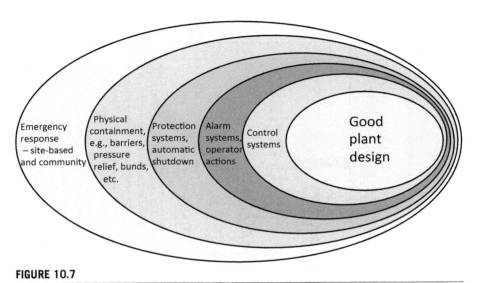

FIGURE 10.7

Layers of protection analysis (LOPA).

where it is important to have diverse routes between any point on the platform and the emergency lifeboats (totally enclosed motor propelled survival craft, TEMPSC).

Quantitative and semi-quantitative methods are summarized in Fig. 10.8.

Safety integrity level (SIL) assessment is used during front end engineering design to develop the reliability requirements for control and protection systems. This is normally done in accordance with international standards such as IEC 61508. This topic was discussed in Chapter 2.

Fault trees, failure modes and effects analysis (FMEA), failure modes effects and criticality analysis (FMECA) and *event trees* use logic, reliability data (component failure rates), and assessed system failure rates, combined with human error failure rates (using methodologies such as HEART or THERP) and other methodologies such as software reliability assessment, to develop estimates of system failure frequencies, and hence plant accident frequencies.

Accident *consequence analysis* is used to assess just how bad a particular postulated accident might be. Typically, complex software models are used to determine, e.g.,:

- Thermal modeling and rate of fire progression, including jet fires and plume fires.
- Explosion overpressures and damage arising.
- Plume dispersal, which will depend on, e.g., meteorological data, local topography, and plume temperature (which may cause the plume to rise).
- Toxicity data for toxic plumes.

The output from a consequence analysis will be an indicative assessment of the likely mortality and physical damage arising from the postulated accident. There

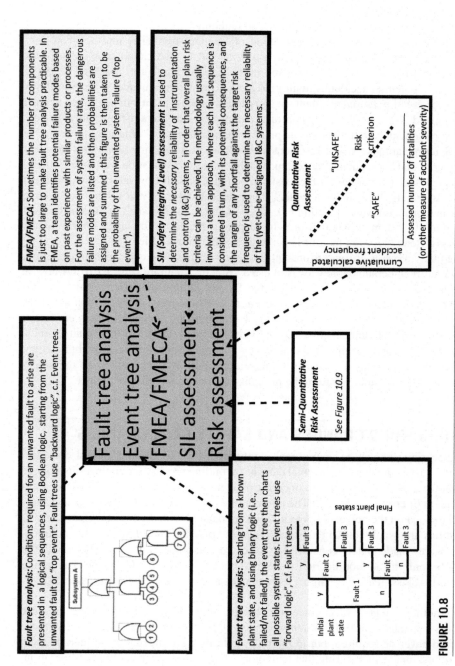

Fault tree analysis: Conditions required for an unwanted fault to arise are presented in a logical sequences, using Boolean logic, starting from the unwanted fault or "top event". Fault trees use "backward logic", c.f. Event trees.

Event tree analysis: Starting from a known plant state, and using binary logic (i.e., failed/not failed), the event tree then charts all possible system states. Event trees use "forward logic", c.f. Fault trees.

Fault tree analysis
Event tree analysis
FMEA/FMECA
SIL assessment
Risk assessment

FMEA/FMECA: Sometimes the number of components is just too large to make fault tree analysis practicable. In FMEA, a team identifies potential failure modes based on past experience with similar products or processes. For the assessment of system failure rate, the dangerous failure modes are listed and then probabilities are assigned and summed - this figure is then taken to be the probability of the unwanted system failure ("top event").

SIL (Safety Integrity Level) assessment is used to determine the necessary reliability of instrumentation and control (I&C) systems, in order that overall plant risk criteria can be achieved. The methodology usually involves a team approach, where each fault sequence is considered in turn, with its potential consequences, and the margin of any shortfall against the target risk frequency is used to determine the necessary reliability of the (yet-to-be-designed) I&C systems.

Semi-Quantitative Risk Assessment

See Figure 10.9

Quantitative Risk Assessment

FIGURE 10.8

Quantitative and semi-quantitative methods used in technical safety and risk analysis.

may be large sensitivities in the analysis, in which case very detailed analysis may be performed on cases with a particularly bad outcome.

In technical safety, risk is a function of accident frequency and accident consequences. "Pure" *quantified risk assessment* (QRA, also called probabilistic risk assessment or PRA) typically uses mortality (i.e., the number of assessed deaths in any given potential accident situation) as a measure of consequences. The overall risk for a variety of potential accidents at a particular process plant can be plotted against recognized risk criteria in a graph of accident frequency against mortality, to help make judgments about whether the plant is "acceptably safe" or not.

Semi-quantitative risk assessment methods are less exacting about how accident frequency and consequences are calculated. For a large organization with many hazardous facilities, it is normal for their risks to be judged against overall industry accident records. Figure 10.9 shows a typical (hypothetical) table of risk criteria for a large multinational company operating hazardous facilities. A table such as this can be used as an aid to judgment when prioritizing capital expenditure for safety improvements across different types of facilities operating in different countries.

Finally, a very useful way to present a large amount of safety assessment information, in a readily understandable way, is to use a "Bowtie" diagram (Fig. 10.10). A bowtie diagram is a combination of a fault tree for the failure of barriers, which should prevent an incident, followed by an event tree for recovery or mitigation activities after the incident. Software tools for these diagrams are available, which enable all the safety critical elements (such as individual plant systems and their maintenance, key staff roles and training, and emergency planning aspects) to be brought together. A bowtie diagram is a powerful management tool since it enables all the safety aspects of a particular hazard to be brought together on one piece of paper.

ENGINEERING CHANGES AND SAFETY CASE CHANGES

Engineering changes to operational plants require thorough and careful consideration – just as much as the original design. An outline process for engineering change is shown in Fig. 10.11. (The engineering change process is often also known as management of change or MOC.) The amount of care, attention and independent review will depend on the perceived potential risk associated with the change. Hence, modifications are usually classified according to the risk that would arise if the modification were "inadequately conceived or executed"; in other words, how bad would the risk become if the modification were fundamentally flawed. Where the risk arising from this classification is "high", there may be a requirement for independent assessment.

Another category of engineering change, which is particularly prone to abuse unless managed carefully, is *temporary modifications or overrides*. Ideally, these should never be required, but in the real world they are inevitable, so there have to be robust means of ensuring that temporary changes are given careful consideration, and that the temporary arrangement is regularized at the earliest possible opportunity.

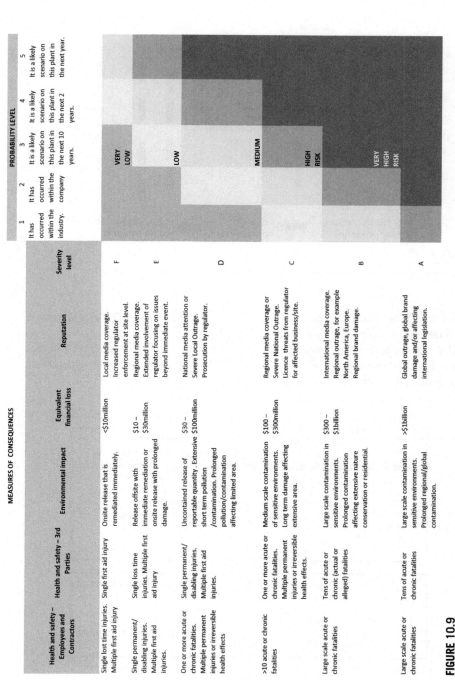

FIGURE 10.9

A typical table of risk criteria for a large multinational company operating hazardous facilities, as might be used in prioritizing safety-related capital spending across different facilities in various countries.

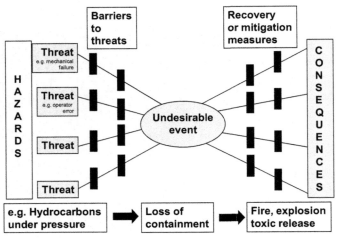

FIGURE 10.10

A bowtie diagram.

Figure 10.12 shows some of the main factors that have to be taken into consideration when designing a temporary modification process.

An important concept here is *configuration management*; it is essential that operations management knows the exact plant state at any given time. Operations management must always be in full control of the plant state.

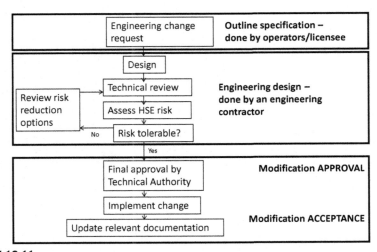

FIGURE 10.11

Typical process for engineering change.

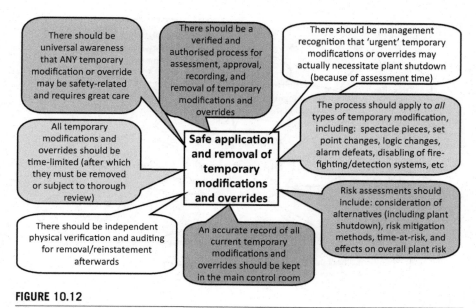

There should be universal awareness that ANY temporary modification or override may be safety-related and requires great care

There should be a verified and authorised process for assessment, approval, recording, and removal of temporary modifications and overrides

There should be management recognition that 'urgent' temporary modifications or overrides may actually necessitate plant shutdown (because of assessment time)

The process should apply to *all* types of temporary modification, including: spectacle pieces, set point changes, logic changes, alarm defeats, disabling of fire-fighting/detection systems, etc

All temporary modifications and overrides should be time-limited (after which they must be removed or subject to thorough review)

Safe application and removal of temporary modifications and overrides

Risk assessments should include: consideration of alternatives (including plant shutdown), risk mitigation methods, time-at-risk, and effects on overall plant risk

There should be independent physical verification and auditing for removal/reinstatement afterwards

An accurate record of all current temporary modifications and overrides should be kept in the main control room

FIGURE 10.12

Some key requirements in a temporary modification process.

It is particularly important that plant operators keep track of all changes that have an effect on the overall plant safety justification or safety case. Major sources of changes to the safety case for an operating plant are presented in Fig. 10.13. Any change, which affects the safety justification should be processed via the engineering change/management of change procedure.

ACCIDENT AND INCIDENT INVESTIGATIONS

Data on accidents and incidents show that, for every major accident, there may be dozens or even hundreds of minor incidents, which might have escalated into a worse situation. It is important that the organization tries to learn from minor incidents and anomalies to ensure no repetition and to avoid possible escalation into something worse. Hence, the management of high-hazard plant should have processes in place to review incidents, learn the important lessons, identify the root causes, implement the necessary changes promptly, and educate all the relevant people about the changes.

A typical flowchart for a root cause analysis process is shown in Fig. 10.14.

One of the best ways to think about root cause analysis is to consider the question, "To when would I have to travel back in a time machine to prevent this accident happening?" In complex accident sequences (such as the Whatcom Park accident described in Chapter 12) there may be more than one answer to this question.

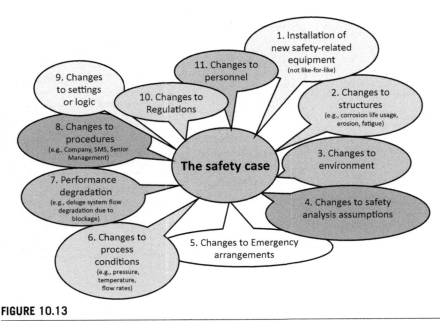

FIGURE 10.13

Major sources of changes to the safety case (or safety justification) for an operating plant.

EMERGENCY PLANNING

Typical outline scopes and objectives of the emergency planning arrangements for hazardous facilities are given below. Note that these are not in priority order – in a real accident, all of these objectives may have to be achieved in parallel.

Nuclear power stations, petrochemical and chemical plants

Basic objectives of emergency plan:

1. Muster and headcount to identify any missing persons.
2. Recovery and treatment of injured personnel.
3. Evacuation of non-essential station staff.
4. Termination or mitigation of the incident.
5. Surveys of surrounding area to establish extent of contamination.
6. Minimize radiation exposure/toxin uptake to the general public by (where appropriate):
   a. Evacuation within a defined emergency planning zone.
   b. Issuing potassium iodate tablets (nuclear power stations only, for iodine-131 release)
   c. Other prophylactic medication as may be appropriate
7. Advice to regional authorities. (police, fire, health authorities, etc.)
8. Receipt of casualties. (who may be contaminated)
9. Communications to the media.

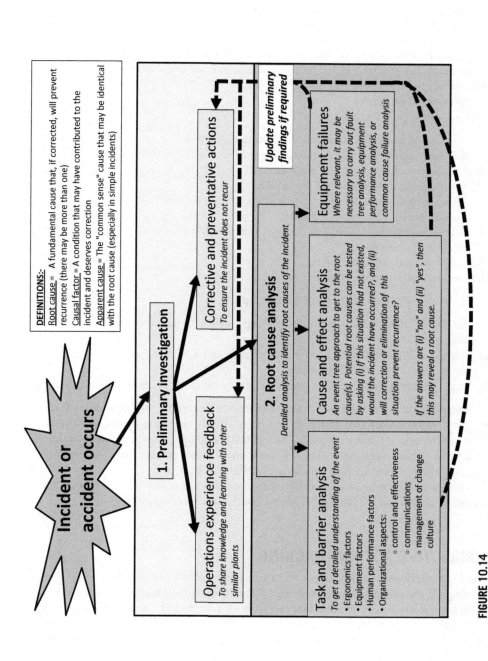

DEFINITIONS:-

Root cause = A fundamental cause that, if corrected, will prevent recurrence (there may be more than one)

Causal factor = A condition that may have contributed to the incident and deserves correction

Apparent cause = The "common sense" cause that may be identical with the root cause (especially in simple incidents)

Incident or accident occurs

1. Preliminary investigation

Corrective and preventative actions
To ensure the incident does not recur

Operations experience feedback
To share knowledge and learning with other similar plants

Update preliminary findings if required

2. Root cause analysis
Detailed analysis to identify root causes of the incident

Cause and effect analysis
An event tree approach to get to the root cause(s). Potential root causes can be tested by asking (i) if this situation had not existed, would the incident have occurred?, and (ii) will correction or elimination of this situation prevent recurrence?

If the answers are (i) "no" and (ii) "yes", then this may reveal a root cause.

Equipment failures
Where relevant, it may be necessary to carry out fault tree analysis, equipment performance analysis, or common cause failure analysis

Task and barrier analysis
To get a detailed understanding of the event
- Ergonomics factors
- Equipment factors
- Human performance factors
- Organizational aspects:
  ○ control and effectiveness
  ○ communications
  ○ management of change culture

FIGURE 10.14

Accident and incident investigation and root cause analysis.

Gasmasks or BA (breathable air) sets should be available to site personnel where necessary to permit ordered evacuation. There should be regular realistic exercises, which include regional authorities, to ensure personnel are familiar with the arrangements. There will be a well-equipped off-site emergency center to deal with non-plant related aspects, such as media communications and the interface with the civil authorities.

Offshore oil platforms

Basic objectives of emergency plan:

1. Muster and headcount to identify any missing persons.
2. Recovery and treatment of injured personnel.
3. Evacuation of non-essential platform staff.
4. Termination or mitigation of the incident.
5. Monitoring for oil releases.
6. Coordinate oil clean-up operations.
7. Advice to regional authorities. (police, fire, health authorities, coastguard, etc.)
8. Receipt of casualties.
9. Communications to the media.
10. Ensuring appropriate isolations to pipeline network.

There are three means of personnel evacuation:

1. If possible, and if time permits, the preferred evacuation method is by helicopter.
2. Lifeboats (TEMPSCs) are used if helicopters cannot be made available, either due to urgency or adverse weather.
3. Direct escape to sea, e.g., using rope ladders or jumping. (*This is a last resort.*)

There should be at least two possible escape routes to the muster stations from any point on the platform. There should be sufficient immersion suits and smoke hoods or BA sets at muster stations. There should be regular emergency drills to test all aspects of the emergency plan. All personnel should receive basic emergency training, which should include time in a TEMPSC lifeboat.

There will be a well-equipped shore-based emergency center to deal with media communications and the interface with the civil authorities.

CORRECTIVE ACTION TRACKING

One final safety-related process should be in place, which to some extent attempts to bring all the other processes together. "Corrective action tracking" is the name given to an over-arching process that monitors all activities at a hazardous site. This is used as a management tool to ensure that anything found to be unsafe, not fit-for-purpose, worthy of further investigation, or even just a learning point from another site, is dealt with in a timely and effective manner. The process should be based on the right and duty of *any* individual to raise a problem event report to identify

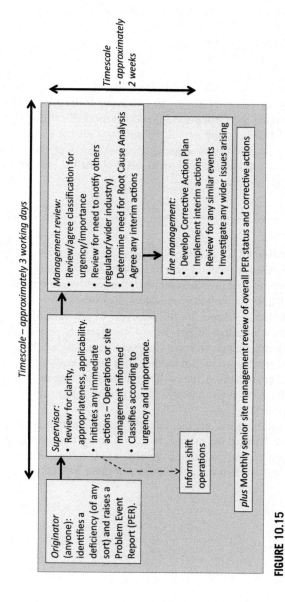

FIGURE 10.15

An overview of a problem event reporting/corrective action tracking process.

an issue (Fig. 10.15). The issues could refer to any equipment failures, inadequate maintenance, procedural failures, other documentation deficiencies (including safety justification), training inadequacies, general plant tidiness, inspection findings, issues arising from emergency exercises, etc. The intent of a good corrective action tracking process should be to ensure that problem event reports are processed with appropriate priority, that problems are fixed, and that root causes are identified and eliminated as far as reasonably practicable so that the problem does not recur.

A healthy corrective action tracking process is also a good indicator of an engaged workforce who are actively seeking to improve the safety of their working environment. Safety regulators often see the condition of the corrective action tracking process as a good key performance indicator for overall safety management and safety culture at a hazardous site – if there are only ever a few active correction actions in the process, it could show a disengaged workforce or else potentially management "chilling", i.e., a management team who "blame the messenger" instead of trying to fix problems.

SYNTHESIS

This chapter has presented an overview of some key safety-related business processes that are important in the safe operation of hazardous plant. Hazardous plant requires robust processes that are implemented thoroughly and conscientiously, and which learn from mistakes – including mistakes elsewhere.

The processes addressed cover the full lifecycle of any hazardous plant, including conceptual design, detailed design, safety justification, personnel training and recruitment, engineering change and risk categorization, safe working arrangements and plant isolations, temporary modifications, root cause analysis, corrective action tracking, and emergency planning. All of these processes come under the umbrella of the site's health, safety and environmental management system (HSE-MS or SEMS) and strategic HSE policy.

For the safe operation of hazardous plant, the senior site management team plays an important role in "setting the correct tone," ensuring that correct responses are given to emerging issues, and maintain a healthy safety culture. A healthy corrective action tracking process is often a good indicator of these aspects.

The Human Factor

11

"The soft stuff is the hard stuff".
Attributed to various people

"Each man calls barbarism whatever is not his own practice".
Michel de Montaigne

"The mind can go either direction under stress – toward positive or toward negative: on or off. Think of it as a spectrum whose extremes are unconsciousness at the negative end and hyperconsciousness at the positive end. The way the mind will lean under stress is strongly influenced by training".
Frank Herbert

INTRODUCTION

Although management systems provide the framework for the safe operation of hazardous activities, it is often the case that accidents happen despite that framework. In particular, even intelligent conscientious people can fail in their jobs when faced with an environment that is either too routine or too stressful.

The two fatal accidents described in this chapter are representative of these two extremes: lack of challenge, and excessive stress. Both accidents defy belief, for various reasons – they just should not have happened.

In the first, a routine piece of basic stress analysis by a reputable design engineering company for a ferry access walkway was botched, and a second reputable company, paid to review the design produced by the first, failed to notice the mistakes. The accident is also notable for the court case that followed, in which the design was described as "inept and incompetent." In the court case, it was found that the company that commissioned the walkway design and operated the harbor also bore a share of the responsibility, even though the faulty design had been produced and reviewed by others. The precedent was thereby established in UK law that the responsible operator is required to understand the basis of the safety of the equipment it operates – i.e., the operator has to be an "intelligent customer."

In the second accident, a fire in the hold of an aircraft, which should have been survivable, nevertheless led to the deaths of all on board even after the plane had landed safely. The stressful situation, combined with a poor working relationship

High Integrity Systems and Safety Management in Hazardous Industries. 978-0-12-801996-2

among the crew (and possibly other cultural factors) led to indecisiveness by the captain and huge loss of life.

THE PORT OF RAMSGATE WALKWAY ACCIDENT, 1994

On 0045 hours on September 14, 1994 at the Port of Ramsgate, Kent, England, a 21 tonne steel walkway onto a cross-Channel ferry collapsed. Six people fell about 10 metres and were killed. Another seven were injured. This accident led to an important court case that helped clarify legal responsibilities [1].

In the early 1990s, Port of Ramsgate had required a new walkway for roll-on/roll-off ferries traveling to Ostend in Belgium. The walkways were to have three spans; the first was between the passenger building and a pontoon, the second was along the pontoon, and the third was from the pontoon to the ferry. The pontoon floated up and down with the tide, so the walkways had to articulate at each end.

On the face of it, this was a fairly straightforward, if slightly novel, piece of mechanical engineering, which would nonetheless require careful design, stress analysis and manufacturing. Since Port of Ramsgate was not a competent design organization, they placed a "design and build" contract with Fartygsentreprenader AB (FEAB), a Swedish company, to build the walkway. FEABs sister organization Fartygskonstruktioner AB (FKAB) did the design, and FEAB did the construction. Both companies had done this sort of work previously.

In a novel design where public safety is important, there will always be a requirement for Independent Assessment. Port of Ramsgate contracted with Lloyd's Register for this part of the project. Lloyd's Register is well known for doing independent reviews of this sort and was considered expert in this field.

The walkway collapsed in weather conditions that were unexceptional. At the time of collapse, the walkway had been in service for four months. Failure was found to have been due to fatigue at one of the walkway bearings (Fig. 11.1).

Later analysis showed that FKABs design stress calculations had neglected to consider any bending moments in the analysis of the walkways. The bending moment

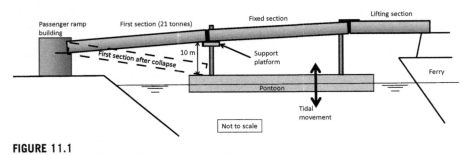

FIGURE 11.1

The Port of Ramsgate walkway collapse.

will have been greater than for a "normal" flat walkway, which would twist readily and therefore minimize the bending moment on the bearings. In this case, because the walkway was fully enclosed, it had a rigidity that meant torsional bending forces were transmitted directly onto the bearings. These design calculations will have been checked within the FKAB organization, so at least two people in FKAB had made the same mistake. An engineer in Lloyd's Register had then carried out independent calculations, which apparently repeated the earlier error by FKAB. So, at least *three* engineers had got it wrong.

In the subsequent prosecution, the design was described by the prosecution as "inherently unsafe" and "inept and incompetent." Inferior steel and bad workmanship were also blamed. *In situ* design changes were made to the pre-fabricated walkway when it was found to be too short.

All four companies – Port of Ramsgate, FEAB, FKAB and Lloyd's Register – were prosecuted under the UK Health and Safety at Work Act. The burden of proof was on the defendants to show that they had done as much as was reasonably practicable to ensure the safety of people using the walkway. This corresponds to the Health and Safety at Work Act's requirement that risk should be "As Low as Reasonable Practicable", which is usually called "the ALARP Principle".

The trial took place during January and February 1997. None of the three contracted parties – FEAB, FKAB or Lloyd's Register – come out of the trial well. All four parties – i.e., including Port of Ramsgate – were found guilty and received fines; FEAB £750,000, FKAB £250,000, Lloyd's Register £500,000 and Port of Ramsgate £200,000. Costs were shared amongst the defendants.

A key issue in this case was whether or not Port of Ramsgate had any responsibility for work that had been done on its behalf by suitable contractors and advisors. Port of Ramsgate had no pretension to being an engineering design organization. However, since the design of the walkway was tailor-made (and not from a catalog), i.e., since Port of Ramsgate had initially prepared some sort of specification, it was argued that Port of Ramsgate was a party to the design process, and had therefore not taken all "reasonably practicable" steps.

Both FKAB and Lloyd's Register made the same mistake in design calculations, which was of course obvious with hindsight. Perhaps a thorough risk analysis might have revealed that the design analysis had not considered all possible failure modes.

Conclusions from this accident are numerous. Perhaps the main lessons for other engineering activities are as follows.

Independent assessors need to "think out of the box" to identify errors of omission, as well as simply checking for errors of commission such as calculation errors. The failure of Lloyd's Register to discover design mistakes undermines the confidence that is placed in Independent Assessors in engineering activities all over the world.

We will never know the circumstances that led highly reputable organizations to fail to notice fundamental flaws in basic engineering design. Perhaps the companies saw the task as fairly mundane so it got little attention, or else it was given to junior staff. Alternatively, perhaps the neglect arose because their offices were very busy at the time. We do not know.

The owner of hazardous plant carries some responsibility for design and manufacturing errors, even if the owner has employed suitably experienced contractors. The owner cannot completely delegate responsibility to contractors. This means that the owner is expected to act like an "intelligent customer" and understand what the contractors are doing.

If we apply the "time machine" test ("If you had a time machine, how far back in time would you have had to go to stop this event from happening?"), we have to conclude that, because the Port had no particular design engineering expertise, it should have retained another company to act on its behalf as architect-engineer, who would manage the initial specification and later be able to review intelligently the design proposals and the independent assessment process.

SAUDIA FLIGHT 163: HUMAN BEHAVIOR DURING CRISIS MANAGEMENT

One of the worst civil aviation accidents of all time, and one of the most bizarre, happened at Riyadh airport on August 19, 1980, when Saudi Airlines (or "Saudia") Flight 163, a Lockheed Tristar, suffered an onboard fire. All 301 passengers and crew died of smoke inhalation while the plane was stationary on the runway (Fig. 11.2). There has been a lot written about this accident, much speculation, and some urban myths, and after more than three decades it is perhaps even less clear exactly what happened, although some new evidence has recently been reported.

All three of the flight crew of Saudi Airlines 163 had decidedly unimpressive training records. Captain Mohammed Ali Khowyter was aged 38 and had worked for Saudi Airlines since 1965. The official accident report [2] notes that he was "slow to learn", needed more training than was normally required, failed recurrent training, and had problems in upgrading to new aircraft.

First Officer Sami Abdullah Hasanain was aged 26 and had worked for Saudi Airlines continuously since 1977, and previously as a trainee in 1974–1975. He had first qualified on Lockheed Tristars only 11 days before the accident, on August 8, 1980. During his initial flying school training in 1975 in Florida the flight school had telexed Saudi Airlines advising of "poor progress" and requesting advice about whether he should continue on the training program. He was then dropped from the training program on October 31, 1975. On March 13, 1977 he was re-instated into pilot training as "a result of committee action"; the exact meaning of this is unclear, but it could have been union action or else some "pulling strings" at a high level.

Flight Engineer Bradley Curtis was aged 42 and had a curious CV. He had worked for Saudi Airlines since 1974. He was a pilot, who had been qualified as a Captain for

FIGURE 11.2

The burnt-out hulk of Saudi Airlines Flight 163 remained at Riyadh airport for several years. The engines were removed after the accident.

(Photo: Michael Busby. Used by permission.)

Douglas DC-3 Dakotas. In 1975 he was assigned to transition training to be a Captain of Boeing 737s, but his training was terminated because of "Progress Unsatisfactory" as either a Captain or a First Officer. Eventually, after further training, he was declared ready for work as First Officer on Boeing 737s, but following a check on March 30, 1978 he was recommended for removal from flying status. He was sent a letter of termination on May 14, 1978. Curtis then offered to pay for his own training to become a Flight Engineer on Boeing 707s. This offer was accepted by the airline, and he began work as a Flight Engineer onboard Boeing 707s on January 24, 1979. He later retrained as a Flight Engineer on Lockheed Tristars, being cleared for duty on May 20, 1980. The Saudi official accident report states that Curtis may have been dyslexic; during the accident, this may have affected his ability to locate the correct emergency procedures.

Flight 163 had flown from Karachi, Pakistan. After a stopover at Riyadh, Saudi Arabia, it took off again at 1807 hours, carrying mostly pilgrims on their way to Mecca for Hajj. The plane was carrying 287 passengers and 14 crew members.

Many of the passengers were poor Pakistanis and Bedouin, who had never flown before. Because they were on pilgrimage, many had brought their own cooking utensils, stoves and gas bottles on board with them. This would have been illegal, so these must somehow have been smuggled aboard, or else the pre-flight checks and security were very lax. Gas bottles were subsequently found in the plane wreckage.

At 1814:53, some seven minutes after takeoff and while still climbing at about 15,000 feet, a smoke detector alarm came up showing smoke in the aft cargo hold, followed at 1815:55 by a second smoke alarm in the same aft hold. Captain Khowyter said at 1815:59, "So, we got to be turning back, right?" but no immediate turn-around was effected. Captain Khowyter asked Flight Engineer Curtis to check the procedure for smoke alarms at 1816:18. Curtis could not find the procedure.

The plane continued its climb to cruise altitude. After more discussion between the Captain and the Flight Engineer, Curtis offered, at 1819:26, to go back in the passenger cabin to see if he could smell anything. Meanwhile, First Officer Hasanain was now trying to find the procedure for smoke alarms, without success. At 1819:44, while Curtis was still in the passenger cabin, Captain Khowyter told First Officer Hasanain "Tell them we're returning back" but Hasanain did not immediately do so. At 1819:58, Captain Khowyter again said, "We better go, go back to Riyadh" but he still didn't turn the plane back.

Meanwhile, Flight Engineer Curtis was still in the passenger cabin. While he was away, the Captain said to First Officer Hasanain, referring to Curtis, "By the way he's a jackass, in the abnormal it is in the checklist" (*sic*), meaning that Curtis had been looking in the wrong place for the procedure.

At 1820:16, Curtis returned to the cockpit, "We've got a fire back there."

Finally, at 1820:27, Riyadh airport was told "One six three, were coming back to Riyadh", and it was only at this point, some five and a half minutes after the first smoke alarm, that the plane was turned round and began its descent back to Riyadh. About 10 seconds later there was a radio message to Riyadh to say there was fire in the cabin. The plane was about 80 miles from Riyadh, at about 22,000 feet.

During the flight, the transcript of the cockpit voice recorder sometimes notes that Captain Khowyter was "singing in Arabic." It has been suggested (without any apparent evidence) that he was praying.

The aircraft descended rapidly to Riyadh. During the descent there was mayhem in the passenger cabin as smoke became thicker. There were warnings to the flight crew of panic in the passenger cabin (1822:08, 1826:42), attempts to fight fire (1825:41, 1826:53), more smoke alarms (1824:16), and requests for passengers to remain seated (or otherwise not to panic) (1824:59, 1827:16, 1827:40, 1828:40, 1830:27, 1830:56, 1833:08, 1834:25, 1834:53). In the cockpit they had concerns about whether or not an emergency had been declared at Riyadh airport (1822:50), and pre-landing checks.

Engine number two developed problems at 1826:53 because apparently, control cables were burned through. It was shutdown and the remainder of the flight was completed on two engines.

At 1831:34, while still airborne, there was a discussion between Flight Engineer Curtis and a member of the cabin crew about whether passengers should be evacuated. "When we are on the ground, yes", Curtis said. Notably, this discussion did not include Captain Khowyter.

At 1831:58, Flight Engineer Curtis asked Captain Khowyter, "Okay, right after landing do you want me to turn off all the fuel valves?" Khowter replied, "No, after we have stopped the aircraft."

Throughout the approach, Captain Khowyter seemed to be flying the plane himself. This was undoubtedly not what should have been happening – he should have asked First Officer Hasanain to fly, and Khowyter should have concentrated on managing the crisis. However, Hasanain had only qualified 11 days previously. Perhaps Khowyter did not trust him to fly the plane in an emergency, in which case this put extra burden on Khowyter.

At 1832:10, while still airborne, Flight Engineer Curtis asked Captain Khowyter, "Do you want us to evacuate passengers, Captain?" Khowyter replied "What?" Curtis repeated, "Do you want us to evacuate the passengers as soon as we stop?" Khowyter did not reply.

Curtis said at 1834:04 that "The girls have demonstrated impact position."

Curtis again said to Khowyter at 1835:17 "The girls wanted to know if you want to evacuate the airplane." Khowter replied apparently in a non-committal way, "Okay, huh." Curtis repeated the question but Khowyter did not answer. Hence, Curtis had asked Khowyter four times about evacuation without receiving a clear response.

At 1835:53, Khowyter is recorded singing in Arabic. At 1835:56, Curtis said "Looking good."

Up until this point, the situation was still potentially going to end without disaster. Suddenly, at 1835:57, Captain Khowyter announced, "Tell them, tell them not to evacuate" for reasons that are unknown.

The last known communication from the passenger cabin was at 1836:09, when a member of the cabin crew warned passengers to adopt the "brace" position for landing. Flight Engineer Curtis commented, "No need for that, we are okay, no problem."

Landing was at 1836:20, and was reported by witnesses to be normal. Some ground staff noted that smoke was trailing behind the aircraft as it decelerated, but others reported they did not see smoke. It will have been close to sunset but visibility was still good.

The cockpit voice record stopped just before landing, for reasons unknown; the official report merely notes that it "ceased to function when the aircraft was about 30 feet in the air and on its landing approach." Thereafter, the only known communications are exchanges with the control tower.

Captain Khowyter did not bring the aircraft to an emergency stop. Instead, the plane taxied for a further 2 minutes and 36 seconds, eventually coming to a halt at 1838:56.

During this time there were exchanges between the cockpit and the control tower, so the crew were still conscious. At 1839:06, the control tower asked if they wanted to continue to the ramp or to shutdown. The aircraft replied "Standby" and then "Okay, we are shutting down the engines now and evacuating."

After a critical further minute and a half, at 1840:33, the aircraft reported to the control tower "Affirmative, we are trying to evacuate now." This was the last transmission received from the aircraft. An eyewitness who had followed the aircraft onto the taxiway in a car later said he had observed fire through the windows on the left hand side of the cabin. He could not see any movement on board. It therefore seems quite likely that, by this time, many or even all passengers and flight attendants will already have been dead or dying from smoke inhalation.

The engines were reported to have been shutdown about 3 minutes after the aircraft stopped, which would be about 1842:00. About 1 minute later, smoke and flames engulfed the aircraft, with a flash fire occurring within the fuselage. Firemen tried to open the doors, at first unsuccessfully, even though cabin pressure was ambient. Eventually, at about 1905 hours, a door was opened. Everyone was dead, and all the bodies of the passengers and the cabin crew were found crowded at the front the passenger cabin, apparently trying to escape the smoke and flames coming from the aft hold.

Autopsy examinations indicated that carbon monoxide poisoning was the cause of death. Some burns had been caused by the flash fire, but these were inflicted *post mortem*.

There had apparently been no attempt made to open the doors from inside the aircraft, indicating that perhaps the passengers and cabin crew were already dead or unconscious at the time of landing, or shortly thereafter.

The exact cause of the fire was not determined, although pilgrims' carriage of gas bottles must be a strong suspect.

For many reasons, this event should never have been the disaster it became:

The fire in the aft hold very probably started (although this was never fully established) because of cooking equipment and gas bottles being carried illegally on board by pilgrims on their way to Mecca.

Captain Khowyter took far too long to turn back after the first smoke alarms went off. The five and a half minute delay between receipt of the smoke alarm and the decision to turn back was critical to survival. (However, it should be noted that the Saudi Airlines emergency procedure recommended that confirmation of fire alarms should be sought before taking emergency action.)

At no point during the crisis was there any suggestion made by the flight crew that passenger oxygen masks could be made available. The official report simply notes that "the flight station oxygen system and the passenger oxygen system were not utilized during the flight."

Preparation should have been made, before landing, for an emergency evacuation. The landing should have been followed immediately by an emergency stop and prompt evacuation. The plane could have stopped moving two minutes earlier than it did so, and this might have had a significant effect on survivability and mortality. The Captain, by not shutting down the engines until three minutes after the aircraft stopped, prevented the cabin crew from initiating an emergency exit. (even if he had not already said they were not to do this)

First Officer Hasanain did not take an active role in events; he should have been helping the Captain, he should have been more assertive, and he should have been actually flying the plane. Flight Engineer Curtis' relationship with the Captain seems to have been poor; the Captain ignored direct questions from him (including repeated requests about passenger evacuation), and called him a "jackass" for failing to find the appropriate emergency procedure quickly.

The question remains: Why did Captain Khowyter not bring the aircraft to a rapid halt, stop the engines and order an emergency evacuation? The official report does not offer any explanation. One eyewitness, Michael Busby, who was living near Riyadh airport and watched in horror from his balcony the events on the runway, published his account on the Internet in 2010 [3]. He says Khowyter's strange decision-making was because King Khalid's Boeing 747 was rolling on the runway at the time Flight 163 landed. (The official report, in witness interview with Nasser Al-Mansour (page 147 of the report), does indeed note that the King's Boeing 747 was taking off at about this time, but no further comment is made.) Mr Busby says Saudi protocol required everything at the airport to stop moving while the King's plane was rolling, regardless of circumstances. He also says that Captain Khowyter will have known the King's plane was moving, and that "a Saudi pilot was not going to risk beheading due to the King's ire." Mr Busby is certain that this will have been a determining factor in Captain Khowyter's behavior – he will have known that the runway would remain in "lock-down" mode after Saudia 163 had taken off, until the King's plane had also departed. Hence Khowyter may have been reluctant to return to Riyadh, and he may not have wanted to perform an emergency stop and evacuation on the runway.

Busby's account does indeed offer a simple explanation for what happened. However, I have some difficulties with his explanation; e.g., he says that ground emergency crew could not move until the King's plane was "wheels up", and yet we are told in the official report that Flight 163 was chased down the runway by emergency vehicles, so these versions seem inconsistent. Also, we are asked to believe that the King was such a tyrant, or that Saudi deference to their King was so strong, that the rules could not be broken even in a genuine emergency. Is that credible?

Another explanation for the long taxiing and the delayed engine shutdown could be that the flight crew were also affected by noxious gases. However, the last transmission from the cockpit at 1840:33 ("Affirmative, we are trying to evacuate now") was one and a half minutes after the aircraft stopped moving.

Flight 163 therefore remains, at least partially, a mystery. Three hundred and one passengers and crew died in a plane that had landed successfully.

If we again apply the "time machine" test ("How far back would one have to go to have stopped this accident happening?"), the answers in this case are as follows.

This accident would have been completely avoided if dangerous flammable material had not been stowed in the aft hold. Ground staff must therefore take a major share of the blame.

Deaths would perhaps have been avoided completely if the Captain had turned back to Riyadh more quickly.

Deaths would perhaps have been avoided completely if the Captain had stopped the aircraft promptly on the runway and ordered emergency evacuation.

Some deaths might have been avoided if oxygen masks had been made available.

Contributory factors must include the poor caliber of the flight crew, and the Captain's failure to delegate the control of the plane to the First Officer.

The significance of the presence of King Khalid's Boeing 747 is unknown, but I find it difficult to believe that Khowyter might have been so afraid of the King's wrath that it made him hesitate about turning back to Riyadh, then carrying out an emergency stop and evacuation.

Also unknown is the significance of the Captain "singing in Arabic" – was he praying when he should have been managing the crisis? Or perhaps it was irrelevant – may be he just sang to himself when he was concentrating.

Captain Khowyter's behavior seems reminiscent of the fictional Captain Queeg, played by Humphrey Bogart in the 1954 movie *The Caine Mutiny*. In the movie,

Queeg/Bogart froze and was indecisive on the bridge of his warship during a typhoon. On the whole, I am inclined to believe that, like the fictional Queeg, Khowyter just froze and became indecisive during the crisis.

Khowyter was clearly not the most able of captains. He was placed under a position of extreme stress, where key decisions had to be taken promptly. He had a sub-standard flight crew who were perhaps less helpful than they should have been. Nevertheless, he was an experienced pilot who should have been able to make those key decisions. He should also have been able to delegate more effectively, and he should have fostered better relationships with his colleagues. He showed signs of wanting too much to be in control of everything, to the point that he seemed to resent, ignore, and even countermand the helpful suggestions about preparing for evacuation made by Flight Engineer Curtis.

CONCLUSIONS

This chapter has discussed two very different aspects of safety management: independent assessment and crisis management. The point if this chapter, the only conclusion possible, is that no matter how good procedures and arrangements might be, it is difficult to foresee the wide variety of ways in which those procedures and arrangements may fail. We can implement arrangements such as independent assessment, which should be a good thing, but in practice the arrangements may be stymied by a variety of factors; in the case of Port of Ramsgate, the designers and assessors, quite simply, did not think sufficiently about the issues they were reviewing.

Similarly, we can develop emergency procedures for all sorts of operational crises, but if the people facing the crisis cannot rise to the occasion, the procedures are worthless. As aviation and other industries have discovered, the best way of helping staff to cope with highly stressful emergency situations is to have frequent, realistic simulator training of emergencies. Aircraft simulators and training have of course improved immeasurably since the Saudia 163 accident, facilitated by the enormous developments in computing since that time.

We develop all sorts of arrangements to try to ensure safety, but any of them can fail because of human frailty and fallibility.

REFERENCES

[1] Joel and Gorton, Health and Safety Laboratory investigation report into Port of Ramsgate walkway failure, undated.
[2] Presidency of Civil Aviation, Jeddah, Saudi Arabia, Aircraft Accident Report, Saudi Arabian Airlines Lockheed L-1011 HZ-AHK, August 19, 1980 January 16, 1982.
[3] Michael Busby's eyewitness account of the Saudia 163 accident is given in http://www.scribd.com/doc/38040625/Death-of-An-Airplane-The-Appalling-Truth-About-Saudia-Airlines-Flight-163. I have corresponded with Mr Busby and I am in no doubt about the sincerity of his belief that the presence of King Khalid's plane was a significant factor.

Hydrocarbon Processing

There are about 700 operational oil refineries in the world. A medium-sized refinery can typically process 100,000 barrels of oil per day. Jamnagar refinery in Gujarat, India, is currently the biggest refinery in the world; it alone processes more than 1% of global output. The top 10 biggest refineries in the world are listed in Table 12.1.

Refineries are high-technology process plants, and the cost of building a large modern refinery is several billion US dollars. The main processes carried out in simple refineries are listed in Table 12.2.

Hydrocarbon processing, transport and storage are undoubtedly very hazardous, high risk activities, and they also feature elsewhere in this book – the Buncefield accident is discussed in Chapter 6, and the Texas City refinery accident is discussed in Chapter 14. The hazards and risks mainly arise because they process hydrocarbons at high temperature and high pressure:

- A leak of high-pressure gaseous hydrocarbon can, if it ignites immediately, produce a jet fire that can impinge on other process plant and then escalate to become a large conflagration, in a similar fashion to the Piper Alpha accident. (Chapter 13)
- An un-ignited leak of high-pressure gaseous hydrocarbon can quickly generate a large, inflammable cloud which may drift until it finds an ignition source, and it can then yield a vapor cloud explosion, as happened at Buncefield. (Chapter 6)
- A spill of liquid hydrocarbon may catch fire and yield a pool fire.

Hence, accidents in refineries can be amongst the most devastating industrial accidents. Marsh Risk Consulting publishes data on major accidents, including refinery accidents, giving the insured losses in each case. (Total losses may, of course, exceed insured losses.) Table 12.3 is a compendium of refinery accident data between 1972 and 2011 from the Marsh reviews for 2001 and 2011, and it shows that at least 53 major refinery accidents occurred during the 40-year period covered. Many of these accidents did cause injuries and deaths, although the table does not list them.

There can be no guarantee of the completeness of these data, in particular because Chinese and pre-1990 Soviet and Eastern European data are not available. It is also noted that data in the period 1972–1979 are strangely absent.

However, the point is made; refineries are almost uniquely difficult places to operate safely. If we allow for some credible under-reporting of accidents, and assuming

High Integrity Systems and Safety Management in Hazardous Industries. 978-0-12-801996-2

Table 12.1 The world's largest refineries

	Location	Operator	Capacity (barrels per day)
1	Jamnagar, Gujarat, India	Reliance Industries	1,240,000 bpd
2	Paraguana, Venezuela	PDVSA	940,000 bpd
3	Ulsan, South Korea	SK Energy	850,000 bpd
4	Yeosu, South Korea	GS-Caltex (Chevron/GS Holdings)	730,000 bpd
5	Ulsan, South Korea	S Oil (Saudi Aramco/Hanjin Group)	669,000 bpd
6	Jurong Island, Singapore	Exxon Mobil	605,000 bpd
7	Baytown, Texas, USA	Exxon Mobil	572,500 bpd
8	Ras Tanura, Saudi Arabia	Saudi Aramco	550,000 bpd
9	Baton Rouge, Louisiana, USA	Exxon Mobil	503,000 bpd
10	Texas City	Marathon (ex-BP)	467,720 bpd

Table 12.2 Principal refinery processes

Refinery process	Function	Products
Crude oil storage	*Self-explanatory*	–
Desalter	Removes impurities from crude oil	Clean crude oil
Crude oil distillation column	Separates crude oil into light and heavy components	Naphtha*, jet fuel, diesel, residue
Hydro-treating	Remove sulfur	Desulphurized products
Catalytic reforming, platforming or isomerization	Converts naphtha to gasoline	Gasoline/petrol
Residual fluid catalytic cracking (RFCC)	Takes residue from distillation column and produces useable product	Gas, LPG, gasoline, diesel, slurry
Thermal conversion and delayed coking plant	Takes heavy oil residue and converts it to coke (batch production only)	Coke
Product storage and blending	*Self-explanatory*	Blended products to meet market needs

*Naphtha is defined as the fraction of hydrocarbons in petroleum boiling between 30 and 200 °C

for the purposes of argument that the number of refineries worldwide has remained broadly constant at about 700 over the timescale (production has increased but bigger plants will have replaced smaller ones), with each plant having something like a 50-year operational life, then we can derive a rough rule-of-thumb that any given refinery has about a one in ten chance of suffering a major accident during its operational lifetime.

Table 12.3 Refinery major accident losses, 1972–2011

	Date	Location	Type of accident (shaded = natural causes)	Value of Insured Losses (2001 USD, except* 2011 USD)
1	July 21, 1979	Texas City, Texas	Vapor cloud explosion	$47 million
2	September 1, 1979	Deer Park, Texas	Explosion	$138 million
3	January 20, 1980	Borger, Texas	Vapor cloud explosion	$65 million
4	August 20, 1981	Shuaiba, Kuwait	Fire	$73 million
5	April 7, 1983	Avon, California	Fire	$73 million
6	July 23, 1984	Romeville, Illinois	Explosion	$275 million
7	August 15, 1984	Las Piedras, Venezuela	Fire	$89 million
8	March 22, 1987	Grangemouth, UK	Explosion	$107 million
9	May 5, 1988	Norco, Louisiana	Vapor cloud explosion	$336 million
10	April 10, 1989	Richmond, California	Fire	$112 million
11	September 5, 1989	Martinez, California	Fire	$62 million
12	September 18, 1989	St Croix, Virgin Islands	Hurricane	$168 million
13	December 24, 1989	Baton Rouge, Louisiana	Vapor cloud explosion	$89 million
14	April 1, 1990	Warren, Pennsylvania	Explosion and fire	$30 million
15	November 3, 1990	Chalmette, Louisiana	Vapor cloud explosion	$25 million
16	November 30, 1990	Ras Tanura, Saudi Arabia	Fire	$40 million
17	January 12, 1991	Port Arthur, Texas	Fire	$31 million
18	November 3, 1991	Beaumont, Texas	Fire	$18 million
19	March 3, 1991	Lake Charles, Louisiana	Explosion and fire	$28 million
20	April 13, 1991	Sweeney, Texas	Explosion	$45 million
21	December 10, 1991	Westphalia, Germany	Explosion and fire	$62 million
22	October 8, 1992	Wilmington, California	Explosion and fire	$96 million
23	October 16, 1992	Sodegaura, Japan	Explosion and fire	$196 million
24	November 9, 1992	La Mede, France	Vapor cloud explosion	$318 million
25	August 2, 1993	Baton Rouge, Louisiana	Fire	$78 million
26	February 25, 1994	Kawasaki, Japan	Fire	$41 million
27	July 24, 1994	Pembroke, UK	Fire	$91 million
28	October 16, 1995	Rouseville, Pennsylvania	Fire	$46 million
29	October 24, 1995	Cilacap, Indonesia	Explosion and fire	$38 million
30	January 27, 1997	Martinez, California	Explosion and fire	$22 million
31	September 14, 1997	Visakhapatnam, India	Explosion and fire	$64 million
32	June 9, 1998	St John, New Brunswick	Explosion and fire	$66 million
33	September 26, 1998	Pascagoula, Mississippi	Hurricane	$357 million

(Continued)

Table 12.3 Refinery major accident losses, 1972–2011 *(cont.)*

	Date	Location	Type of accident (shaded = natural causes)	Value of Insured Losses (2001 USD, except* 2011 USD)
34	October 6, 1998	Berre l'Etang, France	Fire	$23 million
35	February 19, 1999	Thessaloniki, Greece	Explosion and fire	$40 million
36	March 25, 1999	Richmond, California	Explosion	$79 million
37	August 17, 1999	Korfez, Turkey	Earthquake	$210 million
38	December 2, 1999	Sri Racha, Thailand	Explosion	$37 million
39	June 25, 2000	Mina Al-Ahmadi, Kuwait	Explosion and fire	$433 million
40	April 9, 2001	Aruba, Caribbean	Fire	$134 million
41	April 16, 2001	Killingholme, UK	Explosion and fire	$82 million
42	April 23, 2001	Carson City, California	Fire	$124 million
43	April 28, 2001	Lemont, Illinois	Fire	$36 million
44	September 21, 2001	Lake Charles, Louisiana	Fire	$52 million
45	November 22, 2002	Mohammedia, Morocco	Explosion and fire	$190 million*
46	January 6, 2003	Fort McMurray, Alberta	Explosion and fire	$170 million*
47	January 4, 2005	Fort McKay, Alberta	Explosion and fire	$150 million*
48	March 23, 2005	Texas City, Texas	Explosion and fire	$250 million*
49	October 12, 2006	Mazeikiu, Lithuania	Explosion and fire	$170 million*
50	August 16, 2007	Pascagoula, Mississippi	Explosion and fire	$230 million*
51	February 18, 2008	Big Spring, Texas	Explosion and fire	$410 million*
52	September 12, 2008	Galveston, Texas	Hurricane^	$540 million*
53	January 6, 2011	Fort McKay, Alberta	Explosion and fire	$600 million*

^Refers to Hurricane Ike, which affected six refineries in the Galveston area. The loss data relate to the worst affected single refinery.

(Source: Marsh [1])

PIPELINE RUPTURE AND FIRE, WASHINGTON STATE, USA, JUNE 10, 1999

This accident involved the rupture of a buried gasoline (petrol) pipeline. The basic facts are simple yet tragic. The buried pipeline was 16 inches in diameter (40 centimetres), and the rupture released about 237,000 gallons (more than a million liters) of gasoline into a river creek flowing through Whatcom Falls Park in Bellingham, Washington. The park is a woodland area in an otherwise mostly residential suburb, a few miles south of the US–Canadian border.

The gasoline from the pipe flowed down the river creek for about 90 minutes before finding a source of ignition and catching fire. One and a half miles of the river creek burned, and two 10-year-old boys and an 18-year-old man were killed, with a further eight people injured. Property damage was estimated at $45 million.

FIGURE 12.1

Whatcom Falls Park, June 1999, after the fire in the river creek.

(Source: NTSB)

The NTSB report [2], although very detailed, does not name any of the actors in the long sequence of events which led to the accident, probably because legal action was still pending (Fig. 12.1).

As stated above, the facts are simple and tragic. The accident was undoubtedly unusual – a massive fire in a public park on a Thursday afternoon in summer. The surprising things about this accident in the current context are, however, as follows: the number of missed opportunities to prevent the accident; the number of different people and organizational entities that failed in their responsibilities; and the length of time between the first failings and the eventual accident.

The pipeline was owned and operated by the Olympic Pipeline Company, but operation was subcontracted to another company, Equilon, although the 2002 NTSB report says that Equilon disputed this assertion – Equilon said it was not responsible for pipeline operations and that it only loaned employees to Olympic.

The Olympic pipeline system consisted of a network of some 400 miles of pipelines transporting refined petroleum products from refineries in northwest Washington State to various locations as far as Portland, Oregon. The pipeline system included a length of pipeline between Ferndale and Bayview, which passed through Whatcom Falls Park; this was a section of a longer pipeline used for transferring gasoline from a refinery at Cherry Point in northern Washington southwards toward a large storage depot at Renton near Seattle.

The section of pipeline in Whatcom Falls Park where the accident occurred was first installed in 1964, but was re-routed in 1966 because of the construction of a water treatment plant (owned by Bellingham city council) that is situated in the middle of the park. The pipeline was made of 0.312 inches thick steel. It was hydrostatically tested to 1820 pounds per square inch (124 bar). Maximum operating pressure was 80% of the hydrostatic test pressure.

As part of further improvements to the water treatment plant in the early 1990s, Bellingham city council personnel were required to confirm the exact location of the gasoline pipeline using a process called "potholing". This consisted of, first, identifying the rough location of the pipeline with a magnetic detector, then probing with a steel bar, and finally excavation by hand. Olympic's own personnel were present when the council crew carried out this work on different occasions in 1993.

Bellingham council placed contracts with two companies (IMCO General Construction Inc and Barrett Consulting Group, subsequently known as Earth Tech) to implement the modifications at the water treatment plant.

The arrangements were supposed to be that Olympic personnel would be present whenever excavation happened within "10–15" feet of the pipeline, and that all excavation within two feet of the pipeline would be only by hand; i.e., no mechanical diggers were to be used close to the pipeline. Olympic inspectors also made unannounced visits to the work site on a regular basis, more than once per week.

However, the NTSB report concluded that excavation around the Olympic pipeline occurred in August 1994, without Olympic inspectors being present, in order to lay a new water pipeline which crossed over the gasoline pipeline. A subcontractor to IMCO told the accident investigators that he heard the gasoline pipeline being struck by a backhoe – a powered mechanical excavator – during the project, probably on August 11, 1994, when an Olympic representative was not present, and that IMCO personnel decided not to notify Barrett or Olympic. The subcontractor said that IMCO personnel coated the damaged area of pipeline with a mastic coating before backfilling over it. A laborer working for IMCO also recalled IMCO hitting the pipeline. All other IMCO employees denied this account.

Next, Olympic had placed a contract with Tuboscope Linalog Inc to carry out 5-yearly remote inspections of the pipeline along its entire length using a magnetic flux inspection tool, which determines steel thickness of the pipeline wall. (Such a device is called a "pig" in the oil industry – it is passed along the pipeline and records data, which can be analyzed later offline.) An inspection in 1991 showed no defects in the area of the later failure. An inspection on March 18, 1996, however, did show anomalies in that area. These anomalies were assessed by Tuboscope and judged to be small enough not to matter; however, the assessment methodology they used was developed by the American Society of Mechanical Engineers (ASME) specifically for corrosion damage and not mechanical damage. For mechanical damage, the geometry of the defect may be much more precisely defined, and the resolution of magnetic flux inspection may underestimate the severity of the damage.

A separate incident on an Olympic pipeline led, on September 17, 1996, to the Washington Department of Ecology ordering Olympic to carry out further remote inspection work, including the use of a diverse (mechanical) measurement technique called "caliper tools". This led, on January 15, 1997, to another inspection of the pipeline done by a different company, Enduro, using the caliper tool inspection technique that was specifically aimed at identifying mechanical damage. Near the site of the future pipeline rupture, a 0.45-inch "sharp defect" was reported at the same point where Tuboscope had detected an anomaly. The "sharp defect" at the future rupture site was reported as being 23% of wall thickness.

In May 1997, Olympic began exposing locations where anomalies had been measured that might be more than 20% of pipewall thickness.

The future rupture site was declared by an employee of Olympic's construction supervisor to be "too wet" to permit excavation at that time. This was reported to a junior engineer in Olympic, who says he was told they would "go back and try again when the area was dry." Tragically, no further action was taken before the accident.

(After the accident, inspection of the failed section of pipeline showed "numerous gouges and dents." The pipeline was about 10 feet (3 m) underground at the failure location. The failed section of pipe had a tear-like fracture some 27 inches (68 cm) long with a maximum separation of 7 inches (17 cm).)

Meanwhile, several miles away at Bayview, in December 1998, Olympic completed construction of a new terminal with a storage capacity of 500,000 barrels (82 million liters). The contractor was Jacobs Engineering Inc. The gasoline flowed southwards along the pipeline, through Whatcom Falls Park, to the new terminal.

The new Bayview Terminal had a design operating pressure of 740 pounds per square inch (50 atmospheres or 50 bar) and the accident pipeline was designed for much higher pressures, so it was necessary to install pressure-reduction equipment where the pipeline joined the new terminal. There were three layers of control and safety devices: (i) a control valve to throttle the incoming flow, (ii) a pilot-operated spring-loaded relief valve, and (iii) three motor-operated isolation valves. This triple-barrier approach is sound design – a control system, backed-up by two protection systems (the relief valve and the isolation valves). In the event of a problem with the control valve, the relief valve should operate quickly, and only in the event of a fault with the relief valve should the isolation valves ever have been required to operate.

The three safety-critical motor-operated isolation valves operated as follows: two isolated the pipeline from the new terminal, and the third opened the pipeline into a receiver vessel to divert the source of high pressure.

During commissioning of the new terminal, on the night of December 16–17, 1998, it was discovered that the pilot-operated spring-loaded relief valve had been wrongly specified – it opened at 100 pounds per square inch (6.9 bar) instead of the intended 700 pounds per square inch (48 bar). The springs in the pilot valve were replaced with spares, but the replacement springs were actually identical. Furthermore, in the process of replacing the springs, the pilot-operated spring-loaded relief valve was actually rendered unreliable. Finally, to make things even worse, no proper re-testing was carried out to see if the valve was working properly and at its correct relief pressure.

There still should have been opportunities to diagnose and rectify the problems with the relief valve. After the new Bayview Terminal went into service on December 17, 1998, there were operational difficulties. Pressure control continued to be a problem, and the motor-operated isolation valves operated 41 times because of high-pressure within the terminal. Each time this happened, a pressure pulse was sent back along the pipeline toward the future accident site. Some of these pressure pulses exceeded 1300 psig. Because the isolation valves were classified as "safety devices", the closure of these valves should have triggered management concerns – their closures clearly implied that something was wrong with both the pressure control valve and the relief valve. However, no concerted effort was made to find out what was wrong to cause the isolation valves to keep operating, or to find out why with the relief valve was not opening.

The final stage in the sequence of events happened on the day of the accident, June 10, 1999.

The Olympic pipeline was controlled using a supervisory control and data acquisition (SCADA) system – in other words, a digital control system. This system used plant sensors and actuators connected to two DEC VAX computers, with one primary computer, and one backup. The computers received data from the sensors and the actuators every 3–7 seconds. As is normal for systems like this, the operators used screens with pre-programed formats on which the data were displayed, and the operator could interact with the system by mouse click. The system also recorded and stored all data and commands made.

Among its other functions, the SCADA system sent control signals to the control valve, which throttled flow incoming from the north.

At about 1500 hours in the afternoon of the accident, June 10, 1999, the SCADA system's response time slowed significantly following an upload of plant data records into the SCADA historical database. This data upload was done by the computer system administrator. The SCADA system's problems grew more pronounced over a period of about 20 minutes, and for a while the system became completely unresponsive; it was unable to send control signals to any of the equipment on the pipeline system. This meant that the pressure control valve and the isolation valves were not operational.

Although the SCADA fault that caused the system to be unresponsive could not be replicated or explained subsequently, the computer system administrator had been carrying out development work on the live system at the time. This is extremely bad practice. Olympic personnel were using the *operational* pipeline SCADA system as the test bed to develop improvements to its database; normal good practice should be to first test changes on a separate offline system.

Finally, at the same time as the computer system administrator's online development work caused the SCADA system to stop working, the pipeline system controller switched delivery points, which led to a significant pressure pulse through the system.

This pressure pulse went through the system; the SCADA system was unresponsive because of the development work, so the isolating valves did not operate; the spring-loaded relief valve was inoperative because of a specification error; and the pipe ruptured at 1528 hours, at the point where a defect remained from the botched 1994 excavation work.

So, the complete timeline of missed opportunities and mistakes that led to this accident can be summarized as follows.

1. The pipeline was damaged during excavation work carried out IMCO probably on August 11, 1994. This damage was not reported to Olympic.
2. A magnetic inspection carried out on March 18, 1996 did show anomalies in the region of the eventual pipe rupture. The anomalies were assessed to be insignificant, but the assessment technique was intended for corrosion damage. (and not for mechanical damage)
3. Another inspection was carried out on January 15, 1997 using a different inspection technique. A "sharp defect" was reported near the eventual failure site. However, the ground was too wet to allow excavation and it was agreed to excavate later when it was drier. No subsequent excavation took place.
4. A new gasoline terminal was installed some miles downstream and commissioned on December 17, 1997. The pressure relief valve was not set up properly so it did not function, and it was not tested so the improper setup was not revealed.
5. Because the relief valve was not opening when required, there were 41 separate occasions when the isolation valves operated because of overpressure. None of these was investigated.
6. On the day of the accident, June 10, 1999, the computer system administrator had doing software development work on the live SCADA system, which controlled the entire pipeline system, including the isolation valves. For reasons unknown, this development work caused the whole SCADA control system to become unresponsive. This occurred at the same time as the pipeline system controller was switching delivery points, which led to a pressure pulse going through the system.

The result of all the above was that a pressure pulse occurred, but the relief valve had never worked, and the control system was unresponsive, so the isolation valves did not operate. The pipeline pressure rose and the pipe ruptured at the defect caused by the excavation damage in August 1994, allowing 1 million liters of gasoline to flow down a river creek on a summer afternoon, where it caught fire and killed three people. Fig. 12.2 summarizes the situation.

Suddenly, all the holes in the slices of Swiss cheese had become exactly aligned.

Root cause analysis was described in Chapter 10, and the "time machine" question was introduced. "To when would I have to travel back in a time machine to prevent this accident happening?" By this rule, all of the above six points count as root causes.

Probable cause of the accident was ascribed to the following.

1. Damage done to the pipe by IMCO General Construction Inc during the 1999 water treatment plant modification project.
2. Olympic Pipeline Company's inadequate evaluation of online pipeline inspection results.

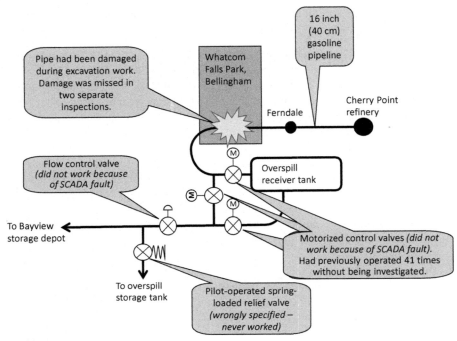

FIGURE 12.2

A schematic diagram summarizing failures that led to the gasoline pipeline rupture and fire in Whatcom Falls Park. This accident was due to a combination of pipeline damage (caused by poorly supervised excavation work around the pipeline several years earlier), inadequate non-destructive inspections of the pipeline, a wrongly specified pressure relief valve, inadequate investigation of valve control problems, and uncontrolled software upgrades ("SCADA" was the computer-based control system).

3. Olympic Pipeline Company's failure to test, under approximate operating conditions, all safety devices associated with the new Bayview Terminal.
4. Olympic Pipeline Company's failure to investigate and correct the conditions leading to the repeated unintended closing of the Bayview inlet isolation valve.
5. Olympic Pipeline Company's practice of performing database development work on the SCADA system while the system was being used to operate the pipeline, which led to the system becoming non-responsive at a critical time during pipeline operations.

This was a long and sorry tale of many people, over five years, all of whom failed to carry out their duties properly. It is difficult from the report to be sure exactly how many people were involved in unsafe acts and poor decisions, but there must have been at least 12 individuals involved. Mostly, their failings were simply laziness, indifference, or cynicism. There was no evidence to suggest time pressure, or financial pressure. It was just that people *couldn't be bothered*. No one cared about what they were doing, and as a result two 10-year-old boys and an 18-year-old man were killed.

Perhaps this is too harsh. Perhaps the people involved simply did not recognize that this sort of accident could happen. Maybe, instead, all those involved thought the only risk was that there might be a pipeline leak – and not a gross rupture – and that the worst that could happen was a little localized environmental damage, instead of fatalities. After all, this was a very unusual accident; it just so happened that the pipeline rupture was large, and that it ruptured near a river creek, and that the gasoline was able to flow un-ignited for over an hour, so that the size of the area affected (when it eventually did find an ignition source) included one and a half miles of the river creek. In which case, their blame can be ascribed to ignorance, or at least lack of imagination, and not indifference.

There is perhaps a parallel with the legal principle that "ignorance of the law is no excuse." In safety, there should perhaps be a principle that "inability to imagine the consequences of your actions (or inactions) is no excuse."

EQUILON ANACORTES REFINERY COKING PLANT ACCIDENT, NOVEMBER 25, 1998

This accident is an unusual multiple fatality refinery accident, which was actually relatively minor in terms of financial losses so it does not even appear on Table 12.3.

It is an unusual accident because it does not involve hydrocarbon liquids or gases under pressure. The accident occurred at one of the coking drums, where coke is produced from long-chain hydrocarbon residues in a thermal cracking process.

It is also unusual as an example of "group-think", where a number of experienced people in a slightly unusual situation persuaded themselves to do something that, with perfect hindsight, seems to be completely crazy.

The delayed coking unit is where heavy oil residues are converted into coke, which can then be used, say, as fuel for electricity generation. The heavy oil residues are passed through a furnace – a "thermal conversion unit" where the long chain molecules are "cracked" – and into large coke drums where the coke formation actually takes place. The term "delayed" is used to indicate the coke formation does not take place in the furnace (which would lead to a plant shutdown) but, instead, the coke crystallizes in the large coke drums after the furnace.

The coke drums are filled and emptied daily in a batch process, although the rest of the refinery operates more-or-less continuously. Like most plants of this sort, the Anacortes refinery had two large coke drums, Drums A and B, stainless steel drums each about 20 m tall and each with adequate capacity for one day's coke production. The process conditions in the operational coke drum are 450–500 °C and 20–30 bar pressure. Only one coke drum is online at any time; the other is offline, being emptied or standing by. Vapor passes from the top of the operational coke drum to a fractionating column, where the gaseous products are separated into the desired fractions. The residue remains in the coke drum to crack further until only the coke is left.

This description of the accident is taken from two sources: a NASA presentation which is available online, and a file on the www.historylink.com website [3].

A powerful storm hit western Washington State on November 23, 1998. It caused widespread damage and also interrupted the electricity supplies to the Equilon Anacortes refinery in Puget Sound for about 2 hours. This meant, in particular, that the delayed coking unit had to be re-started. Drum A was about one hour into a routine charging cycle when the power interruption occurred.

Under normal conditions, at the end of the cycle, the drum would be cooled with steam first and then water. Once the temperature was low enough, a permit to work would be issued and the top of the vessel would be unbolted and removed. The mass of coke in the vessel is then cut up using high-pressure hoses; the coke and water then can be discharged out of the bottom the vessel.

In this case, however, because of the power interruption, the Charge Line at the bottom of the vessel, through which the coke would normally flow out of the drum at the end of the cycle, was clogged with coke that had formed during the power cut. This meant that the operators were unable to put steam or water into the drum to cool the coke. There were some 46,000 gallons (about 200 m³) of hot coke remaining in the drum (Fig. 12.3).

Electric power was restored and at about 1000 hours on November 24 steam was also available. The operators made attempts to clear the clogged charge line. They believed – although there was no evidence to confirm this had happened – that steam

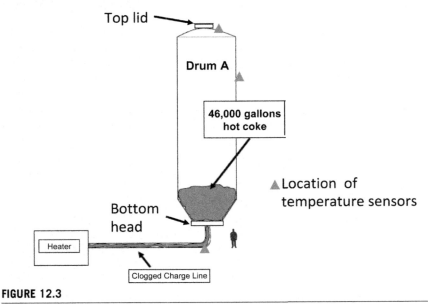

FIGURE 12.3

Drum A in the coking plant at Equilon Anacortes refinery, November 25, 1998. Six people were standing under the bottom head when it was unbolted and removed. They were killed by hot coke and burning heavy oil residue.

had made its way into Drum A. Instead, it is likely that pressure relief valves were actually diverting the steam into a blowdown system.

At the afternoon meeting to discuss the instructions to the nightshift, it was agreed to tell nightshift that the drum was cooling without the need for water, and that dayshift would remove the head from the drum the following day, Wednesday November 25.

There were no credible indications of the temperature of the coke inside the vessel. The only temperature sensors were on the outside of the drum. Although the operators believed that steam had passed into the drum, no steam had actually been admitted to the drum. Furthermore, no attempt had been made to put water into the drum. Nevertheless, on the morning of November 25, the foreman and operators reviewed the drum temperature sensors and concluded that the drum contents had cooled sufficiently. It appears that Equilon plant management were also involved in the decision to go ahead, and a Permit to Work was issued for a specialist contractor, Western Plant Services, to open up the drum. Wearing oxygen masks, workers removed the top lid from the drum safely.

At about 1330 h, only 37 hours after the loss of power which interrupted the normal coke batch production sequence, and without any further effort to determine the temperature of the coke (even after removing the top lid from the drum), the bolts holding the bottom head in place were removed and a hydraulic lift began to lower the head from the bottom of the coking drum. Six men were standing directly under the drum; they expected to find a congealed mass of crude oil residue, but the unit was far hotter than anyone thought.

Hot heavy oil broke through the crust of cooled residue and poured from the drum. The oil was above its auto-ignition temperature so, when exposed to air, it burst into flames engulfing the two refinery workers operating the hydraulic lift and pouring onto four more workers below. Witnesses said they heard an explosion, and saw a large plume of black smoke rise up from the refinery followed immediately by a ball of fire several stories high. The blast was felt several blocks from the refinery. A few minutes later, a site emergency was declared.

The six workers killed in the explosion were two Equilon employees and four Western Plant Services employees. They died from smoke inhalation and burns.

Later analysis showed that, without effective steam or water cooling, it would have taken about 200 days for the 200 m^3 of coke in the drum to cool to a safe temperature.

The Washington State Department of Labor and Industries set up an investigation that lasted 8 months. Their report criticized the Equilon plant managers for allowing the coking drum to be opened when it had only air-cooled for some 37 hours, instead of the normal steam- and water-cooling process.

Equilon subsequently installed a remotely controlled system for removing the drum lids, and installed a gas-fired backup system to maintain steam supplies in the event of a power failure.

On January 19, 2001, a $45 million settlement was reached between Equilon and the families of the six men, and Equilon accepted responsibility for the accident.

What should have happened? First, the situation was very significantly different from the normal plant operating procedures, so this should have triggered all sorts of concerns for the managers and operators.

- Some attempt to measure the temperature of the material in the drum could perhaps have been made, possibly by lowering a temperature measurement device from the top of the vessel. Specialist contractors would almost certainly have been needed to do this.
- Specialist advisers might have proposed some further ways to cool the drum.
- The plant managers should also have contacted specialist technical support who would have tried to calculate temperatures inside the drum.

All these steps would undoubtedly have required offsite specialist assistance, and the offsite experts would undoubtedly have taken a long time, possibly as much as several weeks. During that time the coking plant could only been used at best at half-capacity, using the other drum.

What did happen? It seems that there are two possibilities.

1. By a process of "group-think" the plant managers and operators somehow convinced themselves that there was no problem, that the coke drum would not contain dangerously hot oil and coke. Possibly they also felt commercial pressure to make the plant operational again, which might have affected their judgment.
2. The managers and operators were completely unaware of the nature of the hazard posed by an extremely hot mixture of heavy oil residue and coke being exposed to the air.

Option 2 is difficult to believe. These were people with many, many years of collective experience. In this example, it appears to me that the collective knowledge of the decision-makers means they must have known about the hazard of the hot oil residue-coke mixture, so ignorance is simply not credible. So, somehow, a group of experienced people persuaded themselves that it would be okay to do something really stupid.

This bizarre accident happened because the plant managers somehow convinced themselves that the situation was safe, despite operating well outside their approved operating procedures.

The two examples described in this chapter are very different types of failures. The pipeline rupture in Whatcom Falls Park was the result of numerous separate failings of safety management arrangements over five years, with many separate individuals involved, who will probably not even have known each other. By comparison, the Anacortes refinery coking plant accident occurred when a group of individuals, working together, failed to realize that they were facing a situation that was completely beyond the scope of their normal operating procedures.

Engineering managers have to *worry* about safety. If they are not worrying about safety, they are not doing their jobs.

REFERENCES

[1] Data on major refinery accidents can be found at http://uk.marsh.com/ProductsServices/MarshRiskConsulting.aspx.

[2] US National Transportation Safety Board, Pipeline rupture and subsequent fire in Bellingham, Washington, June 10, 1999, NTSB/PAR-02/02, October 8, 2002.

[3] The Equilon Anacortes coking plant accident is described in http://www.csb.gov/assets/1/19/moc0828011.pdf and also http://www.historylink.org/_content/printer_friendly/pf_output.cfm?file_id=5618.

Offshore Oil and Gas: Piper Alpha and Mumbai High

13

"Safety is not an intellectual exercise to keep us in work. It is a matter of life and death. It is the sum of our contributions to safety management that determines whether the people we work with live or die".

Sir Brian Appleton

PIPER ALPHA ACCIDENT AND THE CULLEN REPORT

About 8 miles from my home, in Strathclyde Park, Lanarkshire, there is a memorial to the 167 people who were killed on July 6, 1988 when huge explosions and fire engulfed and destroyed Occidental Petroleum's Piper Alpha production platform, about 340 km east of Aberdeen in the North Sea. The first explosion took place at 2200 hours, and further explosions happened at 2220 hours, 2250 hours and a final huge explosion happened at 2320 hours. Piper Alpha remains the world's worst offshore oil accident (Table 13.1, Fig. 13.1).

The subsequent report by Lord Cullen into the accident caused a complete re-baselining of safety management in North Sea oil and gas. Cullen carried out a very detailed analysis of the accident and the management systems in place at the time, and made sweeping recommendations which were adopted by the industry. His report is a model of its kind. Today, it remains the starting point for any discussion of offshore oil and gas safety.

In particular, Cullen highlighted two main areas of weakness: failure of the "permit to work" arrangements, and inadequacy of the emergency evacuation and escape arrangements.

First, the initial fire and explosion was caused by a *failure of the "permit to work" arrangements* (see "Safe Working Arrangements" in Chapter 10).

All hazardous facilities – offshore or onshore, oil and gas, petrochemical, nuclear, pharmaceutical, etc. – must have clear controls over how maintenance, inspection or test work is carried out on the plant. Hazardous equipment needs to be carefully isolated, and work needs to be carried out in a planned, careful way. "Isolation" means ensuring that sources of electricity or high-pressure fluid or other hazards cannot reach the equipment being worked on. After the maintenance work is completed, the equipment needs to be tested, de-isolated and re-commissioned carefully.

High Integrity Systems and Safety Management in Hazardous Industries. 978-0-12-801996-2

Table 13.1 The worst offshore oil platform and exploration rig accidents

Platform or rig	Location and year	Fatalities	Brief cause
Piper Alpha (production platform)	UK North Sea, 1988	167 fatalities	Work control failure
Alexander Kielland (accommodation platform)	Norway North Sea, 1980	123 fatalities	Structural failure
Seacrest (drillship)	Thailand South China Sea, 1989	91 fatalities	Capsized in typhoon
Ocean Ranger (drilling rig)	Off Newfoundland, Canada, 1982	84 fatalities	Capsized in heavy storm
Glomar Java (drillship)	Vietnam South China Sea, 1983	81 fatalities	Capsized in tropical storm
Bohai 2 (jack-up rig)	China Gulf of Bohai, 1979	72 fatalities	Capsized in storm while being towed
Enchova (production platform)	Campos Basin, Brazil, 1984	42 fatalities	Blowout; during evacuation the lifeboat lowering mechanism failed.
Mumbai High North (production platform)	Mumbai Indian Ocean, 2005	22 fatalities	Support vessel collided with gas export riser, fire and explosion.
Usumacinta (jack-up rig)	Gulf of Mexico (Mexico), 2007	22 fatalities	Collision with production platform Kab-101, fire.
CP Baker (drilling barge)	Gulf of Mexico (US), 1964	21 fatalities	Blowout, fire and explosion.

Source: www.offshore-technology.com

FIGURE 13.1

Piper Alpha in flames.

So-called "essential" equipment, such as fire-fighting equipment or smoke or fire detection equipment, or backup electricity supply equipment, can only be taken out of service when suitable alternative arrangements have been agreed and are in place.

Before any maintenance work can take place, therefore, a permit to work is written which defines what work is going to be done, what isolations are necessary, how the job is going to be done, and how the equipment will be returned to service. A robust permit to work system is therefore a cornerstone of ensuring safe means of working. It should be signed (at least) by the operations foreman and by the person doing the maintenance work.

When equipment is isolated, padlocks should be fitted so that it is impossible to restart the equipment. The keys for the locks should be kept in a cabinet which is itself padlocked, and the keys for the cabinet should be held *personally* by the maintenance person(s) who has signed the Permit to Work form. Hence the person(s) doing the work knows that the equipment he is working on cannot be de-isolated.

Sometimes problems arise, such as people going home with isolation keys still in their pockets. Any decision to break into isolation cabinets in order to de-isolate equipment, without the permit to work having been signed off by the person doing the work, should only be taken after very careful management review and inspection of the plant.

Cullen noted many deficiencies in the Permit to Work system on Piper Alpha, including the lack of any consistent practice of physically locking-off isolated equipment, and the absence of a review or discussion of permits at shift handover.

In the Piper Alpha Enquiry [1], Cullen reviewed evidence from survivors and experts and tried to reconstruct events. This was difficult because many of the people most directly involved in events had died. Cullen concluded that the most likely cause of the explosions and fire was because a member of the nightshift operations team, George Vernon, had started up a condensate (light hydrocarbon) pump, here called pump A, when it was still under maintenance with an outstanding Permit to Work, and was therefore not in a safe condition. George Vernon died in the accident.

Lord Cullen's report details at some length the sometimes conflicting survivors' accounts of what happened. A summary of the most likely sequence of events is as follows.

Production from the platform depended on having one condensate pump available, so the platform had two condensate pumps located in the gas compression module. Either pump could supply the required flowrate. Hence, to enable the maintenance on pump A, the standby pump (pump B) was operational, with suitably closed isolation valves for pump A. However, at 2145 hours on July 6, 1988, for reasons unknown, pump B stopped and could not be restarted.

Pump A had been isolated for maintenance. The maintenance team on day shift had removed a pressure relief valve, and the open flange where the relief valve was normally attached had had a temporary cover fitted which was neither leak-tested nor pressure-tested. A Permit to Work was still in force for the job, and the nightshift team were aware that pump A had been undergoing maintenance. However, the nightshift operator, George Vernon, decided it was actually fit to return to service. He apparently did not see the temporary cover on the open flange. (The open flange was some distance away from the pump, so this may be why George Vernon thought the pump looked OK to return to service, i.e., he may be did not think to look beyond the area immediately around the pump.) The permit to work had not been signed off by the day maintenance crew as "work completed". Once the nightshift operators in the Control Room (Mr Vernon and Mr Bollands, the Control Room operator) had decided that they really needed to have pump A working, the day maintenance foreman should have been woken up and asked what had to be done so that pump A could be restored to service. But the nightshift operator did not wake up the day maintenance foreman; he consulted the nightshift maintenance lead hand (Mr Clark) in the Control Room, and they apparently decided that it looked OK to restart pump A. George Vernon left the Control Room and quickly de-isolated it – without any consultation with the day shift people who had been doing the work on pump A.

The above decision-making process sounds cavalier and ill thought-out, of course – that was exactly the problem. George Vernon restarted Pump A at about 2200 hours, only 15 minutes after pump B had tripped; the short time interval shows that no detailed review or inspection of the condition of pump A could possibly have been carried out.

Restarting pump A so quickly would never have occurred if a robust Permit to Work and isolation process had existed. The rapidity with which the decision was made to restart pump A, despite the outstanding Permit to Work, suggests that overriding of permits to work was done on a semi-routine basis on Piper Alpha. It appears that there was a culture of lip service to permit to work. Hence the extent of the blame accorded to George Vernon is a moot point – he was just working within that culture. The sanctity of the permit to work system is something that is nowadays generally taken for granted; the idea that, within 15 minutes, someone could decide that it would be alright to de-isolate and restore to service a piece of safety-critical equipment, part-way through maintenance, is quite frankly amazing.

Hence, let us not be too critical of George Vernon. He was undoubtedly too hasty, but the system was faulty – it should have been far harder to re-instate isolated equipment – and the culture clearly placed production over safety.

When pump A was restarted at 2200 hours, the temporary cover on the open flange (where the pressure relief valve would normally have been) failed, and a large quantity of light hydrocarbon liquid and vapor spewed out. Shortly afterward it found a source of ignition and the initial explosion and fire occurred. The fire then escalated with more explosions at 2220 hours, 2250 hours and a final huge explosion happened at 2320 hours. There were also other factors which made control of the fire difficult or impossible – e.g., the switches to start the firewater system were located too near the initial explosion, and the emergency shutdown (ESD) valves on the oil export line were leaking. (which allowed oil to come backup the export pipe and feed the fire)

Secondly, *the emergency evacuation and escape arrangements were inadequate.*
Today, all North Sea oil platforms have a "Temporary Refuge" – i.e., a designated enclosed area of the platform, which should be secure from smoke and should have a minimum assessed time of fire resistance. In the event of an emergency, all personnel muster in the Temporary Refuge and await further instructions.

Nowadays, offshore oil platforms in UK waters normally have designated primary, secondary, and tertiary means of escape. *Primary escape* means the preferred means of evacuating the platform if time allows, which generally means lots of helicopters. Aberdeen is currently the busiest heliport in the world. In an emergency, all available helicopters will be diverted from routine operations and proceed to assist with evacuation of a platform in trouble. This process has been triggered a few times over the years, and is generally used if there is an incident (such as a gas leak), which might possibly escalate into something bigger.

If a fire or explosion has already occurred, or if there is bad weather, such that helicopter operations are not possible, then the crew onboard the platform may be directed instead to go to the lifeboats (called TEMPSCs in North Sea jargon, which stands for totally enclosed motor-propelled safety craft). TEMPSCs are the *secondary means of escape*. It is normally expected that, from any point on an offshore oil platform, there will be two possible routes for personnel to reach the TEMPSCs, so that an alternate route should always be available if the primary route is blocked by fire.

Tertiary escape constitutes "direct escape to sea", i.e., if people cannot find their ways to a TEMPSC, they climb down to sea level, or even jump from high level, and hope for the best. Tertiary escape is clearly not advised in the North Sea where sea temperature is low all the year round, and death from hypothermia may occur quickly, even for good swimmers.

On Piper Alpha, primary (helicopter) escape was simply not an option because of the fire. Smoke had quickly engulfed the landing pad. Many of the crew mustered in the dining room of the Accommodation Module as the emergency plan required, but no clear further instructions were ever given. No secondary escape was even attempted, and no lifeboats were launched. When the final huge explosion occurred at 2320 hours the Accommodation Module fell into the sea, but by that time the 81 people inside the module had already died from smoke inhalation.

The 61 survivors of the Piper Alpha accident survived *because they disobeyed the standing instructions*; they did not wait for an order to evacuate, they opted for tertiary escape and jumped into the sea where they were rescued by support vessels and fast rescue boats from nearly platforms.

Why were no further instructions issued to personnel mustered in the Temporary Refuge? Cullen was extremely critical of Occidental Petroleum's emergency arrangements on Piper Alpha. "...the system was almost entirely inoperative and little command or control was exercised over the movements of personnel" (Paragraph 8.8, page 152). Apart from asking the Radio Operator to issue a "Mayday" message at 2203 hours, the Offshore Installation Manager (OIM), Colin Seaton, did not give any other clear instructions, and he certainly gave no instructions to abandon the platform.

Survivors from the Accommodation Module reported that there was confusion, and that no one seemed to be in charge or giving instructions or advice. Personnel received no further instruction than to wait for a helicopter to take them off. Colin Seaton apparently tried to calm people by saying that a Mayday signal had been put out and that help would arrive, but it should already have been obvious that helicopters could not land on the platform because of fire and smoke. Cullen quotes one survivor: "He did not know whether the OIM was in shock or not, but he did not seem able to come up with any answer." Colin Seaton's failure of leadership was judged by Cullen to be a major factor in the high death toll. It should have been clear to him that anything was better than staying in the Accommodation Module. Fire and smoke meant that helicopters were simply not going to be able to land on the platform, but fast recovery boats and the semi-submersible support vessel *Tharos* were nearby. Hence, the only alternative was, somehow or other, to find a way onto the water.

Personally, I have sympathy for Colin Seaton's predicament. As Offshore Installation Manager, he was on call 24 hours a day, seven days per week while offshore. The sudden emergency will have been as much a surprise to him as it was to everyone else – he will almost certainly not have known about the condensate pump problems or the decision to restart pump A. I imagine the first thing he knew was when the first explosion occurred at about 2200 hours. He should have been trained in dealing with emergencies and the need for clear leadership in crises. (Nowadays, that would be the norm.) I imagine that, in the few minutes after 2200, when something could have perhaps been done to save more lives, Colin Seaton was struggling to appreciate the scale of the disaster, and to accept that his platform was already a write-off. Also, evacuation to the sea will always be a high-risk option, and in such a scenario it was likely that some people would die. Seaton had, very quickly, to make a judgment on the balance of risk: Was it better to risk some deaths in evacuating to the sea, or to stay where they were and hope for rescue to arrive? He would not immediately have realized that the fire and explosions were going to cascade out of control. Because of these internal conflicts, he perhaps failed to understand quickly enough that his *only* focus had to be on getting as many personnel as possible, as quickly as possible, onto the water.

The question we all have to ask ourselves in these scenarios is this: "In the same circumstances, would I have done any better?" Given the state of Occidental Petroleum's emergency planning and training arrangements, I think my answer would be, "I am really not sure."

Smoke began filling the dining room and people had to crouch below the tables to try to keep under the smoke. By 2220 hours – only 20 minutes after the initial explosion – some personnel had already decided to take matters into their own hands. One survivor, JM McDonald, reported to Cullen, "I just said to myself 'get yourself off'. I got my pal Francis, and I got him as far as the reception, but he would not go down the stairs because he says 'We have done our muster job; they'll send the choppers in'. I said to Francis, "...There is something drastically wrong on this rig. We'll have to get off". Francis would not go, and he just slumped down. ...That was as far as I could get him."

JM McDonald found the wind was blowing from the south. He used his knowledge of the platform to find his way to the southwest corner, where he climbed down a hose before dropping into the sea.

Twenty-seven others survived who, like JM McDonald, left the Accommodation Module. However, the rest apparently made no attempt to leave. There was no systematic attempt to lead men to an escape route. The 81 who remained, including Colin Seaton, all died from smoke inhalation.

The other 33 survivors (i.e., those who had never been able to reach the muster point in the Accommodation Module) jumped into the sea from various heights, some extremely high indeed. Fifteen survivors jumped from a deck at the 133 feet level, and five jumped from the helideck at 175 feet (53 m!) above the sea.

Of the 167 fatalities, only 13 were attributed to drowning. The vast majority of fatalities were due to smoke inhalation. Thirty bodies were never recovered.

The accident happened in calm summer weather. There would undoubtedly have been even fewer survivors if it had happened during a winter storm.

Lord Cullen made very extensive recommendations for improving safety in the offshore oil and gas industry. His recommendations were implemented in full, and they addressed the root causes of the accident, as follows.

His first 16 recommendations concerned the requirement for, and content of, safety cases that were to be presented by operators to the safety regulator. The safety cases must address the Safety Management System, the potential major hazards at the platform and the risk to personnel, and emergency evacuation, escape and rescue.

He then made recommendations about legislation and the safety regulatory body. Prior to his report, the regulator for offshore safety was the Department of Energy, which therefore had a perceived conflict of interest between advocating the development of North Sea oil, and managing its safe exploitation. Hence, safety regulation was moved to the Health and Safety Executive.

He made seven specific recommendations regarding the permit to work arrangements, including training and harmonization of permit to work practices within the industry, and the need for physical locking-off of isolation valves.

He made eight recommendations about fire and gas detection, emergency shutdown, and fire and explosion protection. These included specific recommendations regarding the vulnerability of the emergency shutdown valves, and also the firewater deluge system, to damage caused by severe accident conditions.

Finally, he made a further 51 recommendations about emergency escape and evacuation, command and control, and emergency training.

As a result of the accident and the Cullen report, responsibility for offshore safety regulation was moved from the Department of Energy to the UK Health and Safety Executive, which set up an Offshore Safety Directorate [2].

Some of these lessons were re-learned after the Macondo-Deepwater Horizon accident (see Chapter 14).

OFFSHORE SAFETY TRAINING

Cullen also concluded that there was not enough qualified and trained personnel onboard at the time of the accident. Temporary promotions had allowed available people to fulfill critical functions.

Since the Cullen report, standardized training programs have been introduced for the entire North Sea oil workforce. As a minimum, all offshore personnel from management to kitchen staff must do BOSIET training every four years. BOSIET stands for basic offshore safety induction and emergency training.

The training takes place over several days. It includes time onboard a totally enclosed motor-propelled safety craft (TEMPSC) in the sea – albeit within the confines of a sheltered harbor. There are no windows except for a Perspex dome at the top for the helmsman, so there is no visual reference for the horizon. This makes any rocking of the boat feel like it is amplified, even in a calm harbor; it is difficult to imagine a full TEMPSC, in darkness, in open water, in a storm. In rough seas, seasickness would begin quickly, and the smell of sickness means that, very soon, many or indeed all onboard would begin throwing up.

One of the principal safety hazards in the North Sea is helicopter ditching. There is a real risk that the helicopter may turn upside down, because helicopters are not stable on water in an upright position, especially in rough seas. Hence, a significant part of the training is about emergency escape from a ditched, inverted helicopter. The training facilities have dummy full-size helicopter fuselages hanging above a swimming pool, supported in a cradle from an overhead crane.

The trainees all change into *immersion suits*, and also wear lifejackets and *re-breathers*.

An *immersion suit* is a one-piece, loose-fitting rubber suit with watertight seals around the neck and with waterproof seals over the zips. Its role is to improve survival time in open water.

A person's survival time in seawater depends on a wide range of factors, as follows.

- Weather (sea state, water temperature, wind speed, air temperature);
- availability and effectiveness of personal protective equipment such as a lifejacket and immersion suit;
- physical and medical fitness of the individual;
- physical characteristics (age, sex, body fat, diet, behavior);
- training in procedures and discipline;
- effect of coincidental injury, if any;
- psychological condition; and
- type and number of layers of clothing worn beneath the immersion suit.

Death may occur due to "cold shock" within 2 or 3 minutes of immersion if the survivor is not adequately protected. The powerful stimulus of cold water on the skin can cause a loss of control of breathing, a reduction of breath-hold duration and a sharp rise of blood pressure. As a consequence, drowning may occur if wave action is severe while the person is at the surface of the sea. Cold shock exposure has been responsible for many sudden deaths, even among good swimmers. The sharp rise of blood pressure is a significant hazard for those with a potential cardiac problem. Experience has also shown that some individuals, fully conscious and lucid in the water, collapse and die whilst climbing out or lifted from the water, or shortly after.

A *re-breather* is a rubber bag, a bit like a hot water bottle, with a hose and a mouthpiece, and a nose-clip attached. The bag lies on your chest over the lifejacket.

In helicopters, in the event of a ditching, the emergency escape route is via the windows. Unlike airliners, helicopters are unpressurized, so the windows do not have to be pressure-resistant and are designed that they can be quickly knocked out with a sharp blow from the elbow.

The re-breather is for use during a helicopter ditching. The idea is that each person onboard quickly takes a deep breath, fits the mouthpiece, breathes out into his or her re-breather, and then fits the nose-clip. The re-breathed air in the bag should then enable more-or-less normal breathing for at least one minute.

In the training exercise, the trainees swim across the pool wearing their immersion suits and lifejackets, with re-breathers around their necks, and climb into the dummy helicopter fuselage, where each trainee takes a seat and fits a safety belt. The dummy fuselage is then lifted by the crane into the air. The tutor tells the trainees to fit their mouthpieces, exhale into their re-breathers and fit their nose-clips, and then a pretend "ditching" takes place; the dummy fuselage is lowered quickly back into the water with a splash. After a brief pause, the motorized cradle holding the dummy fuselage suddenly rotates so that, within a second or so, the fuselage and all its occupants are underwater and upside down.

The effect of the sudden rotation is extremely disorientating. From sitting upright and more-or-less comfortable, one is suddenly hanging upside down, still strapped into one's seat, with one's head about 2 m down in cold water, with the re-breather mouthpiece in one's mouth and the nose-clip fitted. Since goggles are not worn, vision underwater is extremely blurred. In these circumstances, one must retain one's composure and knock out the "helicopter" window with one's elbow. Then, still upside down, one undoes the seatbelt and climbs through the window while wearing the cumbersome (and awkwardly buoyant) lifejacket and re-breather.

Even non-swimmers must do this training if they wish to work offshore in the North Sea oil industry.

The exercises are deliberately uncomfortable, but if it were a genuine accident, the reality could be much worse. The North Sea is a much, much nastier place than any training facility can ever replicate.

MUMBAI HIGH, JULY 2005

In May 2005, I carried out a "third-party" safety audit of two platforms in the Mumbai High oilfield owned and operated by the Indian state-owned oil company ONGC (Oil and Natural Gas Corporation).

The Mumbai High oilfield complex is about 150 kilometers west of Mumbai, in the Indian Ocean. Mumbai High is actually a number of separate oil fields, which together supply more than 10% of India's oil consumption, so it is very important to their economy. I visited two platforms out of the many in the oilfield.

In general, my safety audit report was satisfactory. They seemed to know what they were doing.

Two months later on July 27, 2005, one of the ONGC platforms in Mumbai High had a devastating accident.

The accident happened at the large Mumbai High North platform complex (Fig. 13.2), which consisted of four adjacent platforms connected by bridges [3, 4]. (I had not visited this particular platform complex.)

The four platforms consisted of a small wellhead platform called NA, a residential platform called MHF, the main processing platform called MHN, and a relatively new additional process platform called MHW. At the time of the accident, a further temporary ("jack-up") platform called Noble Charlie Yester was alongside the complex working on the NA platform.

It was during the monsoon season, so the weather was bad. The wind speed was 35 knots, there was a 5-m swell, and the sea current was three knots. The conditions meant that helicopter operations were not possible. A diving support vessel (multipurpose support vessel or MSV) called the *Samundra Suraksha*, about 100 m long, was working nearby in the oilfield, in support of saturation diving operations.

The accident was strange, as most accidents are. A cook in the galley of the MSV *Samundra Suraksha* had an accident and cut off the tips of two fingers. The bridge of the MSV *Samundra Suraksha* contacted the Mumbai High North platform by radio, asking to transfer the cook to the platform for medical treatment. (ONGC generally

FIGURE 13.2

Mumbai High North, July 27, 2005.

employed qualified doctors onboard each platform; in the North Sea, the normal practice is to have a paramedic only.)

Because of the weather, the only way to achieve this would be by pulling alongside MHN and transferring the cook on a man-riding basket. Normally, this would be done from the leeward (i.e., downwind) side of the platform, which would be more sheltered, but the leeward crane on MHN was not working. The vessel therefore came onto the windward (upwind) side.

Offshore support vessels are normally capable of very fine positioning, with thrusters that give precise control. However, the vessel had problems with its normal control thrusters, so it came alongside using its emergency thrusters. The casualty was successfully transferred to MHN.

At this point the MSV *Samundra Suraksha* experienced a strong heave from the swell, and its helideck collided with the gas riser (the main pipe bringing gas onto the platform complex).

The gas riser failed below its Emergency Shutdown Valve, so the resulting leak could not be readily isolated. The leak ignited very quickly and fire engulfed MHN (the process platform) and MHF (the accommodation platform), while NA and the jack-up Noble Charlie Yester were severely damaged.

In addition the MSV *Samundra Suraksha* also caught fire. Several divers were in decompression chambers, and it was 15 hours later before it was safe to take them off the vessel. It subsequently sank (Fig. 13.3).

Evacuation from the platform complex was difficult, because the fire prevented six of the eight lifeboats being launched from the complex, and only one life raft was launched. Helicopters were not involved because the weather ruled them out. Despite these difficulties, 362 people were rescued over the next 15 hours, although 22 died. The rescue operation should therefore be considered successful.

Safety issues arising from the accident are many but the main issues include, first, the positioning of and lack of protection for the main riser pipes meant they were

FIGURE 13.3

MSV *Samundra Suraksha* on fire after the accident. It sank four days later.

vulnerable to collision and, second there was poor managerial control over the conditions in which it was acceptable for a vessel to come alongside.

The Indian Oil Industry Safety Directorate (OISD) had existed since 1986, although its function was non-regulatory. The Indian Government's response to the accident included giving OISD the mandate for the regulation of offshore safety.

ONGC opened a new Mumbai High North platform in October 2012.

The accident left me wondering whether I missed anything during my brief visit to Mumbai High. My remit had been to consider helicopter safety, platform housekeeping, emergency arrangements and safety training and equipment, plant condition, environmental controls, safety management systems, and accident data. Of these, the emergency evacuation and rescue arrangements during the accident appeared to have worked well. Otherwise, the only item relevant to the Mumbai High North accident appeared to be the safety management systems. I was perhaps remiss in not reviewing the ONGC procedures for vessels coming alongside, but frankly I doubted whether the procedures would have shown anything out of order anyway – it wasn't a question of what the procedures said, rather it was a question of whether anyone had actually paid any attention to the procedures in the urgent situation they faced (i.e. the cook with two fingertips missing).

The accident had happened because, in a rush to transfer the injured cook to the platform, the vessel had approached the platform in an unsafe manner in severe weather. The crew of the MSV Samundra Suraksha were in too much of a hurry to "do the right thing" and help their colleague to reach the medical facilities on board the Mumbai High North platform complex. And, in a hurry to do the right thing, they caused a disaster.

PIPER ALPHA AND MUMBAI HIGH: COMMON LESSONS?

The common issue in both accidents was impulsive behavior by relatively non-senior people when faced with a (minor) crisis. In the case of Piper Alpha, it was the rapid decision to re-instate the standby condensate pump without checking its status properly. In the case of Mumbai High, it was the rapid decision to transfer the injured cook from MSV *Samundra Suraksha* to the Mumbai High North platform, despite adverse weather, the unavailability of the leeward crane, and the exposed design of the gas riser on the windward side.

On any hazardous operating plant, any change of circumstances must be assessed carefully, including careful risk assessment, before action is taken. Staff must be clear about the limits of their decision-making responsibilities. Operations management needs to keep acutely aware of the overall plant situation ("situational awareness", as discussed elsewhere) while also remaining sufficiently detached that they can make calm decisions about the best way forward when circumstances change. This is a difficult balance to achieve in practice – it requires "hands-on detachment", i.e., full awareness with critical faculties intact.

The culture to support this form of calm detached decision-making must be encouraged and supported at senior levels in the organization.

REFERENCES

[1] The public enquiry into the Piper Alpha disaster, Department of Energy, HMSO, (1990).

[2] M. Elisabeth Pate-Cornell, Learning from the Piper Alpha accident: a post-mortem analysis of technical and organisational factors, Risk Analysis 13 (2) (1993).

[3] J.B. Verma, Mumbai High incident and regulatory progress since, International Regulators Forum Global Offshore Safety, Miami, December 2007.

[4] S. Walker, Mumbai High North Accident. HSE Presentation to the Marine Safety Forum; 2006.

BP

"Responsibility is a unique concept... You may share it with others, but your portion is not diminished. You may delegate it, but it is still with you... If responsibility is rightfully yours, no evasion, or ignorance or passing the blame can shift the burden to someone else. Unless you can point your finger at the man who is responsible when something goes wrong, then you have never had anyone really responsible".
Hyman G. Rickover

"The big accidents are just waiting for the little ones to get out of the way".
Carolyn Merritt

"It should not be necessary for each generation to rediscover principles of process safety which the generation before discovered. We must learn from the experience of others rather than learn the hard way. We must pass on to the next generation a record of what we have learned".
Jesse C. Ducommun

INTRODUCTION

BP suffered two enormous accidents in the United States and its territorial waters within a few years of each other – the Texas City refinery explosion and fire on March 23, 2005, and the Macondo-*Deepwater Horizon* blowout and fire in the Gulf of Mexico on April 20, 2010. These were among the first major industrial accidents to happen in the Internet age, and incurred a massive amount of publicity. There was also litigation, some of which is still continuing. The reputation damage suffered by BP was probably the worst any public company has ever had to endure, and the company was vilified by, amongst others, President Barack Obama.

Each of these accidents, in its own way, was uniquely bad.

The Texas City refinery accident killed 15 people and injured a further 180. The accident had a financial cost exceeding $1.5 billion, making it the most expensive refinery accident in history. A highly critical report was published by the US Chemical Safety and Hazard Investigation Board, who also produced an excellent 55-minute video about the accident.

A further independent report was commissioned by BP itself, chaired by James Baker. James Baker had a very long and distinguished career in US politics and government. He had been Chief of Staff in President Reagan's first administration

1981–1985, Secretary of the Treasury in Reagan's second administration 1985–1989, and Secretary of State and then Chief of Staff again in the administration of President George HW Bush 1989–1993. Baker's report was highly critical of the safety culture and the management of safety within BP, as we shall see. BP's ploy of asking Baker to chair the independent report into the Texas City accident was intended to show, very publicly indeed and especially for a US audience, that they were going to learn their lessons and change their ways.

The Texas City accident and the public reports of management shortcomings were followed by much public breast-beating by BP, with BP representatives giving presentations about all that had been wrong at Texas City, and how things were going to change radically within the worldwide BP organization to make things better – in effect, a very public, high profile repentance of their sins. Technical presentations were given in public seminars in various locations.

The dust had just begun to settle on the Texas City accident when the Macondo-*Deepwater Horizon*[1] blowout and explosion occurred. This accident, and the subsequent oil contamination, therefore happened at the worst possible time for BP, which was already standing accused in an extremely high-profile way of poor attitudes to safety. In addition to causing 11 deaths on the day of the accident, April 20, 2010, the Macondo blowout remained uncontained for 87 days, causing oil pollution along the coasts of various southern states including Louisiana, Mississippi, and Florida. The blowout caused the world's biggest oilspill, about 5 million barrels, exceeding the *Ixtoc 1* blowout in 1979 (about 3 million barrels) and dwarfing the likes of the *Torrey Canyon* tanker accident (UK, 1967, less than one million barrels) and the *Exxon Valdez* tanker grounding in Prince William Sound, Alaska, in 1989, which had been the previous biggest oilspill in US waters (see Table 14.2). Throughout the period of the uncontained blowout and beyond, BP remained front page news. Eventually, on October 1, 2010, the publicity became so bad that BP's CEO, Tony Hayward, was forced to resign. His cause was not helped when he was quoted as saying that he "wanted his life back", which was ill-advised given the actual loss of life; this comment was jumped on by the news media.

The Macondo-*Deepwater Horizon* accident became, of course, ensnarled in US politics, partially because New Orleans in Louisiana was affected. New Orleans was widely considered to have been neglected by the previous US Administration and the Federal Emergency Management Administration (FEMA) after Hurricane Katrina had caused massive damage in August 2005. President Obama was therefore keen (and quite rightly so) to be seen to be reacting in a firm and high-profile way toward BP, and also keen to ensure that BP paid for both the clean-up operations from the Macondo oil pollution and also damages for loss of income incurred through loss of fishing or lost income from tourism. However, Obama went so far as to play the "xenophobia" card in some of his pronouncements: Suddenly, BP became again "British Petroleum", a name it had not used for many years.

---

[1]"Macondo" was the name of the oil field, and "Deepwater Horizon" was the name of the drilling rig. The name "Macondo" was taken from Gabriel Garcia Marquez's novel *One Hundred Years of Solitude*, which features a town called Macondo that is a metaphor for Marquez's home country of Colombia.

Estimates of the total cost to BP vary but a figure in the region of $30–40 billion seems likely, making this one of the most expensive accidents in history. BP has said, it has paid $24 billion in expenses related to the oilspill, and it faces fines that may exceed $10 billion. Only a company of BP's size and profitability could withstand such costs without being taken over or going bankrupt.

The Macondo-*Deepwater Horizon* accident also generated a veritable flood of official investigation reports, as well as legislation and regulation changes.

So, two accidents and one company, within just over 5 years: the most expensive accident of any sort ever, and one of the most expensive refinery accidents ever. However, as we shall see further on, the differences between the two accidents greatly outweigh any similarities. The Texas City accident was at an old refinery, carrying out routine production operations. The Macondo-*Deepwater Horizon* accident occurred on a state-of-the-art drilling rig while exploring for oil in very deep seawater– right at the boundaries of current technology.

The history of BP dates back to an entrepreneur, William D'Arcy, who obtained exclusive rights to explore for oil in southwest Persia (modern Iran) in 1901. He eventually found oil in 1908. The company Anglo-Iranian Oil changed its name to British Petroleum in 1954. Major oil finds in the North Sea and Alaska followed in the 1960s and 1970s, and had a transformational effect on the company, making it a truly global enterprise.

The UK Government sold its remaining shares in BP in 1987. Between 1988 and 2003, the company went through a further period of massive transformation; it grew rapidly by a series of mergers and acquisitions, with large increases in revenue and profitability. The companies involved included Britoil, Sohio, Amoco, Burmah-Castrol, and Atlantic Richfield.

At time of writing, BP is the third largest company by market capitalization in the UK Stock Exchange. BP's tax payments are a major contribution to the UK Government's finances, and many large pension funds rely on BP's dividend payments (Fig. 14.1).

BP TEXAS CITY REFINERY – PRELUDE TO THE ACCIDENT

Refinery operations at Texas City refinery had been taking place since 1934. The refinery had been owned by Amoco until it merged with BP in 1998. Under Amoco's ownership, at least three opportunities had been missed to carry our modifications that would have prevented the accident:

1991	The Amoco refining planning department proposed eliminating blowdown systems that vented to the atmosphere, but funding for this plan was not available.
1993	A project was proposed to eliminate atmospheric blowdown systems but funding was not approved.
1997	Despite Amoco's process safety standard prohibiting new atmospheric blowdown systems and calling for the phasing out of existing ones, Amoco replaced the 1950s-era blowdown drum/vent stack that served the raffinate Splitter Tower with an identical system, instead of upgrading to recommended alternatives that were safer.

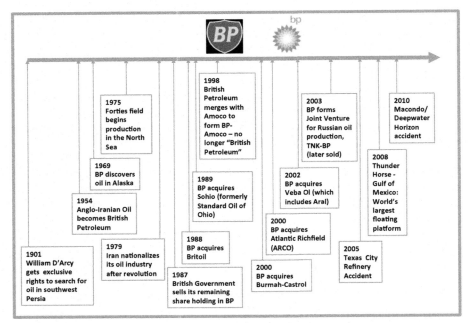

FIGURE 14.1

Timeline of major events in the history of BP.

Accident time-lines almost always read like comedies of errors, instead of the tragedies that they are. When the events are laid out in chronological order, it is hard not to read the stages of the accident unfolding without thinking "They did WHAT?" I have put my comments on the timeline below in italics.

With hindsight, in the years preceding the March 2005 accident, there had been a number of significant indicators that all was not well regarding plant safety at Texas City. In the preceding 30 years, 23 people had been killed in separate accidents on the plant. *This number seems incredibly high – it should have created a major focus for both senior management and safety regulators.*

Budget cuts of 25% were made at all refineries after the BP-Amoco merger in 1998 without any apparent review for their effects on process safety.

Mergers and acquisitions create difficult problems for the management of safety. Responsibilities and reporting routes change. Management communications based on personal relationships can be disrupted. The expectations of the new people in charge may not be clear. The balance between the needs of safety and the needs of production – in which safety should always come first – may become upset, either because plant operators misperceive the expectations of their new senior management, or else because senior management fail to communicate their requirements clearly to their new staff. Senior managers such as VP's and Directors may be unsure of their new CEO – is he really concerned about safety, or is he just paying lip service? If it comes to a tough decision about safety versus production and revenue, senior managers may be thinking, "Where will the new CEO stand?"

Further down the corporate food chain, middle managers may be anxious about their new senior managers, because downsizing normally follows acquisition and they may be worried about their jobs. There may be issues of different company cultures – e.g., the way in which things are done and the way in which concerns are communicated – that can affect safety.

There may even be new paradigms introduced about safety; e.g., there was a vogue in the late 1990s for the rate of industrial safety accidents (i.e., accidents arising from "slips, trips and falls" in the workplace) to be used by senior management as a surrogate key performance indicator (KPI) for the safety of a hazardous industrial process, i.e., the risk of a major process accident such as a fire or explosion. *Clearly, these are two almost entirely different issues.*

So, at Texas City, a KPI for Lost Time Accidents was used as a surrogate measure of process safety – and then the Lost Time Accident rate was, somehow, declared to be a record low in 2004 – the very same year that they had three fatal accidents. *Simply unbelievable.*

BP's own reports during the years immediately before the accident reported multiple safety system deficiencies, and included the following comments and statements (as detailed in the report by the US Chemical Safety and Hazard Investigation Board):

2002	"Infrastructure at Texas City was in complete decline".
	"Serious concerns about potential for major site incident".
	There were 80 hydrocarbon releases at Texas City in a 2-year period.
	A further proposal to replace the blowdown drum/vent system was cut from the budget.
2003	"Current condition of infrastructure and assets is poor at Texas City".
	Maintenance spending was limited by a "chequebook mentality".
2004	"Widespread tolerance of non-compliance with basic HSE rules"
	"Poor implementation of safety management systems".
	"Production and budget compliance gets recognized and rewarded above anything else".

There was a high leadership turnover rate.

The refinery had three major accidents in 2004, including 3 fatalities and $30m damage, but its lowest ever rate of Lost Time Accidents (LTAs). *These two facts, juxtaposed like that, do not ring true – but that is what we are told. The only possible reconciliation is that there was significant under-reporting of Lost Time Accidents. A "punitive culture" with regards to incident reporting was one of the contributory factors cited in the investigation reports.*

2005	The isomerization unit Splitter Tower high-level alarm had been reported as not functioning several times in the 2 years prior to accident – but maintenance work orders were closed without repairs being carried out.

One month before the accident, an internal BP memo said, "I truly believe we are on the verge of something bigger happening."

THE ACCIDENT AT BP TEXAS CITY

On the morning of March 23, 2005, there were lots of contractors on site for maintenance projects. Mostly, they were housed in temporary trailers near hazardous plant, including the isomerization unit. (The isomerization unit's function was to improve the octane rating of raw gasoline.) Start-up of the isomerization unit had commenced during nightshift.

At 0215, operators had started to introduce raffinate into the raffinate Splitter Tower, which is used to distil and separate gasoline components. (The word "raffinate" means a product in the refining process. In this case the raffinate was naphtha – raw gasoline – from the crude distillation column.) The tower was more than 30 m tall. A single instrument (shown as LT) was available for liquid level indication at the bottom of tower which had a maximum indicated level 9 feet (about 3 m). Above this level the instrument just indicated "9 feet". However, operators routinely filled above this level during start-ups to avoid the possibility of low level causing furnace damage.

At 0309, a high-level alarm (shown as LAH) actuated. Another alarm, designated "Hi-hi", failed to actuate.

At 0330, the level indication showed its maximum – 9 feet – and feed was stopped by operators. (The actual level was probably about 13 feet at that point.)

At 0500, the lead operator in the satellite control room for the isomerization unit gave a briefing to the central control room and left to go home early.

At 0600, a new central control room operator arrived to start his thirtieth consecutive day doing 12 hour shifts, because of staff shortages.

Thirty consecutive 12 hour days would obviously be exhausting. In the European Union it would also be illegal under the Working Time Directive, enacted in 2003.

The shift log left by the nightshift was unclear about the level in the raffinate Splitter Tower and the general state of start-up. All that was recorded was "ISOM *(Isomerization Unit)*: brought in some raff to unit."

At 0715, the day supervisor arrived late, so he missed the shift handover.

At 0951, the start-up was resumed. The day shift began to put more feed into the already over-filled Splitter Tower. An Auto-level control valve on the raffinate feed to the Splitter Tower was disabled because of "conflicting instructions". *So the Splitter Tower just kept filling up higher and higher...*

At 1000, the furnace under the raffinate Splitter Tower was lit to start feed heating. Raffinate feed was still going on, although the only level instrument still showed its maximum of about 9 feet.

At 1050, the day supervisor left the site to deal with a family medical emergency. This left no supervisor in the central control room, contrary to the operating rules. A single control room operator, very tired from 30 consecutive 12 hour shifts, was now running three operating units, including the isomerization unit as it went through its start-up procedure.

In 1999, after the BP-Amoco merger, a second operator position had been eliminated.

By about 1200, the level in the Splitter Tower level reached 98 feet (15 times its normal level) but the level instrument showed 8.4 feet and gradually falling. Screen displays in the control room did not show "flowrate in" and "flowrate out" on the same screen (so the control room operator had to toggle between two separate displays, if this was checked at all), nor was there any computer calculation of the total amount of liquid in the tower.

At about 1200, maintenance contractors left their temporary trailers near the isomerization unit for a lunch to celebrate 1 month without lost-time injury.

The irony of this stretches belief. It also says something about the working environment at Texas City refinery that a mere one month *without a lost-time injury was considered sufficient to merit a celebratory lunch. Also, I do not understand how this is consistent with the claim of "zero lost time accidents" in 2004, unless the celebratory lunch was something that had happened every month for a long time....*

At 1241, an alarm appeared in the control room to say there was high pressure at the top of the Splitter Tower. *(This was caused by compression of gases as the liquid raffinate level rose. The Splitter Tower – a distillation column – was now almost completely full of liquid raffinate.)* The control room operator instructed plant operators to respond to this alarm as follows:

- A plant operator opened a manual valve to vent gases into the relief system (which vented unflared gas into atmosphere via the blowdown drum).
- A plant operator also turned off two burners in the furnace at the bottom of the Splitter Tower (thinking this would reduce the pressure).
- A plant operator opened a valve to allow liquid to go from the bottom of the Splitter Tower to storage tanks. This liquid was very hot and flowed through a heat exchanger with liquid entering the Splitter Tower, raising temperature of liquid entering tower by about 141°F (about 90°C).

At 1300, the contract workers returned from their celebratory lunch to their temporary trailers which were located near the blowdown drum.

At 1314, the hot feed raffinate caused boiling, so the level rose until the Splitter Tower was filled completely. Hot liquid gasoline then spilled into the vapor line, which caused pressure relief valves in the vapor line to open (see Fig. 14.2). 52,000 gallons (236,000 litres) of liquid gasoline thereby vented to the blowdown drum, where it overflowed and drained into a process sewer, setting off control room alarms. The high-level alarm in the blowdown drum (shown as LAH) failed to actuate. A geyser of liquid and vapor gasoline erupted from the vent above the blowdown drum, and the hot gasoline formed a large vapor cloud, which was ignited by a running truck engine nearby. An explosion and fire ensued, causing 15 deaths and 180 injuries. The temporary trailers housing the contractors were destroyed in the blast.

To recap: the accident involved the Splitter Tower becoming completely filled with hot liquid raffinate (naphtha or gasoline), when it should have been less than one-tenth full. Hot raffinate then overflowed into the blowdown drum and out through its vent. The Splitter Tower had been receiving raffinate feed for several hours without any apparent concern that it might be overfilling. The ineffective level instrumentation at the

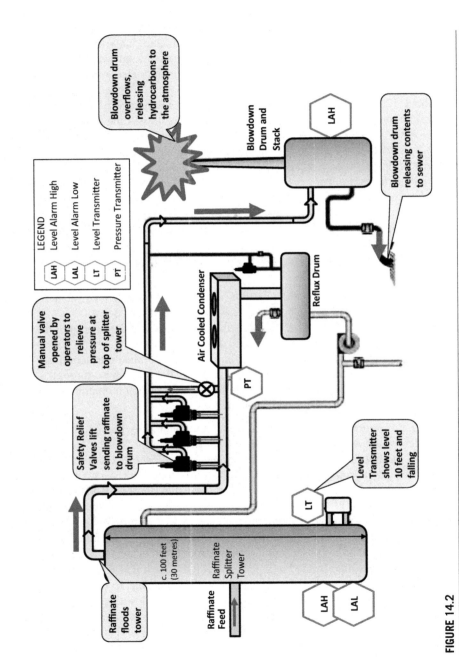

FIGURE 14.2

Schematic diagram of the isomerization unit, BP Texas City.

bottom of the tower never recorded any values above about 9 feet. After the overflow, the blowdown drum level alarm failed to work.

The control room day supervisor had missed the handover from nightshift. He then had to go home because of a family medical emergency. The only remaining control room operator was on his thirtieth consecutive 12 hour shift.

Some of the extremely damning root causes and contributory factors noted in the accident reports are listed in the following Table 14.1.

There was one very significant recommendation, which should apply to all hazardous process plant: "All hazardous chemical operations should be required to review the safety impact of major organizational changes." When the organization is changed – as happens in all companies on a regular basis – the implications for safety have to be considered carefully.

The Baker Report was aimed, in particular, at "the effectiveness of BP's corporate oversight of safety management systems at its five US refineries and its corporate safety culture." Amongst its findings were the following.

- BP's leadership should set the process safety tone at the top of the organization and establish appropriate expectations regarding process safety performance.
- BP had emphasized personal safety in recent years but did not emphasize process safety.

Table 14.1 Root causes and contributory factors to the Texas City accident

	Root cause or contributory factor
1	There was a lack of open event reporting – a "punitive culture".
2	There was de-centralized management which impaired learning from incidents elsewhere.
3	There was a failure to investigate near misses in previous isomerization unit start-ups.
4	There was a lack of modern design for key safety systems (e.g., level instrumentation, blowdown system).
5	There were occupied trailers near the isomerization unit. This neglected industry siting guidelines, and personnel inside the trailers were not advised of start-up operations. BP's own management of change guidelines were not heeded in considering the positions of the trailers.
6	There was serious worker fatigue and no fatigue prevention policy.
7	Inadequate training: the training program had been downsized.
8	There was lack of procedural adherence and, in any case, the procedures were out-of-date.
9	There was "no accurate and functional measure of level in tower" which led to incorrect decisions.
10	There was poor communication during shift handover.
11	There was a lack of robust, enforceable, external independent auditing.
12	There was tolerance of serious deviations from safe operating practices, and apparent complacency toward serious process safety risks.
13	Restructuring following the BP-Amoco merger had resulted in a significant loss of people, expertize and experience.

- BP had not established a positive, trusting, and open environment with effective lines of communication between management and the workforce.
- BP's corporate management had overloaded personnel at BP's US refineries.
- BP had a short-term focus, and its decentralized management system and entrepreneurial culture had delegated substantial discretion to US refinery plant managers without clearly defining process safety expectations, responsibilities, or accountabilities.
- BP's system for ensuring an appropriate level of process safety awareness, knowledge, and competence in its refineries organization was criticized.
- BP had sometimes failed to deal promptly with process safety deficiencies identified during hazard assessments, audits, inspections, and incident investigations.

The Chemical Safety and Hazards Investigation Board (known as the CSB) drew lessons in safety system deficiencies, incident investigation deficiencies, maintenance, management of change and safety culture. From the accident, the CSB drew eight Key Lessons for operators of hazardous plant. These are absolutely generic in nature – they apply to any hazardous plant:

1. Track KPIs for monitoring safety performance
2. Maintain adequate resources for safe operation and maintenance
3. Nurture and maintain a proper safety culture
4. Non-essential personnel should be remote from hazardous process areas
5. Equipment and procedures should be kept up-to-date
6. Manage organizational changes to ensure safety is not compromised
7. Analyze and correct the underlying causes of human errors
8. Directors must exercise their duties regarding safety standards

To paraphrase: the accident happened because the plant was not maintained adequately, the operators were over-stretched and did not adequately understand the plant they were operating, and the management were not paying enough attention.

After repairs and a further period of operation, BP sold the Texas City refinery to Marathon Oil in October 2012. This follows a worldwide trend of major oil companies exiting refining – there is an excess of refining capacity, and refining has become less profitable than oil production.

ABERDEEN 2007 TO 2009

In UK North Sea oil developments, BP has been in the vanguard. In the 1970s, BP developed and put into production the large Forties field, amongst others. As their early fields' outputs reduced, BP sold older platforms to smaller producers, such as Apache and Fairfield. This has become a standard practice in the North Sea – the oil "majors" such as BP and Shell pioneer new developments and, as the fields go past maturity with falling output, the assets are sold to smaller companies to squeeze the remaining oil from the fields. Smaller companies have undoubtedly a good record of running older, less productive fields profitably.

In the first decade of the twenty-first century, BP was again in the vanguard of UK Continental Shelf oilfield development as they developed new large fields in deepwater west of Shetland – the Schiehallion, Clair and Foinaven fields. These were truly impressive cutting-edge developments, in extremely hostile environments, which were completed with admirable safety records.

Between 2007 and 2009 (i.e., post-Texas City, but pre-Macondo/Deepwater Horizon), I ran the Aberdeen office of an international safety consultancy. In that time I met people from all the main oil companies in Aberdeen. I also visited five different North Sea oil platforms operated by four different companies. I had quite a lot of dealings with BP in their Dyce offices, although I never won any work from the company and I never visited any of their offshore installations (although I did visit an old platform that BP had recently sold to another operating company).

Also at that time, senior BP representatives were traveling around the UK and other countries to give seminar presentations about Texas City. I attended one such presentation, which was full of apologies and self-criticism. We were assured BP was improving its safety culture and organization.

Nevertheless, BP's corporate hubris could be astonishing. They had just built very smart, new offices – probably the best of any oil company in Aberdeen – next to their old offices. The "old" offices had been built in the nineteen-seventies, so they were only 30 years old. The new offices were smaller than the old ones; they downsized the organization and those who were remaining in BP employment moved into the new offices, while they closed down the old ones. This seemed an extravagant way of doing things.

There seemed to be an arrogance with many Aberdeen BP personnel that was absent in other oil companies. It felt like they were sure they always knew best, and contractors were expected to be in awe of them. One of the BP engineers said something to me privately that seemed inappropriate, and perhaps smug: "Of course, there are not really any lessons from Texas City relevant to North Sea operations." He said this even as the post-Texas City BP corporate bandwagon was rolling, with seminars declaring publicly that BP had been guilty of a bad safety culture, and promising improvements.

BP is a massive multinational organization – it currently employs about 80,000 people worldwide. It has a number of totally different businesses, including onshore and offshore exploration, onshore and offshore oil and gas production, oil refining, pipelines, shipping, global transportation, and sales. It operates in 28 countries. The diversity of these activities – their completely different technologies, their different political and cultural environments, and their differing relative profitability (and therefore level of re-investment) – mean that, actually, a BP person working on cutting-edge North Sea developments will have next to nothing in common with a BP (former Amoco) person working on an aging refinery in Texas. Hence, maybe the engineer who told me there were no lessons for BP Aberdeen from the Taxes City accident was right: there were no lessons for BP in Aberdeen from Texas City. Perhaps the corporate *mea culpa*s were just a bit of essential public relations management.

THE MACONDO-*DEEPWATER HORIZON* BLOWOUT, FIRE AND OILSPILL, APRIL TO JULY 2010

Like all oil companies, BP had a company risk matrix for assessing projects. Their highest level of project risk was "category A", which was applicable if a project had potential large hazards. "Large hazards" were defined as those which could lead to more than 50 fatalities, or large-scale oil pollution more than 100,000 barrels in sensitive coastal waters, or equivalent financial loss greater than $1 billion, or serious reputation damage. (For clarity, these descriptions of "category A" hazards relates only to potential accident *consequences*. To be acceptable, the BP project risk assessment for any new project would have had to show that the *likelihood* of any of these accidents occurring was very low.) Although there were "only" 11 deaths in the Macondo-*Deepwater Horizon* accident, these descriptions of "category A" corporate risk otherwise describe exactly the accident with regards to Environmental Impact, Equivalent Financial Loss and Reputation Damage.

The last category, "serious reputation damage", was described as "global outrage, global brand damage and/or affecting international legislation" and included the following possible outcomes:

1. Change to group strategy
2. Intervention from major western Government – US, UK or EU
3. Unplanned change of CEO
4. Public outrage in major western markets – US, UK or EU
5. External Auditors qualify accounts

These descriptions are strangely prescient of the actual reputation damage that arose following the Macondo-*Deepwater Horizon* accident.

BP Exploration and Production Inc was the lease operator of Mississippi Canyon Block 252 in the Gulf of Mexico, which contains the Macondo prospect. Geological data indicated that there might be a significant oil and gas reservoir deep below the seabed in this block. BP had minority partners in this venture, Anadarko (25%) and MOEX (10%).

BP had hired the drilling rig *Deepwater Horizon*, owned by Transocean Ltd, based in Vernier, Switzerland, to carry out exploratory drilling at Macondo. Transocean is listed on both the New York Stock Exchange and the Swiss Stock Exchange. It is a very large company in its own right: in 2010 (the year of the accident), it had 18000 employees, its market capitalization was about $21 billion, revenues almost $10 billion, and it declared a profit of $961 million.

Transocean was responsible for operation of the drilling rig and for the safety of operations, and to maintain well control equipment. Transocean purchased, maintained and operated a range of subsea equipment, including the blowout preventer (BOP) and associated control systems.

The lease of *Deepwater Horizon* from Transocean cost BP about $1 million per day. On the day of the accident, April 20, 2010, BP and the Macondo prospect were almost six weeks behind schedule and more than $58 million over budget.

Deepwater Horizon was classified as an "ultra-deepwater" drilling rig. In the offshore oil business, "deepwater" is defined as more than 1500 feet (400 m), and "ultra-deepwater" is more than 6500 feet (2000 m). Transocean had a lot of expertise in this work, and the company owned some 30 drilling ships and "semi-submersibles" for carrying out this work. The rig was originally contracted until 2013 to BP, for work at a variety of sites. *Deepwater Horizon* was a semi-submersible that cost about $350,000,000 to build in 2001, and it represented the cutting edge of drilling technology. It was a floating rig, capable of working in up to 10,000 feet (3 km) water depth, although the Macondo site was "only" 4992 feet deep – or about 1.5 km. The rig did not need to be moored; rather, a triple-redundant computer system used satellite positioning to control powerful thrusters that kept the rig on station within a few feet of its intended location, at all times.

The third major player in this drama was Cameron International, a major supplier of equipment for the oil and gas and process industries, who had designed and supplied the blowout preventer (BOP), which failed to operate in the accident. Cameron International is based in Houston, Texas, and has 20,000 employees and annual revenue of $8.5 billion (2012).

The fourth major player in the drama of this accident was Halliburton, one of the world's largest oilfield services companies, who provided engineering services, materials, testing, mixing and pumping for cementing operations. This included onshore engineering support, offshore equipment, and personnel based on *Deepwater Horizon*. Halliburton provided technical advice regarding the design, modeling, and testing of the cement that was pumped into place behind the casing string and in the shoe track to isolate the hydrocarbon zone from the wellbore at the Macondo well site. Halliburton has dual headquarters in Houston and Dubai. It has 68,000 employees and its 2011 revenue was $24.8 billion.

The fifth and final major player was the US government agency responsible for safety regulation, the Minerals Management Service (MMS). BP had paid some $34 million to the MMS for an exclusive lease to drill in Mississippi Canyon Block 252, which was nine miles square. An exploration plan was submitted by BP to MMS before exploratory drilling began in Mississippi Canyon Block 252, and an Application for Permit to Drill was submitted for approval before drilling began at the Macondo well. Macondo would be the first well in the Mississippi Canyon Block 252 lease.

MMS was one of the first organizational casualties of the accident. Regulation of offshore oil exploration in the US at the time of the Macondo-*Deepwater Horizon* accident had largely developed in response to the Santa Barbara oilspill in 1968, which occurred at Platform Alpha operated by the Union Oil Company of California (later Unocal, which was taken over by Chevron). The Santa Barbara oilspill caused a public outcry because of environmental damage, and this led ultimately to the formation of the Environmental Protection Agency in 1971. However, offshore exploration was regulated by MMS, whose remit was wider than safety, and which perhaps led to conflicts of interest.

After the Macondo-*Deepwater Horizon* accident, MMS's role in safety regulation was transferred to a more dedicated agency, the Bureau of Ocean Energy Management,

Regulation and Enforcement (BOEMRE) on June 18, 2010. A subdivision, the Bureau of Safety and Environmental Enforcement (BSEE), was given particular responsibility for safety.

THE MACONDO-*DEEPWATER HORIZON* ACCIDENT, APRIL 20, 2010

On the day of the accident, April 20, 2010, the intention was to seal temporarily the Macondo well so that *Deepwater Horizon* could move away – "temporary abandonment". The plan was that, at some later point, another "completion" rig would be put into place to install hydrocarbon production equipment. This rig would in turn be replaced by a production facility that would connect to the subsea wellhead, left by the completion rig, so that production could be started.

Conditions at the deep ocean seabed are always just a few degrees above freezing. The oil reservoir itself – probably more than a billion barrels – was a further 4.5 km below the seabed, where the temperature would be about 100°C. Figure 14.3 gives a schematic representation of the layout, comprising *Deepwater Horizon* on the surface, the blowout preventer (BOP) on the seabed nearly 5000 feet – some 1500 m – below the surface, and the hydrocarbon reservoir 18,360 feet (nearly 6000 m) below the surface.

Pressures at well depth were immense, and this presented the major challenge for all parties. Macondo required a path to be drilled that controlled the huge pressure without fracturing the geologic formation that contained the oil. The drillers are required to keep the reservoir pressure balanced with counter pressure inside the wellbore. Too much counter pressure can fracture the rock formation around the well, and lead to oil losses into the surrounding rock. Too little pressure can cause oil and gas to rush out of the well in an uncontrolled manner – in other words, a blowout. A blowout is a difficult problem if the wellhead is at ground level. It is a more difficult problem if the wellhead is in shallow water, such as the North Sea. It becomes extremely difficult indeed if the wellhead is 1500m below sea level, as was the case with Macondo.

Deepwater Horizon had placed a blowout preventer (BOP) on the wellhead at the seabed. Everything below the seabed level – drilling pipe, drill bits, casing and mud – had passed through the BOP. Transocean personnel tested the BOP before and during drilling operations. The BOP weighed about 400 tonnes, and could seal the well in various ways. "Annular preventers" could seal off the annular space around the drill pipe. A "blind shear ram" was designed to cut through the drill pipe inside the BOP to seal the well in an emergency, and could be activated manually from *Deepwater Horizon*, or remotely using a Remotely Operated Vehicle (ROV), or using a "deadman system". A casing shear ram was designed to cut through the casing.

Drilling mud is used for two purposes; it lubricates and cools the drill bit and, also, the weight of the column of mud within the well bore counterbalances the reservoir pressure.

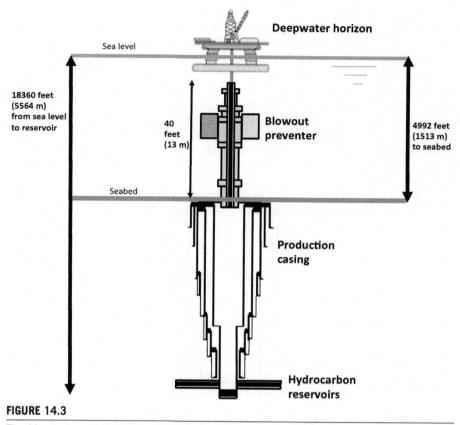

FIGURE 14.3

The Macondo well, the blowout preventer and the well casing assembly.

The well bore is also lined with casing strings which line the well as drilling progresses. The space between the casing strings and the well bore is sealed with cement. Once a prospect has been deemed suitable for production, holes are punched through the casing and its surrounding cement to allow oil and gas to flow into the well.

BP was the legal operator and two BP Well Site Leaders were always on board *Deepwater Horizon*. These two BP personnel remained in contact with BP engineers onshore.

Macondo had been a difficult well for the drillers. At 18,360 feet, the drillers were struggling to keep the balance between reservoir pressure, drilling fluid, and rock formation integrity. It had been intended to drill to over 20,000 feet but it was decided to stop further drilling because of well integrity concerns. Between April 11 and 15, measurements were taken down the hole which led BP to conclude the reservoir would be economic without going any deeper. The next step was to install a final production casing, to leave the well ready for future production. The design of the production casing was discussed extensively between BP and Halliburton, led by BP

Wells Team leader John Guide. The production casing needed a number of centralis-
ers down the bore to keep it straight within the well bore. However, insufficient cen-
tralisers were available so a decision was made to go ahead with a reduced number.

In the morning of April 18, the production casing began to be lowered into place
down the well. The leading piece of the casing is called the "shoe track". This was
completed in the afternoon of April 19. To prepare for cementing, mud fluids were
pumped down through the casing string, out through the shoe track and up through
the annulus. However, flow of mud could not be established. BP Well Site Leader
Bob Kaluza, after consulting with people onshore, increased the pressure until it fell
dramatically as mud began to flow – although the circulation pressure (to keep it
flowing) then seemed suspiciously low. It was concluded that the pressure gage was
faulty (Fig. 14.4).

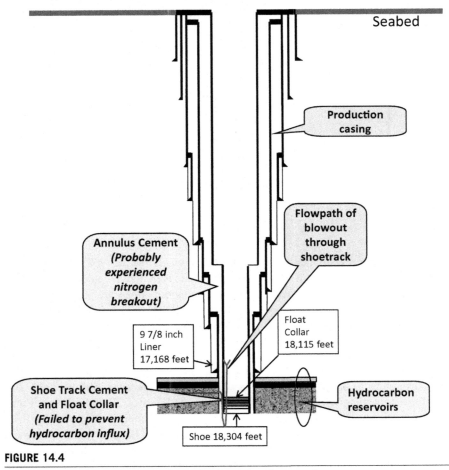

FIGURE 14.4

The Production Casing between the seabed and the oil reservoir.

In order to put cement into the annulus, a slug of cement is pumped down the casing, followed by more mud, until it goes through the shoe track and enters the annulus. If the cement is pumped too far or not far enough, the casing will not be sealed. If the drilling mud contaminates the cement, it may not set properly. A further possibility is that the cement may not fill the annulus fully, leaving a channel for hydrocarbons to flow through. The difficulty is this: The crew on the drilling rig have to try to infer the state of the cement based on pressure and flow measurements that are read miles above the reservoir.

The design of the cement – meaning its volume, density, chemistry and positioning – was nominally the job of Halliburton, but BP overrode them on some points. BP's main concern appears to have been that of too much pressure, leading to fracture of the rock formation and losses of future production. Normally a full flush of the well bore with drilling mud would be expected, but BP was worried about damage to the rock formation, so only about 15% of the mud was circulated. BP also reduced the speed at which the cement was pumped into the well, again to keep the pressure on the rock formation down. BP then limited the volume of cement that Halliburton would pump down the well, again because they were concerned that too much cement could exert too much pressure on the rock formation; it was decided that the annular cement column should extend only 500 feet above the uppermost hydrocarbon zone. This was actually a breach of BP's own internal guidelines which specified the top of the annular cement should be 1000 feet above the uppermost hydrocarbon zone. Only 60 barrels of cement would be pumped down the well.

BP and Halliburton together chose nitrogen foam cement, which provides lighter cement, again because of concerns about damage to the rock formation. A sample of this cement formula had been sent for laboratory testing by Halliburton in February, with the results received on March 8. Further tests were done in early April. Both sets of tests actually showed the nitrogen foam cement was unstable (i.e., the nitrogen bubbles separated from the cement before it set) although the results of the second set of tests were not sent by Halliburton to BP until after the accident.

On the morning of April 20 the cement had been pumped down the well, and at 0545 hours Halliburton's Nathaniel Chaisson sent an email saying "We have completed the job and it went well." BP engineer Brian Morel sent an email saying "… the Halliburton cement team…did a great job." This conclusion was on the basis that there were no fluid losses into the rock formation. A team from Schlumberger (a company specializing in oil well logging and instrumentation), who had equipment ready to carry out extra cement evaluation tests, were sent home without doing their tests.

Thereafter, BP issued their "temporary abandonment procedure" – i.e., the preparations needed for *Deepwater Horizon* to move away from the well, leaving it ready for the well completion rig to do its job. Amongst other things, this procedure called for: A positive pressure test to confirm that the casing was sealed; a negative pressure test to confirm the integrity of the cement job against oil entering the well; setting a surface cement plug and installing a lockdown sleeve at the seabed level. This procedure was never completed, because of the blowout.

The first step – the positive pressure test – was completed in the afternoon of April 20. The blowout preventer (BOP) blind shear ram was closed and well pressure was increased to 2500 pounds per square inch (about 167 atmospheres) and held for 30 min without leakage, so it was considered successful. This test only proved the casing was pressure tight. It did not prove the integrity of the cement plug within the casing annulus.

By contrast, the negative pressure test checked both casing integrity and cement plug integrity – and this was in fact the only test done which tested cement plug integrity. If the casing and the cement were installed properly, they should prevent oil entering the well even when the weight of mud was reduced. After abandonment, the plan was that reservoir pressure would exceed the downward force of fluids in the well, so the integrity of the well – including the cement plug – had to be confirmed.

To prepare for the negative pressure test, the drill pipe was run down to 8367 feet below sea level and the drilling crew pumped a liquid "spacer" – a liquid mixture that serves to separate heavy drilling mud from seawater – followed by seawater, to displace 3300 feet of mud. After some problems, the crew were then able to bleed the pressure down to zero pounds per square inch, but the pressure quickly rose again. This was repeated several times, each time with the same outcome – the negative pressure test was apparently unsuccessful.

The Transocean and BP personnel then repeated the test, this time monitoring pressure via another pipe called the "kill line". On this instrument the pressure stayed at zero pounds per square inch and, on the basis of this result (while neglecting the other instruments which showed pressure quickly rising), they concluded that the negative pressure test had been successful. Hence, by carefully selecting which data they used, they concluded erroneously that the negative pressure test had been successful. Hence they got the "successful" test result that they wanted to see – but in a way that gave the wrong result.

Unplanned ingress of oil and gas into the well can be monitored on the drilling rig as "kicks". A small volume of gas in the reservoir will expand enormously as it finds its way up the wellbore, pushing mud ahead of it and accelerating further as it expands. Hence the drill crew are trained to look out for kicks. Kicks become evident because the volume of mud held on the rig will increase – the volume of mud coming back into the rig starts to exceed the volume of mud being pumped into the well, or else mud comes out even when none is being pumped in. It is also possible (though less reliable) to detect a kick from drill-pipe pressure. However, there was an awful lot of mud involved in such a deep reservoir, so it was perhaps difficult for the drill crew to track the amount.

Pressure records show that, at about 2101 hours on April 20, a "kick" began and drill-pipe pressure began to increase, rising by about 100 psi over 7 minutes. This showed itself as a gently rising trend on a screen which followed various parameters – it would hardly have been noticeable unless someone was looking very closely.

The mud flow stopped and was replaced by the 400 barrels of spacer fluid returning to the rig. The pumps were shut off for a test of oil in the spacer fluid – a "sheen test". At that point, there should have been no flow coming back – and this visual check was carried out, apparently successfully.

Between 2108 and 2114 hours, the drill-pipe pressure increased again, although this was unnoticed.

The pumps were turned back on at 2114 hours and at 2118 one of the relief valves on the pumps blew. A repair was necessary.

At 2120, BP Well Site Leader Jason Anderson said, "It's going fine, I've got this." Jason Anderson was one of those killed in the accident.

At 2130 the pumps were shut off to investigate an unexpected pressure difference between the drill-pipe and the kill line. Drill-pipe pressure increased by 550 pounds per square inch over the next five and a half minutes – while the kill line pressure remained significantly lower. (*Remember, low pressure in the kill line had been taken as the reason for confirming the negative pressure test was satisfactory. With hindsight, therefore, it looks like the kill line was blocked.*) This was now clear evidence of a kick, but no further action was taken.

At 2139 hours, the drill-pipe pressure started falling again. This was actually bad news – it meant the expanding hydrocarbon gas had now pushed its way past the heavy drilling mud. Between 2140 and 2143 drilling mud began pouring onto the rig floor – the first time the crew realized a kick was occurring. They could have routed the mud overboard (which might have prevented the accident) but instead they routed the mud to a mud-gas separator – which was rapidly overwhelmed by the large flow-rate. The crew also closed one of the annular preventers on the BOP to shut in the well. However, these attempts were too late – high-pressure gas burst out into the mud-gas separator, and the first explosion occurred at 2149 hours.

In those few minutes 2140–2149, there was perhaps a window of opportunity for the Well Site Leader to activate the BOP and shut down the well. This would have been an admission of failure – the Macondo well had blown, and the BOP would have been isolated on the seabed, the months of work required to reach that point would have been wasted. Perhaps there was a reluctance to activate the BOP until it was too late. There was a conflict of interest: the Well Site Leader, responsible for bringing the well to operational readiness, also had an opportunity to take a decision which might, in effect, have killed the project at a stroke.

After the first explosion, the crew on the bridge tried to activate the BOP using the Emergency Disconnect System, which should have closed the blind shear ram, severed the drill pipe, sealed the well and disconnected *Deepwater Horizon* from the BOP. However, nothing happened – the first explosion had damaged cables to the BOP, preventing the sequence from working.

The BOP also had a "deadman system" which should have operated automatically. This failed too. The BOP was supposed to be a high-integrity system, probably intended to meet SIL 3 requirements or equivalent. The failure of the BOP is discussed further below.

Eleven workers were killed in the initial explosion. The rig was evacuated, with numerous injured workers airlifted to medical facilities. After burning for approximately 36 hours, *Deepwater Horizon* sank on April 22, 2010. The oilspill continued until July 15. See Fig. 14.5.

At the time of the accident, there were 126 personnel on board, of whom 79 were Transocean and 7 were BP. The other 30 were from Halliburton and various other companies.

FIGURE 14.5

Deepwater Horizon ablaze – the helipad has already burned through © PA.

SO WHAT WENT WRONG?

The Macondo-*Deepwater Horizon* accident is the most complex of accidents. It was a combination of: Complicated and highly specialized technical issues about how to seal a difficult well far below the seabed in deepwater; the design, manufacturing and testing of a blowout preventer that was supposed to operate extremely reliably almost 5000 feet (1500 m) below sea level; poor decision making; and unclear contractual divisions of responsibilities between the main parties.

In short, well integrity was not established, because the cement plug had failed. Oil and gas entered the well and well control was lost. A large "kick" of gas overwhelmed the mud/gas separator on *Deepwater Horizon* and then ignited, and the high-integrity blowout preventer (BOP) 5000 feet below on the seabed failed to operate as intended.

The final BOEMRE report gave the central cause of the Macondo-Deepwater Horizon accident as "failure of the cement barrier in the production casing string".

- There had been swapping of cement and drilling mud ("fluid inversion") in the shoe track at the bottom of the casing string.
- The cement had been pumped beyond its target location in the well.
- BP and Transocean personnel missed an opportunity to remedy the cement problems when they misinterpreted the anomalies encountered during the negative pressure test.

The final BOEMRE report also criticized poor risk management, last-minute changes to plans, failure to observe and respond to critical indicators, inadequate well control response, and insufficient emergency bridge response training by the companies and individuals responsible.

If members of the rig crew had detected the hydrocarbon influx into the well earlier, they might have been able to take appropriate actions to control the well.

Contributory factors included project pressures arising from BP's cost and program overruns, and BP personnel changes and conflicts.

A visual presentation of the main findings of the BOEMRE is presented in Fig. 14.6.

The Report to the President by the National Commission on the accident identified nine examples of decision making that had increased risk while potentially saving time, including:

- BP onshore decided not to wait for more centralizers to be available for installation within the production casing.
- Halliburton did not wait for the nitrogen foam cement stability test results.
- BP onshore (approved by MMS) set the surface cement plug 3000 feet below the mudline in seawater.
- BP onshore displaced the mud from the riser before setting the surface cement plug.
- BP onshore chose not to install additional physical barriers during the temporary abandonment procedure.
- BP staff on *Deepwater Horizon* chose not to perform further well integrity diagnostics despite the unexplained negative pressure test results.
- Transocean bypassed some testing and conducted other simultaneous operations during displacement.

The US Academy of Engineering also published a report into the accident, adding some other notable findings:

- The reservoir formation, encompassing multiple zones of varying pore pressure and fracture gradients, posed significant challenges to isolation using casing and cement. The approach chosen for well completion failed to provide adequate margins of safety and led to multiple potential failure mechanisms.
- The BOP system was neither designed nor tested for the dynamic conditions that existed at the time of the accident. Furthermore, the design, test, operation, and maintenance of the BOP system were not consistent with a high-reliability, fail-safe device.
- The actions, policies, and procedures of the corporations involved did not provide an effective systems safety approach commensurate with the risks of the Macondo well. The multiple flawed decisions that led to the blowout indicated a lack of a strong safety culture and a deficient overall systems approach to safety. Industrial management involved with the Macondo-*Deepwater Horizon* disaster failed to appreciate or plan for the safety challenges presented by the Macondo well.

A visual presentation of the main findings of the US Academy of Engineering is presented in Fig. 14.7.

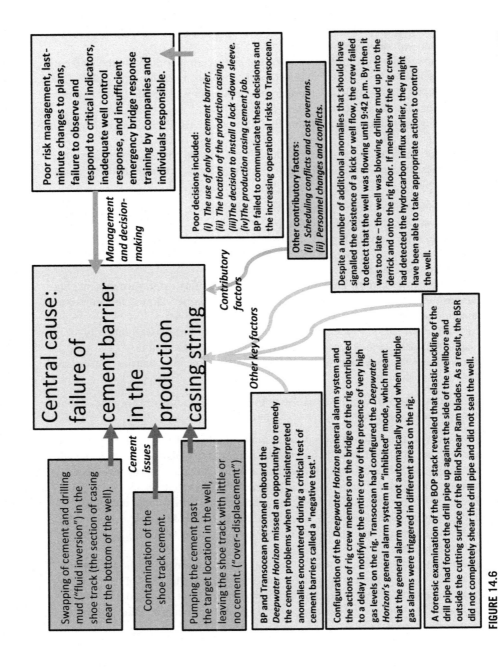

Central cause: failure of cement barrier in the production casing string

Management and decision-making

Poor risk management, last-minute changes to plans, failure to observe and respond to critical indicators, inadequate well control response, and insufficient emergency bridge response training by companies and individuals responsible.

Poor decisions included:
(i) The use of only one cement barrier.
(ii) The location of the production casing.
(iii)The decision to install a lock -down sleeve.
(iv)The production casing cement job.
BP failed to communicate these decisions and the increasing operational risks to Transocean.

Contributory factors

Other contributory factors:
(I) Scheduling conflicts and cost overruns.
(ii) Personnel changes and conflicts.

Despite a number of additional anomalies that should have signalled the existence of a kick or well flow, the crew failed to detect that the well was flowing until 9:42 p.m. By then it was too late – the well was blowing drilling mud up into the derrick and onto the rig floor. If members of the rig crew had detected the hydrocarbon influx earlier, they might have been able to take appropriate actions to control the well.

Other key factors

Cement issues

Swapping of cement and drilling mud ("fluid inversion") in the shoe track (the section of casing near the bottom of the well).

Contamination of the shoe track cement.

Pumping the cement past the target location in the well, leaving the shoe track with little or no cement. ("over-displacement")

BP and Transocean personnel onboard the *Deepwater Horizon* missed an opportunity to remedy the cement problems when they misinterpreted anomalies encountered during a critical test of cement barriers called a "negative test."

Configuration of the *Deepwater Horizon* general alarm system and the actions of rig crew members on the bridge of the rig contributed to a delay in notifying the entire crew of the presence of very high gas levels on the rig. Transocean had configured the *Deepwater Horizon's* general alarm system in "inhibited" mode, which meant that the general alarm would not automatically sound when multiple gas alarms were triggered in different areas on the rig.

A forensic examination of the BOP stack revealed that elastic buckling of the drill pipe had forced the drill pipe up against the side of the wellbore and outside the cutting surface of the Blind Shear Ram blades. As a result, the BSR did not completely shear the drill pipe and did not seal the well.

FIGURE 14.6

A summary of the major findings of the BOEMRE report.

1. The flow of hydrocarbons that led to the blowout of the Macondo well began when drilling mud was displaced by seawater during the temporary abandonment process.

2. The decision to proceed to displacement of the drilling mud by sea water was made despite a failure to demonstrate the integrity of the cement job even after multiple negative pressure tests. This was one of a series of questionable decisions in the days preceding the blowout that had the effect of reducing the margins of safety and that showed poor safety-driven decision making.

3. The reservoir formation, encompassing multiple zones of varying pore pressure and fracture gradients, posed significant challenges to isolation using casing and cement. The approach chosen for well completion failed to provide adequate margins of safety and led to multiple potential failure mechanisms.

4. The loss of well control was not noted until more than 50 min after hydrocarbon flow from the formation started, and attempts to regain control by using the BOP were unsuccessful. The blind shear ram failed to sever the drill pipe and seal the well properly, and the emergency disconnect system failed to separate the lower marine riser and the *Deepwater Horizon* from the well.

5. The BOP system was neither designed nor tested for the dynamic conditions that most likely existed at the time that attempts were made to recapture well control. Furthermore, the design, test, operation, and maintenance of the BOP system were not consistent with a high-reliability, fail-safe device.

6. Once well control was lost, the large quantities of gaseous hydrocarbons released onto the *Deepwater Horizon*, exacerbated by low wind velocity and questionable venting selection, made ignition all but inevitable.

7. The actions, policies, and procedures of the corporations involved did not provide an effective systems safety approach commensurate with the risks of the Macondo well. The lack of a strong safety culture resulting from a deficient overall systems approach to safety is evident in the multiple flawed decisions that led to the blowout. Industrial management involved with the Macondo-*Deepwater Horizon* disaster failed to appreciate or plan for the safety challenges presented by the Macondo well.

FIGURE 14.7

A summary of the major findings of the US Academy of Engineering report.

Importantly, there was a lack of clear division of responsibilities between BP, Halliburton and Transocean. Halliburton were supposed to be the experts recruited to look after well design, but BP's people seemed to overrule them when they felt like it. This behavior may have been exacerbated by the cost and program overruns.

WHY DID THE BLOWOUT PREVENTER FAIL TO WORK?

As previously stated, the blowout preventer (BOP) was designed and manufactured by Cameron International of Houston, Texas. It was owned and maintained by Transocean, the owners of *Deepwater Horizon*. Failure of the BOP was an absolutely key aspect of the accident; this was a multiple-channel high-integrity redundant system, the sole function of which was, *in extremis*, to cut the drill pipe within the well bore and seal off the well, and yet it did not work.

The BOP was eventually recovered from the seabed and taken to NASA's Michoud facility in New Orleans for forensic examination, initially by DNV, a major consultancy. DNV's report was published one year after the accident. Four years after the accident, in June 2014, the US Chemical Safety and Hazard Evaluation Board (the CSB) published another extremely detailed report on the failure of the BOP. There were three main findings.

1. The extremely high well pressures at Macondo had caused the drill pipe to buckle within the well bore such that the blind shear rams in the BOP were unable to cut the pipe properly (Appendix 2A of the CSB report). This buckling was undetectable by the operators. The high pressures inside the drill pipe had caused "effective compression" axially along the length of the drill pipe, which made the drill pipe buckle within the BOP such that it was off-center between the blind shear rams. When the BOP was actuated, the rams were unable to cut the eccentric drill pipe fully (see Fig. 14.8). The damaged, leaking drill pipe remained lodged between the blind shear rams, leaking uncontrollably. Hence this "high-integrity" system was actually *incapable of carrying out its design function*. Cameron had never tested shearing of an off-center drill pipe. The relevant American Petroleum Institute (API) standard did not address this issue. (CSB report Appendix 2A, p34.)

2. Furthermore, the redundant control systems (shown as the "blue control pod" and "yellow control pod" in Fig. 14.9) both had serious failings. Both pods were twin-channel systems, so there were four channels in total. *This supposedly "high-integrity" system had multiple wiring faults and a battery fault*, as follows:

   a. The blue control pod had internal wiring errors which led to its internal batteries being drained, so it was completely ineffective.

   b. The yellow pod also contained (different) wiring errors on one of its two channels which might have completely prevented operation of the dual-coil solenoid valve which triggered operation of the blind shear ram, except

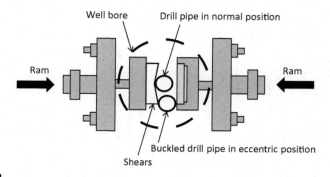

FIGURE 14.8

The blowout preventer blind shear rams from above. The drill pipe had buckled under high internal pressure which caused "effective compression" axially along the pipe. This made the drill pipe eccentric between the blind shear rams. In this position, the blind shear rams were unable to cut the pipe, so the well was not sealed.

that there was another, coincident failure which drained its battery – so the second channel of the yellow pod was effective. (Appendix 2B of the CSB report.)

3. Pre-accident test procedures were incapable of revealing these failures, because the design of the BOP control system did not allow independent functional testing of the redundant systems, neither while the BOP was on the rig nor while it was in subsea service. Without such independent functional testing, latent failures had caused three of the four channels to fail. (Appendix 2B of the CSB report.)

To summarize: the BOP was intended to be a high-integrity system, but it was mechanically ineffective in its main function of shearing the drill pipe, because the drill pipe was eccentric in the well bore. Furthermore, its four-channel control systems were incapable of being independently tested, so latent failures that existed in three of the four channels were unrevealed – and the fourth channel would have failed also, except there were two failures in one channel which serendipitously canceled each other out.

Appendix 2B of the CSB report (Appendix G) contains a review of the Cameron Factory Acceptance Tests (FATs) for the BOP. The tests *should* have been thorough, comprehensive, as realistic as possible, and carried out conscientiously. CSB found numerous deficiencies in the test procedures, in particular:

- Testing of the "Deadman" system was not realistic.
- Procedures were modified during the tests, were not "end-to-end", and did not exactly fit the equipment.

High-integrity systems must be capable of carrying out their functional requirements, and this must be demonstrated under realistic conditions. They must also be testable in-service, as far as is reasonably practicable, to demonstrate that they remain in a healthy condition. To be blunt, the Cameron BOP seems to have been poorly conceived, badly manufactured, and inadequately tested.

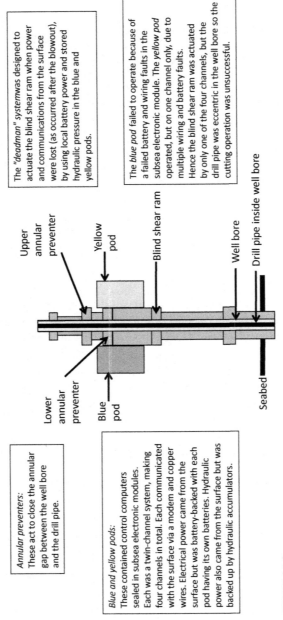

Annular preventers:
These act to close the annular gap between the well bore and the drill pipe.

Blue and yellow pods:
These contained control computers sealed in subsea electronic modules. Each was a twin-channel system, making four channels in total. Each communicated with the surface via a modem and copper wires. Electrical power came from the surface but was battery-backed with each pod having its own batteries. Hydraulic power also came from the surface but was backed up by hydraulic accumulators.

The *"deadman" system* was designed to actuate the blind shear ram when power and communications from the surface were lost (as occurred after the blowout), by using local battery power and stored hydraulic pressure in the blue and yellow pods.

The *blue pod* failed to operate because of a failed battery and wiring faults in the subsea electronic module. The *yellow pod* operated, but on one channel only, due to multiple wiring and battery faults. Hence the blind shear ram was actuated by only one of the four channels, but the drill pipe was eccentric in the well bore so the cutting operation was unsuccessful.

Upper annular preventer

Yellow pod

Blind shear ram

Well bore

Drill pipe inside well bore

Seabed

Lower annular preventer

Blue pod

FIGURE 14.9

The four-channel redundant I&C systems in the BOP had numerous defects. The blind shear rams were actuated by only one of the four channels.

AFTERMATH

As already noted, in addition to 11 deaths and the loss of *Deepwater Horizon*, there was a huge oilspill that continued for 87 days, causing widespread pollution along the coast of the southern United States. BP's CEO, Tony Hayward, was forced to resign on October 1, 2010. Final costs to BP are likely to be in the range $30–40 billion.

The Macondo blowout is the largest ever oilspill. Table 14.2 lists other major spills. BP has recovered some of its costs from the other parties, as follows.

Transocean $1.4 billion
Anadarko $4 billion (Anadarko was a junior non-operating partner in the project with BP)
MOEX $1.07 billion (MOEX was a junior non-operating partner in the project with BP)
Cameron $250 million

However, litigation continues (2014) and these amounts are unlikely be final. BP and Halliburton may share responsibility for the failure of the cement plug, while BP, Transocean and Cameron may share responsibility for failure of the BOP to operate.

The responsible safety regulator – the Minerals Management Service or MMS – was recognized as being fatally flawed. Its mandate included four distinct responsibilities – offshore leasing, revenue collection and auditing, permitting and operational safety, and environmental protection. Each required different skill sets and cultures and presented potential conflicts of interest. A new agency, the Bureau of Ocean Energy Management, Regulation and Enforcement (BOEMRE) was established on June 18, 2010. A subdivision, the Bureau of Safety and Environmental Enforcement (BSEE), was given responsibility for safety and environmental issues only.

Table 14.2 World's largest oilspills

Accident	Date	Approximate oil spillage
Macondo-*Deepwater Horizon*	USA 2010	5 million barrels
Ixtoc 1	Mexico 1979	3.3 million barrels
Nowruz Field	Persian Gulf 1983	2 million barrels
Fergana Valley well	Uzbekistan 1992	2 million barrels
Castillo de Bellver tanker	South Africa 1983	2 million barrels
Amoco Cadiz tanker	France 1978	1.7 million barrels
Odyssey tanker	Canada 1988	1 million barrels
D-103 well	Libya 1980	1 million barrels
Atlantic Empress tanker	Trinidad and Tobago 1979	1 million barrels
Torrey Canyon tanker	UK 1967	900,000 barrels
Exxon Valdez tanker	USA 1989	200,000 barrels

Macondo-*Deepwater Horizon* has demonstrated the sometimes difficult and unpredictable behavior of high-integrity systems operating in extremely hostile environments. As oil exploration and production moves into more and more hostile environments – deepwater, Arctic, Antarctic, and so-called high-pressure, high-temperature (HPHT) fields, and sour fields with high sulfur content – the challenges increase to ensure that safety critical systems can always be relied upon to operate when required.

Deepwater oil exploration and production are well established and accelerating rapidly, but the Macondo accident has illustrated the difficulty of blowout recovery in the ocean depths. Currently (2014) there are deep-water oil developments in West Africa, Southeast Asia, the eastern Mediterranean, west of Shetland (UK), the Gulf of Mexico, and Brazil. The deepest offshore drilling to March 2014 has been in 10,411 feet (3174 m) deepwater by the drilling rig *Dhirubhai Deepwater KG1*, in water off India, by the Indian national oil company ONGC. (Source: *Offshore* magazine.)

TEXAS CITY AND MACONDO-*DEEPWATER HORIZON* – CAN ANY GENERAL CONCLUSIONS BE DRAWN ABOUT BP?

The Texas City refinery accident and the Macondo-*Deepwater Horizon* accident are completely dissimilar – it is very difficult to identify *any* similarities between the accidents:

> The accidents related to different technologies, operating under completely different conditions.
> The plant and equipment involved were at the opposite ends of their lifecycles – Texas City was an old refinery, Macondo was a new prospect, and *Deepwater Horizon* was cutting-edge twenty-first century technology.

Personnel involved in operating Texas City refinery were mainly ex-Amoco employees. Personnel operating *Deepwater Horizon* were mostly from Transocean or Halliburton, with a few BP people and other contractors.

The question that many have asked is whether it is possible to draw any overall conclusions about BP from these two horrendous accidents, which happened five years apart. After all, the only common organizational link between the two accidents is right at the very top of the BP organization.

First, we must remember that other parties than just BP share some of the blame for the Macondo-*Deepwater Horizon* blowout; Halliburton, Transocean and Cameron.

It is tempting to try to draw the conclusion that a penny-pinching attitude was at least a factor in both accidents. For Texas City, maintenance cuts were evident, and the operating budget had been reduced by 25% after BP took over Amoco. For *Deepwater Horizon*, Macondo was a high-profile "prestige" project, and the cost and program overruns for Macondo development will have received corporate-level attention.

Table 14.3 BP profits 2007–2011

	2007	2008	2009	2010	2011
Revenue $M	291,438	367,053	246,138	308,928	386,463
Profit (EBIT) $M	32,352	35,239	26,426	(3702)	39,817

Source: http://www.bp.com/liveassets/bp_internet/globalbp/STAGING/global_assets/downloads/F/ FOI_2007_2011_full_book.pdf

BP has been an extremely profitable company, which generates a great deal of tax revenue for the UK Government. The effect of Macondo-*Deepwater Horizon* on 2010 profitability can be seen in the table below, where large provisions were made in the accounts against eventual costs arising (All values are in $million.) (Table 14.3).

Looking at these figures, and ignoring 2010 because that year's results were affected by the accident, it seems difficult to reconcile them with an organization that *needs* to be penny-pinching. The Macondo project was $58 million over-budget at the time of the accident, which equated to 0.2% of BP's 2009 profit.

So, were the people in the front line feeling pressure, even low-key pressure, from their line managers?

- Did managers at Texas City *really* feel they could not obtain more money from the Company for safety-related maintenance?
- Was there any sort of "chilling" effect where Texas City managers did not ask for more money because they knew such a request would not be received favorably? (Some CEOs will say "I have never turned down a request for safety-related expenditure", but that is because their managers may be too afraid to ask.)
- Did the BP people on board *Deepwater Horizon* and the people they consulted onshore feel either *direct* pressure to "get results" (i.e., line managers telling them to hurry up) or *indirect* pressure (e.g., an implicit recognition that their next personal annual appraisals would be difficult if Macondo was not ready for production)?
- Another issue for both Texas City and Macondo was operator fatigue. At Texas City, the control room operator was on his thirtieth consecutive 12 hour shift. The Macondo well had been a long, difficult project and staff will have been keen to finish the job and get home for a break.

Another general observation relates to the type of engineering managers recruited by BP. My own experience, albeit based on a very limited sample, was that there seems to me to be a "BP type" – a type that does not seem to predominate in other oil companies in anything like the way it did in BP. The "BP type" is a person who is very focussed and determined on an outcome, and very confident about his or her own capabilities. Is there some sort of common-mode problem in their corporate HR recruitment policies, as if they are only looking to recruit one type of engineer? All companies using high-hazard technologies need to have in their midst small but significant populations of gainsayers – people who are prepared to stand up and say "this is not right." If an entire management team is made up of gung-ho high-achievers who lack any self-doubt, well, there may be trouble ahead.

However, Macondo-*Deepwater Horizon* is a complicated accident because several other parties were also involved and share responsibility:

- Did the Halliburton people express concern that the test results for the stability of the nitrogen foam cement were not available? (If the test results were not available, what was the point of doing the test?)
- The design and testing of the blind shear arms in the BOP were inadequate – the BOP was unable to cope with an eccentric drill pipe. The prime responsibility here must lie with Cameron (although the relevant API standard was lacking detail). However, did not Transocean (the owner of the BOP, and a highly experienced company) have any views about BOP design and testing? Did Transocean share some blame also?
- The BOP control systems were defective and were not designed to be fully testable. There were simple wiring faults which show poor quality assurance and inadequate factory testing. Should not Transocean (the rig owner and operator) have witnessed the Factory Acceptance Tests by Cameron (the BOP designer)?

Macondo-*Deepwater Horizon* had another unusual dimension. The poor Federal response to the destruction of New Orleans by Hurricane Katrina in 2005 meant that the US Government was keen to show a firm response to the blowout. BP caught the public relations backlash.

There are no easy general conclusions that can be drawn about BP relating to the two accidents. One could say the same about the two accidents on Equilon plant described in chapter twelve – there were no obvious common factors.

(At time of writing, a further example of a single operating company suffering multiple major accidents has just occurred. Tulsa-based pipeline operator Williams has suffered three large accidents in one year: A Louisiana chemical plant explosion (June 2013, two dead and 80 injured), a liquefied natural gas storage plant explosion in Plymouth, Washington (March 2014) and a fire at a natural gas plant in Wyoming in April 2014 which required the town of Opal to be evacuated.) (Source: www.tulsaworld.com, 17 May 2014.)

Tentative conclusions from BP's two massive accidents, which could be read across to other companies operating high-hazard industries, are as follows.

Companies need to ensure that managers and employees *really* know that delaying a project or reducing output because of safety concerns is alright. Any anxiety over safety should always lead to a pause for thought. Companies need to make examples of such decision making – e.g., internal newsletters should clearly be praising safety decisions which have adversely affected output – so that their front-line engineers really know this sort of behavior is alright, and the company is serious about safety.

Companies need to address responsibly the issue of worker fatigue.
It is not enough for the CEO to keep saying that old cliché, "Safety is (of course) our number one priority." Neither is it enough for the CEO to say "I have never turned down any request for safety-related expenditure." Companies need to have some diversity in the types of people they recruit. If the same type of person is always recruited (i.e. confident, positive, and determined to achieve goals) everyone may think the same way at critical moments.

A final thought: Just suppose there really were no common cause factors causing the accidents which affected the two BP facilities within five years. Macondo-*Deepwater Horizon* was one of the most expensive accidents of any sort ever, and Texas City was one of the most expensive refinery accidents ever. So, if there really were no common cause factors, that would mean BP was just plain unlucky to have these two huge accidents so close together. Is that credible?

REFERENCES

Texas City
[1] US Chemical Safety and Hazard Investigation Board, Investigation report 2005-04-I-TX, March 2007.
[2] Anatomy of a Disaster, http://www.csb.gov/videoroom/detail.aspx?vid=16&F=0&CID=1&pg=1&F_All=y.
[3] The report of the BP US refineries independent safety review panel (The Baker Report), January 2007.

Macondo-*Deepwater Horizon*
This accident is technically very complex. It has produced a flood of reports from various parties. Some of the principal reports are listed here. All of these are available online.
[1] Deep Water – The Gulf oil disaster and the future of offshore drilling, Report to the President, National Commission on the BP Deepwater Horizon Oil Spill and Offshore Drilling, January 2011.
[2] Forensic examination of Deepwater Horizon blowout preventer, Final report for United States Department of the Interior, vols. 1 and 2 and Addendum, Det Norske Veritas (DNV), Report No. EP030842, March 20, 2011.
[3] Deepwater Horizon accident investigation report and appendices, BP, September 8, 2011.
[4] Deepwater Horizon containment and response: harnessing capabilities and lessons learned, BP, September 1, 2010.
[5] Final report on the Investigation of the Macondo Well Blowout, Deepwater Horizon Study Group, March 1, 2011.
[6] K.L. McAndrews, Consequences of Macondo: a summary of recently proposed and enacted changes to U.S. offshore drilling safety and environmental regulation, Society of Petroleum Engineers, SPE-143718-PP, The University of Texas at Austin, March 2011.
[7] Macondo well–Deepwater Horizon blowout: lessons for improving offshore drilling safety, National Academy of Engineering, December 2011.

[8] Investigation Report (Vol. 1 and 2 and Appendices 2A and 2B), Explosion and Fire at the Macondo Well, US Chemical Safety and Hazard Investigation Board, June 2014, downloadable at http://www.csb.gov/macondo-blowout-and-explosion/ which also contains an excellent 11 minute video explaining the failure of the blow-out preventer, accessible at https://www.youtube.com/watch?v=FCVCOWejlag.

[9] Some of the most interesting contemporary commentary on the accident, from an engineering perspective, was written in a series of excellent articles by Ian Fitzsimmons and others in the magazine "Offshore Engineer" during 2010 and 2011, available online at www.oilonline.com.

[10] Consultants ABS have carried out a detailed FMECA of blowout preventers: Blowout preventer (BOP) failure mode and effect criticality analysis (FMECA) for the Bureau of Safety and Environmental Enforcement, Final Report, 2650788-DFMECA-3-D2, June 28, 2013.

Chernobyl and Fukushima 15

"The past is a foreign country: they do things differently there".
LP Hartley

"It is a riddle, wrapped in a mystery, inside an enigma".
Winston Churchill

PRELUDE: TOURISM BEHIND THE IRON CURTAIN, 1984

In April 1984, two years before the Chernobyl accident, my wife and I visited Beijing and Moscow on a package tour holiday. This was an unusual thing to do at that time; the Cold War with the Soviet Union was still going very strongly. Meanwhile, China was still poor and recovering from the Cultural Revolution.

I was working at the time for the UK Atomic Energy Authority. The Head of Security at Dounreay had to brief me before I left. "Don't talk to strangers", he said, like I was a child about to go to school unescorted for the first time.

He then had to de-brief me when I returned. "Did any strangers approach you?" he asked, so I told him about our bizarre holiday while he earnestly took notes.

Beijing was mostly friendly and keen to impress, but the Cultural Revolution was far too recent for people to have got over it. Conditions for the Chinese were still fairly primitive – at that time, apart from the occasional official car, the roads were empty except for buses and bicycles. There were no high-rise buildings. There were no advertisement billboards. Our small party of 15 was followed everywhere by a comic-book secret policeman who wore sunglasses, a hat, and an overcoat with a turned-up collar.

On the way back from Beijing, our strange party of 15 assorted people – what sort of person took a holiday behind the Iron Curtain in 1984? – stayed for two days in Moscow. Seventy-two year old Konstantin Chernenko was briefly the leader of the increasingly gerontocratic Soviet Union. Chernenko was dying of emphysema even before he had become General Secretary, two months before our visit. He was a metaphor for the Soviet Union itself – frail, aging, devoid of energy, unable to recognize the need for change. He died in office after only 13 months. Much of that time he was wheelchair-bound, or in hospital, with Mikhail Gorbachev deputising but not yet in power.

Moscow had opened up a little to Westerners since the Moscow Olympics in 1980. Our visit to Moscow in April 1984 was brief but it was memorable, albeit mostly for strange reasons. The memories are still like snapshots in my head, quite vivid after all the intervening years.

Our room on the sixteenth floor of the Hotel Cosmos had an enormous television that did not work. I left it switched on for 10 minutes but nothing happened. Later, another member of our party told me if you left it switched on for 15 minutes, the vacuum tubes eventually got hot enough and it suddenly sprang to life.

The big department store, GUM, on Red Square – their showplace store – seemed almost empty of any worthwhile products, yet the tourist "Berioska" shops (which only accepted hard currencies such as dollars and pounds, so normal Russians could not shop in them) had a decent range of consumer luxury goods.

All shops were grossly over-staffed: One person would help with selection of the required product, another would wrap it, and a third would take the money.

Western clothes were considered cool. Teenagers in the streets, when they thought no one was looking, offered in poor English to buy my clothes from me. However, they could only offer to pay in roubles, which were useless to me since it was impossible to buy anything with them.

Our tour guide in Moscow was called Rosa, from Intourist, the Soviet travel agency. In the bus we got a running commentary delivered in a loud, intimidating monotone which also seemed strangely bored and indifferent: "On your left...on your right...great Soviet technology..Great Patriotic War..." Their efforts at propaganda were cack-handed and quite unconvincing. It was as if we were expected to be completely overawed by everything when actually it all seemed, well, a bit shabby. Their arrogance seemed misplaced.

Stalin's purges, the slaughter of the Great Patriotic War, the Gulag, the KGB and finally the mindless bureaucracy of the State seemed to have removed most of their humanity. It was as if the entire population of the country had been schooled into not thinking for themselves. There was a widespread cynical indifference. It was the Orwellian dystopia made real. The only spark of real humanity was the teenagers wanting to buy my clothes.

It is difficult to explain the Cold War to my children's generation. It seemed like half the world was imprisoned.

GENERIC TECHNICAL SAFETY REQUIREMENTS FOR ALL NUCLEAR REACTORS

The generic technical requirements for any safe design of nuclear reactor can be stated quite concisely.

The *first requirement* for any nuclear power reactor is that the reactor should behave predictably during operation. The reactor output power should not change unless the operators want it to change, and when the operators do want the power to change, it should do so in a smooth, predictable manner. This is discussed further in the section below.

The *second requirement* is that the reactor must have reliable and quick shutdown systems that will operate under all foreseeable design-basis conditions.

Operating nuclear reactors have within their fuel pins a large amount of highly radioactive and radiotoxic fission products, some of which are volatile at low temperatures. These include in particular iodine-131, and cesium-137. There are also non-volatile but radiotoxic alpha-emitters such as plutonium-239. Hence the *third requirement* is that there must be robust containment of all the fission products under all foreseeable design-basis conditions.

Because of short-lived radioactive isotopes, a recently shutdown reactor continues to generate a lot of "decay heat", the heat produced by the decay of the fission products. The amount of decay heat falls from about 10% of the pre-shutdown power immediately after shutdown, to 0.1% of the pre-shutdown power after about 50 days. The value after 50 days may not sound much, but it still corresponds to about 3 MW of heat for a typical large power reactor. Hence the *fourth requirement* is that reactors must have reliable means of removing the decay heat from recently shutdown reactors under all foreseeable design-basis conditions.

Predictable behavior, reliable shutdown, robust containment and reliable decay heat removal: these four requirements are unambiguously applicable to all reactor designs. All four are necessary for a safe design.

STABILITY AND PREDICTABILITY OF NUCLEAR REACTOR BEHAVIOR

Nuclear engineers talk about *reactivity*, which means, in essence, the instantaneous rate at which the reactor power is increasing (or decreasing). For a reactor to behave predictably and with stability there must be fast-acting negative feedback processes which cause the reactivity to fall if the fuel becomes hotter. This is called *negative thermal feedback*.

There are a number of processes that affect thermal feedback, the most important of which are the *Doppler feedback* coefficient and the *steam voidage* coefficient.

Doppler feedback is a key process in reactor stability that takes place at the level of the atomic nucleus. In nuclear reactors, Doppler feedback is due to thermal vibration of uranium-238 nuclei, which makes the nuclei appear bigger to passing neutrons. Hence neutrons are more readily captured by uranium nuclei when the fuel temperature increases. Nuclear reactor fuel is normally only 2.5–3% fissile (heat-producing) uranium-235. The other 97–97.5% is uranium-238, which is non-fissile – i.e., it does not fission easily to produce heat. Instead, uranium-238 captures neutrons and becomes uranium-239, which with a half-life of just over two days becomes plutonium-239 (which is again fissile). The key point is that the immediate effect of higher fuel temperature is that non-fissile uranium-238 mops up more neutrons within the reactor core without producing more heat.

Hence, Doppler feedback in nuclear reactors works like this. As the reactor fuel becomes hotter, the uranium-238 absorbs more neutrons. This means that, if the reactor fuel becomes hotter, the reactivity reduces. Also, because uranium-235 (which generates the heat of fission) and uranium-238 are mixed at the atomic level, the Doppler Effect is very fast-acting. In essence, the Doppler Effect is the reason we have nuclear reactors – it is a fundamental physical process that makes them stable (Fig. 15.1).

The second fundamental physical process that affects the behavior of water-cooled nuclear reactors is steam voidage. The water in nuclear reactors carries out two functions: first it acts as a coolant, and second, it acts as a neutron moderator – i.e., it slows down the neutrons to enable their capture by uranium-235 nuclei. Water boils inside the reactor of the RBMK, so that means that there is an "absence of water" where steam forms, since steam is much less dense than liquid water. Where steam bubbles form, the neutrons may not (on average) be slowed down so much – and faster neutrons are less likely to cause fission. Also, where steam bubbles form, this means (on average) the neutrons will travel further – so more neutrons may escape the reactor into the shielding and be "lost". These effects due to steam voidage are beneficial – they lead to a reduction in reactivity as more steam bubbles form because of rising reactor temperature. This is called a *negative steam voidage coefficient*.

However, steam voidage can also have another effect because, as well as slowing down neutrons, water also absorbs some neutrons. Hence, if the neutrons are already fully moderated (e.g., by the graphite in the RBMK reactor core) then increased steam voidage will just mean that fewer neutrons are absorbed in the water; this means that increased steam voidage can lead to an increase in reactivity – exactly the effect that we do not want. This is called a *positive steam voidage coefficient*.

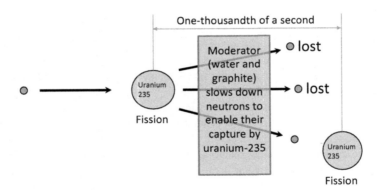

Reactivity is a measure of the instantaneous rate at which the reactor power is increasing (or decreasing), which is proportional to "how many more (or less) neutrons there are in any generation than there were in the previous generation".

The *"lost"* neutrons are absorbed by non-fissionable materials. These include:
- Shielding,
- Structural steel,
- Moderator,
- Uranium-238 – this is most important because of the *Doppler Effect* – higher fuel temperature means that uranium-238 absorbs more neutrons.

FIGURE 15.1

Neutrons and fission in a nuclear reactor.

NUCLEAR REACTOR TECHNOLOGY IN THE SOVIET UNION

Right up until the time of the Chernobyl accident, little was known about Soviet nuclear activities. Their nuclear weapons' program was of course extremely secret, but even their nuclear electricity generation program was veiled in secrecy. Visitors to the Soviet Union were not allowed to travel freely; generally, they were never allowed to go anywhere except a few major cities like Moscow and Leningrad (now called St Petersburg).

In the early 1970s the Soviet Union embarked on an ambitious nuclear power program aimed at commissioning some 15,000 MW of generating capacity by the end of 1980. There were two main designs employed, the VVER and the RBMK. The VVER was a Russian version of the western Pressurized Water Reactor (PWR); several were built in the Soviet Union, and the design was also built in other Eastern European countries. The other design, the RBMK, was only ever built within the Soviet Union, where it provided the larger share of the nuclear program. The RBMK design (*Reaktor Bolshoy Moshchnosti Kanalniy* or High Power Channel-type Reactor) was unique to the Soviet Union; there was nothing like it in the West. It combined vertical water-cooled fuel channels with graphite moderator. The water boiled in the fuel channels and then was delivered, via steam separator drums, direct to steam turbines. The fuel channels could be refueled on-load. It can therefore be thought of as a bit like a combination of the UK's Advanced Gas-cooled Reactors (AGRs) and American Boiling Water Reactors (BWRs).

One reason this design was adopted was due to reasons of transport. The Soviet Union was a vast country, and their preference was for a design where the components could be readily transported by rail. The RBMK did not have a massive reactor pressure vessel like the BWR or PWR designs. Hence, the RBMK could be built wherever there was a railway line.

Almost all that was known outside the Soviet Union about the RBMK prior to the accident was from a visit by British engineers to an RBMK plant in Leningrad in 1975. At the time, the UK was considering building a design called the Steam Generating Heavy Water Reactor (SGHWR), which was vaguely similar to the RBMK except it used heavy water as a moderator instead of graphite. Hence a knowledge exchange visit was arranged, and a report with the findings of the UK team was issued. The report [1] was highly technical and no-one in the West was really too concerned about the safety of the RBMK design at that time – the report was a comparison between the SGHWR and the RBMK – but some of the information buried in the details was prescient.

- The reactivity worth of the control rods – i.e., how much neutron absorption capacity there was in the control rods – was inadequate to meet all possible shutdown requirements. It was possible to conceive of situations where the control rods might not be able to shutdown the reactor.
- The reactor had a positive steam void coefficient at low powers.
- The graphite moderator was blanketed with nitrogen in normal operation, because it operated at temperatures in excess of 700°C. At this temperature,

when exposed to air, the graphite ignites spontaneously. The report said that graphite operating temperatures were "unacceptable".

- The containment structure was less robust than would normally be considered adequate in Western designs. Pressure tube failure could be disruptive.

Why did the RBMK design have a positive void coefficient? This had been a design decision. The RBMK design deliberately used too much graphite – the UK engineers reported that moderation was "beyond the optimum in terms of fuel enrichment and seems to give a positive void coefficient." The explanation for this decision would have been unseen to the UK team at the time, but it appears the Soviet Union was short of uranium enrichment capacity. An over-moderated design enabled the reactor to use only 2% enriched uranium (i.e., 2% uranium-235) instead of a more typical 2.5–3%. The penalty of over-moderating with graphite is that there was a positive void coefficient at low power. Under particular circumstances, the reactor could be unstable and suffer runaway power increases.

In addition to the weaknesses in the RBMK design described above, the detailed implementation of the design also allowed greater freedom of action for the operators than would be normal in a reactor design elsewhere. For example, the operators could override reactor trip systems at the flick of a switch; in Western designs, key interlock systems would have prevented this. Also it was essential for the safe operation of the plant that the control rods should never be withdrawn beyond the point at which the control rod "reactivity margin" became dangerously low, yet this vital aspect was left entirely to the operators, with no automatic trip system.

Also, reactor emergency shutdowns were normally carried out by motoring the control rods into the core, and not (as is common in some other designs) by disconnecting electromagnetic clutches and allowing them to fall quickly into the reactor core under gravity. Because the rods were driven-in instead of being allowed to drop, shutdown could take 20 s to occur (Fig. 15.2).

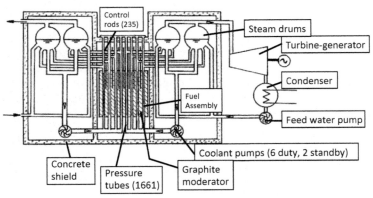

FIGURE 15.2

A schematic diagram of the Chernobyl RBMK nuclear power station design.

THE CHERNOBYL ACCIDENT

At the time of the accident, Mikhail Gorbachev was General Secretary of the Communist Party of the Soviet Union. In the West, he seemed like a breath of fresh air: he was relatively young, pragmatic and seemingly less restricted by Communist dogma. He wanted to improve the Soviet centrally planned economy which was struggling under its inflexible bureaucracy, political dogma, *apparatchiks*, and over-manning. He sacked Andrei Gromyko, the long-standing Minister of Foreign Affairs, in an effort to control old-style thinking in the Politburo. He forced up the price of alcohol to try to reduce alcoholism and its effects on the country.

Just prior to the Chernobyl accident, at the Communist Party Congress in February and March 1986, he announced plans for *perestroika* (restructuring), and he subsequently used the Chernobyl accident as one of his first efforts at *glasnost* (openness). *Perestroika* and *glasnost* were to be Gorbachev's two watchwords in his attempts to salvage the sclerotic Soviet economy but, ultimately, it was too little, too late – for both the Soviet economy and the RBMK reactor design.

The accident at Chernobyl happened shortly after midnight on Saturday April 26, 1986, although no announcement was made by the secretive Soviet authorities. The following Monday morning, airborne radioactivity was detected at a nuclear power station in Sweden, but the Soviet authorities would not initially admit anything had happened. At 9pm on Monday 28, after the Swedish authorities had threatened to notify the International Atomic Energy Agency based in Vienna, the Soviet news agency TASS released a brief message:

> "An accident has occurred at Chernobyl nuclear power station. One of the atomic reactors has been damaged. Measures are being taken to eliminate the consequences of the accident. Aid is being given to the victims. A government commission has been set up".

For the next few weeks the world watched as, gradually, more details came out of the Soviet Union about the accident. Television cameras on board Soviet military helicopters looked straight down into the core of reactor 4, where the burning graphite could be seen. Other helicopters dropped sand and boron into the reactor. The world saw Soviet soldiers receiving large radiation exposures as they tried to mitigate the consequences.

All of this was previously almost unimaginable for two completely different reasons. First, the rest of the world was actually seeing events as they occurred inside the Soviet Union in a way that had never happened throughout its history. The Soviet Union had always been pathologically secretive. Churchill's description of the Soviet Union in 1939 as "a riddle, wrapped in a mystery, inside an enigma" had remained true, yet now, suddenly and for the very first time, we seemed to be seeing through the layers.

The second reason that this seemed unimaginable was that, for years, the nuclear industry around the world had said that such an accident was extremely unlikely. Nuclear reactors in the West were stable, and there were multiple layers of defences.

No one knew for certain the radiological consequences of a severe reactor accident, but hypothetical studies had been carried out. Few in the nuclear industry expected to see an "experiment" done for real in which the consequences of a severe reactor accident would be measurable.

During the week beginning August 25, 1986, Soviet scientists presented information about the accident, in an unprecedentedly candid way, at a Post-Accident Review Meeting held at the International Atomic Energy Agency in Vienna. Following that IAEA meeting, in October 1986 a nuclear industry seminar was held in London where explanations were presented of what had happened, and preliminary estimates of the radiological consequences were given [2].

Ironically, the accident had occurred during a test to check whether the emergency core cooling systems were adequate. The test was to be performed during a planned shutdown for routine maintenance. The main events were as follows.

1. Thermal power was to be reduced in stages from 3200 MW to 700 MW (MW). Operator error caused the power to undershoot to 30 MW, and the consequent increase in xenon poisoning meant that the operators could only manage to raise the power back to 200 MW. This power level was lower than stipulated in the test instructions. The operators should have abandoned the test at this point. ("Xenon poisoning" is a transient behavior of nuclear reactors during power reductions. Xenon-135 is a fission product that absorbs neutrons, and which decays with a half-life of about 9 h. When power is reduced, xenon-135 levels increase, which makes it difficult to increase power until the xenon-135 has decayed.)

2. In violation of the test procedure, the operators started up the standby coolant pumps. This meant that core power was only 7% normal but coolant flow was 120% normal, which made the whole core virtually isothermal.

3. Difficulties in steam drum water level control at this juncture led the operators to override the reactor trip signals generated by low drum level. They then topped up the drum water level under manual control, with relatively cold feedwater. This caused a fall in reactor temperature and therefore a reduction in the amount of steam voidage inside the reactor core; hence (via the positive steam voidage coefficient) the control rods had to be withdrawn yet further to maintain reactor power level at 200 MW.

4. The operators noticed that the "control rod reactivity margin" was too low – the control rods had been raised so far that they would be unable to have any significant and immediate effect to reduce reactor power when required. Indeed, the reactivity margin was at a level where operating rules stipulated that the reactor should have been shutdown. No such action was taken.

5. The test was to be initiated by tripping the turbine. Normally, this would have tripped the reactor also. However, it would appear that the operators wanted to reserve their options in case the test of the emergency core cooling system was unsatisfactory; hence they overrode the reactor trip. This meant (they thought) that they would be able to repeat the test if necessary.

6. At 0123.10 h on the April 26, 1986, the turbine was tripped. The operators had contrived, unwittingly, to put an already unstable reactor design into a highly dangerous condition. The reactor was operating at fairly low power (200 MW), and was almost isothermal due to the high coolant flow-rate. By tripping the turbine the only significant heat sink had been removed. In addition, the control rods were too far out of the core to have any significant immediate effect in the event of a reactor trip.

7. The water coolant temperature now began to rise steadily (because of the heat being generated in the reactor) until it approached the saturation temperature, at which point bulk boiling occurs. Because of the virtually isothermal state of the reactor, the value of the positive void coefficient had been maximized, i.e., bulk boiling began throughout the core more-or-less simultaneously.

8. At 0123.40 h, a rise in power was noted and a reactor trip was initiated manually, by starting to drive the control rods into the core. Because the rods were so far out of the core, however, they had no significant immediate effect on reactivity, which continued to rise.

9. Coolant bulk boiling led to a rapid rise in power, which Doppler feedback could not counteract.

10. The power rose to 530 MW at 0123.43 h, and thereafter extremely rapidly, doubling each fraction of a second; the reactor went "prompt critical", and the fuel (uranium oxide ceramic) shattered due to the thermal shock of the sudden power rise; the fuel cladding melted and the white hot fuel fragments came into contact with the cooling water, causing a steam explosion at 01.23.48 h. This was followed a few seconds later by a hydrogen explosion; the hydrogen was generated by zirconium–water and graphite–water reactions. (There was no nuclear explosion.) The explosions ruptured the containment.

11. The hot graphite moderator caught fire upon exposure to air.

12. Truly heroic efforts by firemen, helicopter pilots and engineers led to the fire being extinguished, and the radioactive release being stopped by May 5,10 days after the accident began. In the interim, it is estimated that 20% of the iodine inventory and 12% of the cesium inventory in the reactor was released to the atmosphere, together with practically all of the noble gas fission products.

The Chernobyl accident was about as bad a nuclear reactor accident as it is possible to imagine. The reactor core burned without any containment for ten days. It is difficult to postulate any accident that could lead to a greater release of radioactivity to the environment. A total of 31 people died either at the time of the accident or within a few weeks from radiation sickness. The nearby town of Pripyat, with a population of 49,000, was evacuated within a few days and an exclusion zone of 30 km radius around the Chernobyl plant was declared.

The RBMK design violated at least three of the four generic technical safety requirements described previously. The reactor power was unstable, the reactor shutdown systems were slow and inadequate, and the containment systems were insufficiently robust.

Also, the operators showed complete disregard for their operating instructions: the initial conditions for the test were not as stipulated, protection systems were overridden, and the control rods were raised too high to retain control of the reactor.

By the time the operators realized that the reactor power was increasing uncontrollably, it was too late.

AFTERMATH – RADIOLOGICAL AND HEALTH CONSEQUENCES

In the days and weeks following the accident, radiation spread across Western Europe, with some unlikely places being worst affected. The contamination happened in discrete locations, according to the vagaries of the weather, with other nearby places sometimes being unaffected. Inside the Soviet Union, there was significant contamination within a 30-mile radius of Chernobyl, straddling the Ukraine/Belarus boundary. There was also significant contamination one hundred miles to the northeast, on the Belarus/Russia boundary. Some 300 miles further east, there was significant contamination in Russia between the towns of Bryansk and Tula. (At the time of the accident, Belarus, Ukraine and Russia were all part of the Soviet Union.)

Further afield, wherever rain fell it caused local contamination. Finland, Sweden, Norway and Austria all acquired large areas of land with significant contamination from cesium-137. Some hill farms in Scotland, Wales and Cumbria also became contaminated [4]. Controls were placed across the European Union on the sale of meat with contamination levels more than 1000 Bq per kilogram. This figure was of course arbitrary, a sop to public opinion, but politics demanded that action had to be taken and some limit had to be imposed.

Few technical subjects are so prone to misrepresentation as the health effects of low-level nuclear radiation. Journalists can seemingly always find a knowledgeable-sounding, self-appointed expert who is prepared to make unreasonable assertions about the consequences of nuclear accidents. The subject of health effects of radiation was controversial, and remains so, perhaps to a lesser extent, to this day. Also, in the months following the accident there were new daily revelations from inside the Soviet Union, which had hitherto been mostly *terra incognita*. The combined effect was to put the news media into a feeding frenzy about an industrial accident, in a way that was probably not matched until the Macondo-*Deepwater Horizon* accident in 2010.

The principal isotopes of concern from the accident were iodine-131 and cesium-137. Iodine-131 is of concern because it concentrates in the thyroid gland and can lead to thyroid cancer, although this is treatable with survivability of about 95 per cent. Iodine-131 has a short half-life of about 8 days, so after two months it has effectively disappeared. Thyroid cancers may then develop within a few years of the exposure. In Pripyat and some other areas, people were given potassium iodate tablets to swamp their thyroids with non-radioactive iodine, thereby preventing uptake of iodine-131.

Cesium-137 has a half-life of about 30 years, so cesium-137 from Chernobyl will remain detectable for more than two hundred years. (An interval of eight half-lives

(240 years) corresponds to a reduction factor of 2 to the power 8, which equals 1024.) It behaves in the human body like common salt, so it spreads evenly throughout the body, without concentrating in any particular organ. Hence its biological effects are similar to those of external gamma or X radiation.

The difficulty in discussions about low-level radiation is that a lot of people die from cancer anyway – typically some 30% of the population, although the exact figure depends on regional variations in life expectancy (if you die young, you are less likely to die of cancer), together with diet, smoking and alcohol consumption. Also, we all receive doses of natural background radiation – typically one or two milliSieverts (mSv) per year. Hence it becomes problematic to separate the "noise" (normal background levels of cancer) from the "signal" (artificial-radiation-caused cancer).

Two United Nations bodies have produced reports about the health effects of Chernobyl – the International Atomic Energy Agency (IAEA) [3], and the United Nations Scientific Committee on the Effects of Atomic Radiation (UNSCEAR) [5]. Their reports seem objective, neutral, unbiased and honest.

The socio-economic impact of Chernobyl has been enormous, but it was made worse by the collapse of the Soviet Union in 1991. It becomes difficult to separate these two issues, as the 2005 IAEA report made clear. It painted a disturbing and depressing picture of the overall effects.

Radiation doses could only be estimated sometime after they occurred by careful evaluation of all available information. The average dose was 17 mSv to Ukrainian evacuees, with doses to individuals ranging from 0.1 to 380 mSv. The average dose to Belarusian evacuees was 31 mSv, with the highest average dose in two villages being about 300 mSv.

The IAEA recognized that the number of deaths attributable to the Chernobyl accident has been of "paramount interest" to all concerned – the general public, scientists, the mass media, and politicians. The IAEA noted that claims had been made that tens or even hundreds of thousands of persons have died as a result of the accident, and stated that these claims were exaggerated: the total number of people that could have died or could die in the future due to Chernobyl-originated exposure over the lifetime of emergency workers and residents of most contaminated areas was estimated to be around 4000. This included some 50 emergency workers who died of acute radiation syndrome in 1986 and other causes in later years, 9 children who died of thyroid cancer, and an estimated 3940 people that could die from cancer contracted as a result of radiation exposure.

The IAEA went on to say that the relatively low dose levels to which the population of the Chernobyl-affected regions was exposed meant that there is no evidence or any likelihood of observing decreased fertility among men or women. The doses were also unlikely to have any effect on the number of stillbirths, adverse pregnancy outcomes, delivery complications or the overall health of children.

The IAEA noted that the Chernobyl nuclear accident, and government policies adopted to cope with its consequences, imposed huge and incalculable costs on the Soviet Union and three successor countries, Belarus, the Russian Federation and the

Ukraine. These costs are impossible to calculate precisely because of the non-market conditions prevailing at the time of the disaster and the high inflation and volatile exchange rates of the transition period that followed the break-up of the Soviet Union in 1991. However, a variety of government estimates from the 1990s put the cost of the accident, over two decades, at hundreds of billions of dollars.

Since the Chernobyl accident, some 350,000 people have been relocated away from the most severely contaminated areas.

There was a significant drop in life expectancy, particularly for men, which in the Russian Federation in 2003 stood at an average of 65 (just 59 years for men). However, "the main causes of death in the Chernobyl-affected region are the same as those nationwide – cardiovascular diseases, injuries and poisonings – rather than any radiation-related illnesses. The most pressing health concerns for the affected areas thus lie in poor diet and lifestyle factors such as alcohol and tobacco use, as well as poverty and limited access to primary health care."

The IAEA regretted the exaggerated or misplaced health fears, and a sense of victimization and dependency created by government social protection policies in the affected areas. The IAEA criticized the extensive system of Chernobyl-related benefits, which had created expectations of long-term direct financial support and entitlement to privileges, and had undermined the capacity of the individuals and communities concerned to tackle their own economic and social problems. "The dependency culture that has developed over the past two decades is a major barrier to the region's recovery," it concluded.

The 2008 UNSCEAR report, which is more tightly focused on the direct health impact of the accident, declines even to offer an estimate of the number of radiation-related deaths attributed to the accident.

"...there is a limit to the epidemiological knowledge that can be used to attribute conclusively an increased incidence to radiation exposure. Therefore, any radiation risk projections in the low dose area should be considered as extremely uncertain, especially when the projection of numbers of cancer deaths is based on trivial individual exposures to large populations experienced over many years." (UNSCEAR)

In other words: We don't know how many people have died or will die because of low-level radiation from the accident. Furthermore, we can't ever know because the data we want to measure are completely overwhelmed by other factors. The answer is almost certainly between zero and 4000, over many decades, spread across the entire population of Europe (about 740 million in 2011). However we don't know for certain and we never will.

CHERNOBYL: INDIRECT CAUSES OF THE ACCIDENT

The accident at Chernobyl was directly attributable to an unsafe design and a complete disregard of operating procedures. Indirectly, however, the accident can be attributed to other factors. The design was compromised by political expediency – shortage of uranium enrichment capacity led to the design being "over-moderated", which led to the positive steam void coefficient. The whole RBMK program, which was considered in the 1976 UK report to have a variety of serious shortcomings, was at least partially constrained by the need to transport all parts through railway tunnels. This need not necessarily have compromised safety but, in the absence of any free and open discussion about nuclear safety in the Soviet Union, it apparently did so.

The Soviet Union undoubtedly produced some very good science. Between 1958 and 2003, ten Nobel Prizes for Physics were awarded for work done in the Soviet Union prior to its collapse. However, the translation from science to engineering did not have a good record.

During the period between 1969 and 1973, the Soviet Union had an especially bad time regarding the safety of its advanced technologies:

- The Tupolev 144 supersonic airliner, dubbed "Concordski", broke up in mid-air in front of the world's TV cameras at the Paris Air Show on June 3, 1973. Six people on board and eight people on the ground were killed.
- The giant N-1 moon rocket, the Soviet equivalent of the American Saturn 5, had four unmanned development launches between February 1969 and November 1972. In each case the rocket launcher blew up shortly after lift-off. The program was canceled in 1974, although the accidents and indeed the whole program remained secret for decades [7].
- On June 30, 1971, the crew of Soyuz 11 was returning from a 22-day, first-ever trip to a manned space station, Salyut 1. Their re-entry capsule depressurized during descent and all three crew asphyxiated.

Throughout the Cold War, the Soviet Union generally tried at least to match US developments in the nuclear arms race – missiles, warheads and nuclear submarines. However, the Soviet record of nuclear submarine safety was not impressive [6]:

- An explosion on board the K-19 submarine July 4, 1961 caused 8 deaths.
- On May 24, 1968 a reactor accident on board the submarine K-27 caused 9 fatalities.
- On June 23, 1983, the K-429 sank in shallow water, killing 16.
- A reactor accident on board the K-431 on August 10, 1985 caused 10 fatalities.
- The K-219 had an explosion and fire in a missile tube on October 3, 1986, causing 4 fatalities and sinking shortly after.
- The K-278 suffered a fire and sank on April 7, 1989. 42 died.

Finally, it is worth recalling that the Soviet Union had previously had a nuclear accident that was even worse than Chernobyl – the 1957 Kyshtym accident [8]. This accident was not publicly known in the West until 1976, when the dissident Soviet scientist Zhores Medvedev published some initial details, which were later confirmed as more information became available. The accident occurred at a plant storing liquid radioactive waste in a fuel reprocessing plant for the Soviet nuclear weapons program. The cooling system of a storage tank, containing some 70 tonnes of liquid waste, had failed and subsequently remained unrepaired; the liquid gradually evaporated until only a solid residue was left. The residue was highly radioactive and it also contained explosive ammonium nitrate. A massive chemical explosion blew large quantities of mostly long-lived radioactivity into the air, which remains detectable to the present day and is known as the East Ural Radioactive Trace. It spreads for some 300 km.

It seems reasonable to conclude that the Soviets were trying to do too many high-technology military and prestige projects, with too few resources, and with too many politically driven targets. There was also a cavalier attitude to safety and too little (if any) independent safety regulation. In particular, the very concept of "independent safety regulation" will always be especially difficult in top-down, centrally planned economies. It is always the role of the independent safety regulator to "speak truth to power," which can be difficult in autocratic regimes.

THE GREAT EAST JAPAN EARTHQUAKE AND TSUNAMI, MARCH 11, 2011

The Indian Ocean earthquake and tsunami on December 26, 2004 was the first tsunami of the Video Age. Before that event, we only really understood tsunamis from eyewitness descriptions, or else drawings made after the event. Videos taken that day showed the world what previously had to be imagined – first, the sea recedes and then, a few minutes later, a remorseless overwhelming wall of water comes ashore, destroying everything in its path. We also saw for the first time "before" and "after" satellite photos of villages obliterated by the tsunami. An estimated 230,000 people were killed.

A little over six years later, on March 11, 2011, the Great East Japan Earthquake struck with an epicenter about 70 km east of the northeast coast of mainland Japan (Honshu). Its magnitude was 9.0, the most powerful earthquake that has struck Japan since the Jogan earthquake of 896 AD, and the fourth biggest anywhere in the world since 1900.

Japan is in a very seismically active area, so it has become a global center of expertise in seismic design. During the earthquake, high-rise buildings swayed but remained intact. Houses shook but by-and-large survived. Infrastructure did not fare so well, however – roads and railways were badly damaged, water and electricity were cut off, and communications by both landline and mobile phone were disrupted. Most of Japan's nuclear generating plants shutdown at, or very shortly after, the earthquake. One dam failed, and there was a major refinery fire. The damage was extensive and expensive but, on the whole, the country survived the initial earthquake in reasonable condition and without huge loss of life.

Aftershocks, some as strong as magnitude seven, continued for days.

After the first earthquake shock at 1446 hours local time on March 11, 2011, the tsunami followed between 25 and 70 minutes later, depending on location. Many people heeded warnings and moved to higher land in the interval between the first earthquake and the tsunami.

In many places on the northeast coast of Honshu and in Hokkaido the tsunami defences were inadequate. 561 square kilometers of mostly densely populated land were inundated – about one-third of the area of Greater London. Initial estimates of mortality exceeded 25,000, but this estimate has been refined and it now appears that the death toll was about 20,000, the vast majority from drowning.

The total number of residential buildings damaged was approximately 475,000 including fully destroyed, half-destroyed, partially destroyed and inundated structures. The number of cases of damage to public buildings and cultural and educational facilities was as many as 18,000. In addition, approximately 460,000 households suffered from gas supply stoppages, approximately 4,000,000 households were cut off from electricity, 1,500,000 were left without water, and 800,000 phone lines were knocked out.

There was a considerable local variability in tsunami damage. Some of the most horrific video and photographs came from Miyako City, in the Taro area, where the population had felt safe behind a 10-m tsunami barrier known locally as the "Great Wall of China". The barrier was over-topped and huge damage ensued (Fig. 15.3).

FIGURE 15.3

The tsunami over-topping the barrier in Miyako City: one of the most powerful images of the disaster (Photo: AFP Getty Images, used by permission). A 15-min video taken from the same position – an upper story of Miyako city hall – can be seen at www.youtube.com/watch?v=VqlQFZZoVR4. It shows the remorseless and overwhelming power of the tsunami. It is spine-tingling and terrifying to watch a city being destroyed in real time. At its high point, the tsunami barrier was completely submerged and invisible beneath the water.

A short distance away, however, in the village of Fudai, the inhabitants did not even get their feet wet. As the (somewhat stilted) English translation of the June 2011 Japanese Government report noted:

> "...the 15.5m embankment was installed in the Ootabe area, Fudai village in Iwate Prefecture following a strong desire of the village chief learning from previous experiences with tsunami. This embankment was able to resist the 15m tsunami and prevented the damage within the embankment zone... These areas are rias type coastlines that have, historically, suffered significantly from giant tsunamis in the 15m range such as the Meiji Sanriku Tsunami (1896) and the Showa Sanriku Tsunami (1933), the lesson of preparation against a 15m-class tsunami has been instructed... Against these tsunamis, there was a sharp contrast between the Ootabe area, which heeded the lessons of the past, and the Taro area."

An AP news report on May 13, 2011 attributed the survival of Fudai village to a stubborn former mayor, Kotaku Wamura, who had experienced a tsunami in 1933. Massive tsunamis had flattened Japan's northeast coast in 1933 and 1896 and, in Fudai, these had destroyed hundreds of homes and killed hundreds of people. "When I saw bodies being dug up from the piles of earth, I did not know what to say. I had no words", Wamura wrote of the 1933 tsunami.

It is in this context of overwhelming and widespread damage, huge mortality, massive destruction of infrastructure, loss of communications, and national shock at the severity of this natural disaster that the response to the emerging Fukushima accident must be judged.

THE ACCIDENTS AT FUKUSHIMA DAIICHI

The Japanese were perceived as world experts on seismic design, because their country is in such a seismically active area. There was a belief that their nuclear plants were well placed to withstand earthquakes.

Many reports have been published about the Fukushima accident, for example [1-7, 9-10]. These include reports by the Japanese Government and its official inquiries, the Japan Nuclear Technology Institute, the Institute of Nuclear Power Operations (in Atlanta, Georgia), and the World Health Organization. In the following a brief synopsis is presented of what happened and some key lessons learned.

Nuclear reactors throughout Japan shutdown automatically as designed in response to the ground accelerations of the earthquake. At Fukushima, the ground accelerations were 0.55 times gravitational acceleration (0.55g) in a horizontal direction, and 0.3g in a vertical direction. These ground accelerations are easily strong enough to knock people off their feet, and to reduce non-seismically designed buildings to piles of rubble – this had been a massive earthquake – but the Japanese nuclear plants apparently survived intact.

The Fukushima Daiichi nuclear power station consists of six Boiling Water Reactors (BWRs) designed by General Electric (GE), built by GE, Toshiba and Hitachi, and all operated by the Tokyo Electric Power Company (TEPCO). The six Fukushima Daiichi reactors were "Mark One" GE BWRs, which began generating electricity in 1970,

1973, 1974, 1978, 1977 and 1979 for reactors 1–6, respectively. Units 1–4 were built in close proximity, whereas units 5 and 6 were built a short distance away, to the north.

Crucially, units 5 and 6 were built with a higher breakwater (or tsunami barrier) than Units 1–4. Hence units 5 and 6 were never flooded to the same catastrophic extent as units 1 to 4, and some standby electricity supplies remained available. In addition, units 4, 5 and 6 were actually shutdown for refueling and maintenance at the time of the earthquake, so their decay heat problems were much less than for the recently operating units 1–3. (Japan has seasonal peaks of electricity demand in winter (for heating) and summer (for air-conditioning). Hence planned shutdowns take place in spring and autumn, and the tsunami occurred in early spring.)

After the tsunami inundated the site, the following situation faced the station staff looking after units 1–4: First, the entire Fukushima Daiichi station, units 1–6, was cut off from all electricity grid connections because of earthquake damage to pylons, cables and substations, and/or flooding. Also, back-up diesel electricity supplies failed on units 1–4, and hence all post-trip cooling was inoperative on those units. Even if power had been restored, the cooling pumps relied on high-voltage electric motors, which had been immersed in seawater and were therefore useless.

In addition, roads were impassable and communications were poor, so power station staff must have been extremely concerned about their families and their homes. There were also on-going major aftershocks, even after the tsunami had stopped, which continued for many days.

The power stations had emergency batteries, which were able to provide a few hours' supply for essential instrumentation and lighting only. After the batteries ran out, the power plants were literally blacked-out. Staff could only find their way round the plant using torches. There were no longer any control room screens or other indications of plant state.

The plant operators will have been keenly aware of the time pressures to try to restore post-trip cooling – they would know from their training that core melt would begin shortly after complete loss of cooling, but the circumstances meant that there was little the operators could do but watch, in the hours succeeding the tsunami, as the reactors overheated and the cores began to melt.

(We now have an insight into the enormous pressures faced by the operators, from the testimony of the Plant Manager of Fukushima, Masao Yoshida. His account tells us of the impotence of the engineers in the situation they faced ("Nobody came to help"). His account also tells us of the unhelpful advice and instructions coming from the Prime Minister's office, who wanted him to delay the use of seawater cooling directly into the reactors – which were already write-offs – because this would mean "the reactors would have to be decommissioned". Extracts of Masao Yoshida's account are available in English at http://www.asahi.com/special/yoshida_report/en/ (downloaded September 15, 2014). Masao Yoshida died from unrelated illness in July 2013.)

At 1636 hours on March 11, 2 hours after the earthquake, the Emergency Core Cooling Systems were declared to have failed on units 1 and 2. Once the core began to overheat, the water level fell inside the reactors as the remaining water boiled. Where the fuel was exposed, the temperature of the zirconium-alloy metal cladding the uranium oxide fuel rose rapidly to much more than its normal operating temperature and the zirconium began to react with steam to generate hydrogen gas.

Operators were able to monitor the pressures within the reactor containment structures with old-fashioned mechanical gages and battery torches: they could see the pressures rising due to the combination of high temperatures and hydrogen production. On March 12 and 13, they had no choice but to vent gases, including hydrogen, from the reactor containments into the reactor buildings. (The purpose of the reactor building – i.e., the externally visible part – in this BWR design is basically just to keep the reactors out of the weather. It has no function for containment of radioactivity.)

Some source of ignition caused hydrogen explosions within the reactor buildings on March 12, 14 and 15, and these explosions were captured by television crews operating several miles away with zoom lenses.

The unit 3 hydrogen explosion caused damage to the integrity of the fuel storage ponds on units 3 and 4, with the risk of spent fuel becoming uncooled and uncontained, so water cannons were aimed at the fuel ponds on March 17. Helicopters were also seen trying to dump water onto the reactor buildings.

Efforts were made to start some reactor cooling with seawater on March 12–14. It was April 3 before proper reactor cooling using off-site electricity supplies was restored.

The Japanese government acted promptly to announce evacuation orders. On the evening of March 11, a 3-km evacuation area was ordered, and this was extended to 10 km and then 20 km on March 12 (Figs. 15.4 and 15.5).

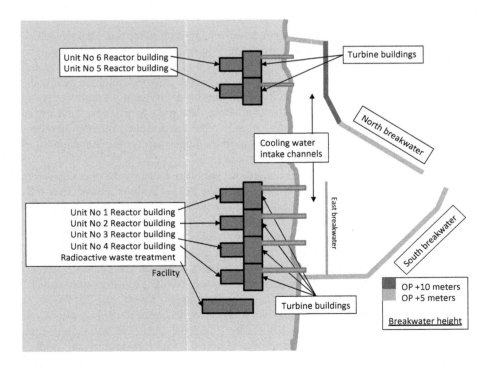

FIGURE 15.4

Plan view of Fukushima Daiichi units 1–6.

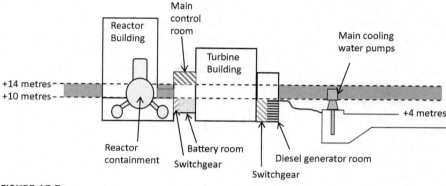

FIGURE 15.5

Section through units 1–4, showing the level of maximum inundation. The diesel generators, batteries and switchgear were all below the maximum water level.

MAJOR TECHNICAL ISSUES

The *first technical issue* arising from the accident was the inadequate design of the tsunami barrier.

The Japan Society of Civil Engineers (JSCE) had published revised guidelines on tsunami barrier design in 2002, "Tsunami Assessment Methods for Nuclear Power Plants in Japan." TEPCO voluntarily reassessed its tsunami design basis against these guidelines. Using these new deterministic evaluation techniques, TEPCO determined that the design basis tsunami would result in a maximum water level of 5.7m. (The previously assessed design basis tsunami was 5.5 m.) In accordance with this evaluation, the elevation of the Unit 6 seawater pump motor for the emergency diesel generator was raised by 20 cm and the seawater pump motor for high pressure core spray was raised by 22 cm. These changes were meant to ensure all vital seawater motors were installed higher than the new inundation level of 5.7m. The 5.5 m breakwater was not modified when the new tsunami height was implemented because it was not intended to provide tsunami protection, but rather to minimize wave action in the harbor.

With hindsight, changes in height of only 20 cm or so seem ridiculous when the actual tsunami was some 14 m, i.e., 8.3 m above the design basis.

After the accident, TEPCO was quick to point the finger of blame at the Japan Society of Civil Engineers, making it clear that Fukushima Daiichi had been designed in accordance with the 2002 JSCE guidelines. Responsibility here is a moot point – see the discussion in Chapter 11 of the Port of Ramsgate accident regarding the operator's responsibility for design decisions. (A revision to the JSCE guidelines had been due for publication at the end of 2011, after the accident.)

So, TEPCO's application of the 2002 JSCE guidelines apparently got this issue very wrong indeed. The methodology of the 2002 JSCE guidelines was to use a database of historical tsunami in order to calculate a design basis tsunami height, which exceeded all the recorded historical tsunami heights at each nuclear plant site. What is not clear is how TEPCO applied the JSCE guidelines.

Press reports have actually suggested that TEPCO knew by the time of the accident that the "design basis tsunami" was insufficient. The Japan Times for July 3, 2011 reported that TEPCO was aware as early as 2008 that a 10-m-plus tsunami could hit the Fukushima Daiichi nuclear power station, and that the results of new analysis work had been reported for the first time to the safety regulator (the Nuclear and Industrial Safety Agency or NISA) on March 7 – only four days before the tsunami occurred.

The role of the safety regulator NISA should have been central in this issue. It should not have been up to TEPCO to "interpret" the JSCE guidelines. Either (i) NISA should have given detailed prescriptive requirements to TEPCO, or otherwise (ii) TEPCO should have had to prepare a detailed justification, for NISA approval, why their chosen height for "design basis tsunami" was justified. (The first is the US regulatory model and the second is the UK regulatory model.)

Very soon after the accident, the role of NISA began to be called into question. TEPCO had reported their concerns about the size of the design-basis tsunami to NISA on March 7, 2011, only four days before the tsunami struck. NISA officials had replied to TEPCO that "countermeasures are urgently needed", calling for modifications, but it was too late.

In a similar way to the changes to oil and gas safety regulation in the USA following the Macondo-*Deepwater Horizon* accident, NISA was one of the first organizational victims of the tsunami. NISA was a division within MITI, the Japanese Ministry for Trade and Industry (MITI). One of MITI's roles was the promotion of the nuclear industry, so there was a clear conflict of interest within MITI. NISA was seen to be insufficiently challenging to the Japanese nuclear industry, and insufficiently pro-active in monitoring threats and the industry's responses to those threats.

NISA was therefore seen as a key factor in the failure to prevent the accident. In August 2011 it was announced that the Japanese government intended to bring NISA under the Environment Agency, which was seen as relatively untainted by close ties with the nuclear industry and therefore better able to regulate in a "hands-off" manner.

A *second technical issue* related to the ability of the Fukushima BWR's to cope with hydrogen production in the reactor core in fault conditions. If the reactors had been able to disperse the hydrogen produced by the overheating fuel safely, the explosions would have been avoided and the release of fission products would have been much reduced.

A *third technical issue* was the inability to top-up from a remote location the water level in the fuel storage ponds.

A *fourth technical issue* was the positioning of the diesels generators – they were at a low elevation in a room that was not watertight.

Three BWRs of almost identical design and vintage to Fukushima Daiichi are operated by the Tennessee Valley Authority (TVA) at Browns Ferry, Alabama. The interesting issue, regarding the second, third and fourth technical issues described above, is that TVA had already implemented improvements to address these issues.

TVA has stated that the Browns Ferry plants already had explosion-resistant pipes to vent hydrogen from the containment, there were already pre-placed fire-hoses in place to add water to the fuel storage ponds, and the diesel-generators were situated in hardened rooms with watertight doors [8].

These design shortfalls at Fukushima Daiichi, when improvements had already been made to similar plants elsewhere, pose very difficult questions for both TEPCO and NISA. Both organizations could reasonably be accused of sleeping on the job. Engineers in TEPCO will have known of the improvements made at Browns Ferry; nuclear operators share this sort of information openly.

OTHER LESSONS LEARNED, CULTURAL ISSUES, AND CONSEQUENCES FOR THE NUCLEAR INDUSTRY ELSEWHERE

A summary of the key "Lessons Learned" taken from the June 2011 interim report is shown in the Fig. 15.6. These have been color-coded according to "prior risk estimation", "accident response", "engineering design" and "offsite response". In addition, the four major technical issues described in the above section are grouped together at

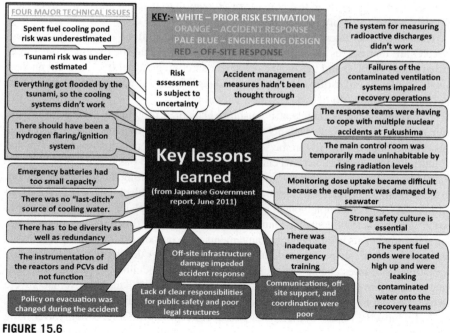

FIGURE 15.6

Key lessons learned from the Fukushima accident.

the top left. The four major technical issues can be thought of as root causes: if any one of these had been addressed, there would either have been no accident, or the consequences would have been significantly improved.

The other items in Fig. 15.6 by-and-large reflect the chaos of trying to manage a nuclear emergency affecting four reactors simultaneously, while at the same time the national infrastructure (especially communications and roads) had suffered massive damage. To put it simply, the emergency response arrangements were predicated upon the situation of "large nuclear accident coincident with failure of national infrastructure" not arising, which seems surprising in a country where seismic events are regular occurrences.

Cultural issues also feature as root causes for the Fukushima accidents. One of the many reports on the accident was published by the Japanese Diet (parliament), which concluded that the accident was entirely "Made in Japan". An underlying cause was apparent unwillingness by TEPCO and the Japanese regulators to take advice from other countries or from international bodies such as the International Atomic Energy Agency (IAEA) and the World Association of Nuclear Operators (WANO) – because advice was seen as criticism, and was therefore considered insulting. Another relevant cultural aspect was traditional Japanese deference to those in senior positions. This manifested itself as unwillingness (or inability) to report bad news upwards through the management chain.

A key issue in the safety management of all hazardous installations is that there must be a willingness at all levels in the organization to listen to and consider advice. Nobody has a monopoly on wisdom.

(However, the international nuclear safety organizations such as IAEA and WANO are themselves not immune to some criticism. Had they pointed out forcefully to both TEPCO and the Japanese nuclear safety regulator NISA that US nuclear plants similar to Fukushima Daiichi had had significant safety upgrades carried out, and TEPCO should do likewise? If not, why not?)

In reaction to the accidents at Fukushima Daiichi, a program of "stress tests" was implemented across nuclear plants in Europe, in order to re-assess the design basis threat conditions considered in the safety justifications for operating reactors. This led to further safety analysis, such as enhanced analysis of flood risks. Some design (hardware) improvements have been identified and, at time of writing, largely implemented. In particular, better resistance against the risk of "station black-out" (i.e., loss of all grid power) has been implemented on many nuclear power plants by enabling portable power generators to be connected into the stations" electrical distribution systems. Also, some improvements in emergency arrangements were introduced for fault conditions affecting more than one reactor at multi-reactor sites. Further development of emergency operating procedures and severe accident management guidelines, to provide clear advice to operators in emergencies, was also carried out. Some plant improvements were also carried out to prevent hydrogen explosions in loss-of-cooling faults.

Politically, the impact of the Fukushima accidents was variable. Notably, Germany announced immediate closure of several nuclear plants, with an intention to shutdown the others by 2023.

RADIOLOGICAL AND HEALTH CONSEQUENCES

A large amount of radioactivity was released from the Fukushima accidents, but a lot of the radioactivity drifted out to sea. The area of seriously contaminated land was much less than for Chernobyl – probably less than one-tenth (Fig. 15.7).

Initially, about half a million people living within 20–30 km of Fukushima were evacuated. These were people who had survived the trauma of the earthquake and tsunami and were now required to leave their homes. The psychological impact of the accident is commonly recognized as creating the most serious health consequences. The Japanese government has announced its intention to try to decontaminate some of the worst affected areas, but this is likely to take many years. A further major effort for temporary re-housing has been necessary in addition to the effort already necessary for people made homeless by the earthquake and tsunami.

Long-term non-psychological health consequences are subject to the same uncertainties as the assessments from the Chernobyl accident, as discussed in "Chernobyl: Indirect Causes of the Accident" above. Any long-term additional cancer mortality from the accident will not be discernible against "normal" cancer mortality.

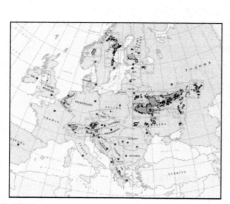
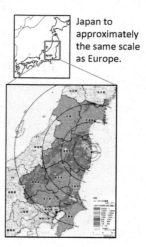

Japan to approximately the same scale as Europe.

FIGURE 15.7

Maps showing caesium-137 contamination from Chernobyl and Fukushima: The bright blue areas in the Japanese map (60–100 kBq/m²) can be compared with the pink areas in the European map (>40 kBq/m²). Hence the area affected by the Fukushima accidents appears to be less than one-tenth of that affected by Chernobyl.

Sources:http://radioactivity.mext.go.jp/ja/1910/2011/09/1910_092917_1.pdf (Recovered September 2011).(MEXT = Japan Ministry of Education, Culture, Sports, Science & Technology).http://rem.jrc.ec.europa. eu/RemWeb/pastprojects/atlasfiles/TEXT/ENGLISH.PDF.Atlas of caesium deposition on Europe after the Chernobyl accident, Office for Official Publications of the European Communities, 1998.

The Fukushima accidents were initiated by, and compounded, a natural disaster, which by itself had caused massive loss of life, damage and hardship. The Great East Japan earthquake and tsunami was the most expensive natural disaster in history, estimated by the World Bank to cost $235 billion.

The comparative magnitudes of the four major reactor accidents to date (Windscale 1957, Three Mile Island 1979, Chernobyl 1986, and Fukushima 2011) are shown in Fig. 15.8. This is a log–log graph where iodine 131 release has been used as a surrogate measure for the total radiological release, or "source term" (which in practice is a mixture of a large number of different isotopes.) The estimation of the quantity of radioactive material released is made from measurements of radioactivity on the ground and in the air. The population health impact is assumed to be proportional to the collective effective dose equivalent (CEDE), i.e., the integrated amount of radiation (measured in Sieverts) for the whole exposed population. This is assessed by means of measured radioactivity uptake as well as doses inferred from calculations.

Both the CEDE and the source term are assessed sometime after the accidents. CEDE has been used in the past to estimate potential cancer mortality in the general population, but this is now considered to be a meaningless figure since any "extra" cancer deaths arising from the accidental exposure are completely subsumed by normal rates of cancer mortality. (The International Commission on Radiological Protection stated in its Publication 103 (2007) that CEDE "is not intended as a tool for

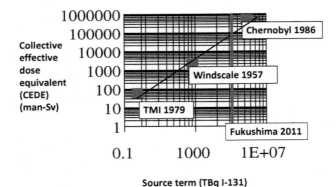

FIGURE 15.8

A log–log graph showing the relative magnitudes of the radiological releases from four major reactor accidents.

Notes:

(1) The above graph uses iodine 131 as a surrogate measure of radiological release. Other isotopes (notably cesium and plutonium) will also have been significant. Iodine 131 is used for simplicity as a common single measure of the magnitude of radioactive release.

(2) CEDE estimates are taken from the relevant recognized "definitive" reports (Kemeny, NRPB, IAEA).

epidemiological risk assessment and it is inappropriate to use it in risk projections".)
Hence, no definitive CEDE for Fukushima has yet been published, although (from
Fig. 15.8) this is likely to be of the order of a few tens of thousands of man-Sieverts.

COMMON THREADS BETWEEN CHERNOBYL AND FUKUSHIMA: NATIONAL CULTURAL ISSUES AND INDEPENDENT NUCLEAR SAFETY REGULATION

If there is one single lesson that can be drawn from the Chernobyl and Fukushima
accidents, it is the absolute requirement that any nuclear power program should have
robust and independent oversight of its safety management arrangements. "Independent oversight" means that the regulator must be independent of the nuclear industry,
and also independent from political interference.

In the case of Chernobyl, the nature of the government in the Soviet Union meant
that independent oversight was not really viable; the requirements of central planning seemed to take precedence. The RBMK design was unsafe by international
standards.

In the case of Fukushima, the Japanese nuclear industry regulator was at the time
also charged with promoting the nuclear industry, which led to conflicts of interest.
The nuclear safety regulator had missed altogether (until too late) that the tsunami
barrier was inadequate. The regulator also failed to notice that important safety improvements to enable hydrogen venting had been made on similar plants elsewhere.
Clearly the operator TEPCO bears the prime responsibility, but a strong independent
regulator would have picked up on these matters earlier. It appears that the regulator
had too close a relationship with TEPCO. The regulator may have been "captured"
by the licensee. "Regulatory capture" is the term used when a regulatory body behaves in the interests of the industry it is regulating, instead of working primarily
for protecting public safety. This can occur when the regulator is highly dependent
on information from the regulated companies, or when the regulator has a symbiotic
relationship with regulated companies (Other [2]).

Finally, the "soft" issue of national culture – and how it relates to safety culture –
seems to be important in both accidents. The Soviet Union was autocratic and centrally planned, and people doing work did not get thanked for challenging their
instructions. In Japan, there is a centuries-old culture of deference to authority, po-
liteness, and a reluctance to criticize publicly, because this may be seen as humiliating. In both countries, therefore (and for different reasons), there was a "bad news
non-return valve", where instructions could be transmitted downwards but it was
difficult for junior staff to transmit their concerns back upwards through the organization. This is anathema to effective safety management – people at all levels must
be able to voice their safety concerns.

National culture, and how it affects people's behavior in safety management, is
an interesting area. Geert Hofstede, a Dutch social psychologist, developed in the
1980s his theory of cultural dimensions which characterizes national cultures in four

dimensions: "power distance", "individualism versus collectivism", "uncertainty avoidance", and "masculinity versus femininity" (Other [1]). Both the Soviet Union and Japan would score high for "power distance"; in such cultures, less powerful members of organizations and institutions accept and expect that power is distributed unequally. Their societies would also score high for "collectivism", in which unquestioning loyalty is a valued behavior trait. One could conclude that, in cultures where there is an acceptance that power resides elsewhere and there is also widespread and unquestioning loyalty, serious safety management errors may occur. Individuals with valid concerns may not want to "rock the boat".

Large multinational corporations have to work hard to achieve a common corporate culture which over-rides national characteristics. As part of that process, it is important to inculcate the correct challenging behaviors required for safety management.

REFERENCES

Chernobyl

[1] The Russian Graphite Moderated Channel Tube Reactor, NPC(R) 1275, Nuclear Power Company Ltd, March 1976.(This report does not appear to be available on-line. It gave an early Western view of the RBMK design and noted some of its major weaknesses.).

[2] Chernobyl, A Technical Appraisal: Proceedings of the seminar organised by the British Nuclear Energy Society held in London on October 3, 1986, BNES, 1987.

[3] Chernobyl's legacy, Health, environmental and socio-economic impacts, IAEA, Vienna, 2005.

[4] Sources and effects of ionizing radiation, Vol. 2, Annex D, Health effects due to radiation from the Chernobyl accident, UNSCEAR, New York, 2008.

[5] J. Thomson, Sample calculations of risk from Chernobyl fall-out in the UK, Nucl Eng 27 (no 5), 1986.

[6] For information on Soviet submarine accidents, see: en.wikipedia.org/wiki/Category:Soviet_submarine_accidents.

[7] The N-1 rocket saga is now well-documented. See for example www.space.com/10763-soviet-moon-rocket-infographic.html and en.wikipedia.org/wiki/N1_.(rocket).

[8] Information on the 1957 Kyshtym accident began to be declassified in about 1990. Wikipedia presents a good balanced account. See en.wikipedia.org/wiki/Kyshtym_disaster.

Fukushima

[1] Report of the Japanese Government to the IAEA Ministerial Conference on Nuclear Safety, June 2011.

[2] Interim report of the investigation committee on the accidents at Fukushima nuclear power stations of the Tokyo Electric Power Company, December 26, 2011.

[3] Fukushima Daiichi accident – technical causal factor analysis, EPRI, Final Report, March 2012.

[4] Review of accident at Tokyo Electric Power Company Incorporated's Fukushima Daiichi nuclear power station and proposed countermeasures (Draft), Japan Nuclear Technology Institute (JANTI), October 2011.

[5] Special report on the nuclear accident at the Fukushima Daiichi nuclear power station, INPO, November 2011.

[6] Health risk assessment from the nuclear accident after the 2011 Great East Japan Earthquake and Tsunami, World Health Organisation (WHO), 2013.

[7] The official report of the Fukushima nuclear accident independent investigation commission (NAIIC), The National Diet of Japan, 2012.

[8] Technical lessons learned from the Fukushima-Daiichi accident and possible corrective actions for the nuclear industry: an initial evaluation, MIT-NSP-TR-025 Rev 1, July 26, 2011.

[9] J. Thomson, Fukushima and its consequences, Nucl Future 8 (2), 2012.

[10] J. Thomson, Emergency planning after Fukushima, Nucl Future 8 (2), 2012.

Others

[1] G. Hofstede, Culture's consequences, SAGE Publications, 1984.

[2] K.S. Choi, Y.E. Lee, H.S. Chang, S.J. Jung, Developments of checklist for self-assessment of regulatory capture in nuclear safety regulation, Trans Korean Nucl Soc, 2011.

Toxic Releases

16

INTRODUCTION: SEVESO, BHOPAL, MISSISSAUGA, SANDOZ

The risk of toxic release accidents into the atmosphere or watercourses is very real, and there have been many examples. Four very notable and well-known examples – Seveso, Bhopal, Mississauga and the Sandoz chemical spill into the Rhine – are briefly outlined below, together with a more detailed description of a less well-known incident.

The range and diversity of toxic materials used in process industries are large, but common toxic and volatile chemicals include, e.g., chlorine, bromine and phosgene. Large release of any of these could present lethal risk many kilometers downwind. The hazard arising from toxic releases is a function of the chemical's toxicity (obviously), the discharge rate (which will affect airborne concentration), the chemical volatility, whether fire or other heat source are present (since this may induce buoyancy and reduce ground-level concentrations), local population density, and local weather at the time of the release.

Taken from a very high-level view, the safety challenges of the chemical industries and the oil and gas industries have some similarities. In particular, the main safety requirement for both is *containment* – i.e., keeping the hazardous materials within their process boundaries. However, whereas refineries are mostly large scale continuous operations, many chemical manufacturing activities can be small-scale operations, sometimes operating in batch mode, with a variety of different activities taking place on one site. Each different process may be technically challenging and each may have unique safety and operational requirements, and hence each small-scale operation may require its own specific safety measures, such as protective equipment and sensor systems.

Also, containment of systems that may operate under pressure means there has to be some form of overpressure relief – and that is often where toxic release accidents arise. When dealing with toxic chemicals, any overpressure relief needs to be channeled via gas scrubbers which can remove the toxic substances before discharge.

The *Seveso* accident at the ICMESA plant north of Milan in Italy on July 10, 1976 – which may have killed no one, although many people were subsequently affected by disfiguring chloracne – remains a very important event (see, e.g., [2]). As a result of this accident, 1800 hectares of land became contaminated with the carcinogen 2,3,7,8-tetrachlorodibenzo-*p*-dioxin (or dioxin for short). This accident led ultimately to the European Union's three "Seveso Directives" which have been aimed at reducing the risk of major accident hazards in Europe. The latest Directive, Seveso III issued in 2012 [3], includes the following:

- It identifies (in Appendix 1) the maximum quantities of dangerous substances that may be stored at hazardous installations.
- Article 5 places the responsibility clearly on the operator: "…the operator is obliged to take all necessary measures to prevent major accidents and to limit their consequences for human health and the environment."
- Article 7 requires the operator to inform the relevant safety regulator (or "competent authority") about the site activities and quantities of hazardous substances involved.
- Article 10 requires the operator to produce a safety report, demonstrating that safety management systems have been implemented, that major accident hazards and scenarios have been identified, and that the design is safe and reliable.
- Article 12 requires the operator to draw up emergency plans.
- Article 13 requires the operator and EU Member States to ensure safe distances between hazardous installations and public areas.
- Article 20 requires hazardous installations to be inspected by the relevant safety regulator.

Similar legislation applies in other countries, for example OSHA regulation 29CFR part 1910 in the United States.

The most appalling industrial accident ever remains the *Bhopal* tragedy at the Union Carbide pesticide plant in Madhya Pradesh, India, on December 2–3, 1984, when water was accidentally allowed to pass into a tank containing 30 tonnes of methyl isocyanate (MIC), which caused exothermic decomposition, thereby blowing a bursting disc and venting the MIC into the atmosphere. The accident happened at night, and the surrounding area was densely populated. Several thousand people were killed and several thousand more suffered blindness and lung damage. The death toll assessments vary – officially the figure is less than 4000 but other estimates suggest the toll may exceed 10,000. Union Carbide is now owned by Dow Chemical Company.

The *Mississauga* accident is an important example of successful emergency planning [2]. An extremely brief account is as follows. Just before midnight on November 10, 1979, a freight train with 106 cars derailed in Mississauga, Ontario, Canada. The rail cars included propane tankers, toluene, caustic soda, styrene and (notably) one car was a tanker containing 90 tonnes highly toxic chlorine. Initially, one propane tanker caught fire and exploded. Literally within minutes, police, firemen and ambulances were present. The train conductor had a manifest with him and consulted with police. Police and firemen thought they smelled chlorine, and the disaster plan was initiated. The first official evacuation involving residents began at 0147 h. Propane tanker explosions continued for many hours, with a third propane car exploding at 0950 h. The evacuation area widened until 250,000 people had been evacuated from an area of 117 square kilometers. The fires burnt themselves out after about 48 h, when it was discovered that all 90 tonnes of chlorine had escaped. No one died in this accident, partly due to the successful evacuation, and also due to the thermal updraft from the fires, which helped dilute and disperse the chlorine. The city of Mississauga was re-opened on November 16.

On November 1, 1986 a chemical warehouse owned by Sandoz (now part of Novartis) in Basle, Switzerland, caught fire. A large amount of various chemicals subsequently entered the Rhine, which caused widespread destruction of wildlife within the river, notably salmon which were subsequently absent until 1997. The pollutants were washed into the river by water used by fire fighters to tackle the fire. Chemicals introduced to the Rhine included some 30 tonnes of pesticides, as well as mercury and agrichemicals. The Rhine flows through Switzerland, Germany, France and Holland.

TOXIC RELEASES AT DUPONT BELLE, WEST VIRGINIA, JANUARY 22–23, 2010

DuPont is a company with a long history dating back to 1802, and the company has been in the forefront of many developments in chemicals manufacturing and also in the management of safety in process industries. They began with the manufacture of gunpowder and other explosives in the nineteenth and early twentieth centuries. Later they were one of the leaders in developments with polymers, with inventions such as nylon, neoprene and Teflon to their credit.

DuPont has had a very good record indeed for operational safety, and their approach to safety has been modeled and copied by others. They are very highly regarded worldwide for process safety management. DuPont actually has a consulting division which sells operational risk management consulting services to other companies. Their website says "DuPont is a world-class safety leader. We have provided workplace safety training to over three million people worldwide, helping our clients meet and exceed regulatory standards."

Prior to January 2010, the DuPont Belle plant in West Virginia, about 8 miles east of Charleston on the Kanawha River, had the best safety record of any DuPont production facility. It was therefore something of a surprise when this plant had three separate incidents within about 33 hours on January 22–23, 2010. The three incidents involved, respectively, the release of methyl chloride, oleum (or fuming sulfuric acid), and phosgene. The last incident (phosgene) caused the death of an operator [1].

The *first (methyl chloride) incident* was discovered at 0502 hours on the morning of January 22. About 1 tonne had been released over the preceding five days, implying an average leakage rate of 2–3 g/s. Methyl chloride is toxic, flammable and volatile. Its LC_{50} is about 1800 mg/kg. (LC_{50} is the concentration that will cause a lethal dose to 50% of those exposed.) Although no one was harmed, this was a serious incident.

The incident occurred at a plant which makes an intermediate used in the production of a herbicide. Fumes from a chemical reactor vessel exit via a vent line to a scrubber, and then a thermal oxidizer used for emission control. (The thermal oxidizer is used to decompose hazardous gases.) To avoid the possibility of overpressure upstream of the thermal oxidizer, a bursting disc was installed. It transpired that the bursting disc had been ruptured during a nitrogen purge activity before reactor start-up, and this went unnoticed for five days until the methyl chloride was detected by chemical sensors.

A bursting disc is a thin membrane blanking a pipe, which will fail at a pre-determined pressure. The bursting disc had a sensor (thin metal wires in the membrane) to advise operators if it had ruptured. This sensor had been in alarm for five days, ever since the nitrogen purge during start-up. This alarm had a history of unreliability so operators had learned to ignore it – it "was in and out of alarm every three minutes." This problem had been experienced from 2005 until the incident. An internal audit in 2007 had commented upon the number of standing alarms.

Root causes were assigned to a failure of the Engineering Change process, called Management of Change or MOC. The MOC process had approved for use an unreliable alarm system for detecting bursting disc failure. A second root cause was failure to resolve the nuisance alarms in a timely manner.

The *second incident*, the least significant of the three incidents, involved oleum or fuming sulfuric acid which is a solution of sulfur trioxide in sulfuric acid. Oleum fumes are toxic and of course corrosive. At the Belle plant, oleum had been used in the production of methacrylic acid which is used for making acrylic polymers.

The incident was discovered at 0745 hours on January 23. About 10 kg of oleum had been released via a corrosion hole in a one-inch (2.54 cm) sample line. The sample line was made from 304L stainless steel which should have been robust against the oleum. It appears an unknown defect, possibly from manufacturing and undetectable with routine non-destructive examination, had allowed corrosion on the pipe to start.

This oleum leak was a minor event – its only real significance was that it happened to occur between two more major events.

The *third event* was the most significant, since it involved phosgene (which was used in World War One as a poison gas) and it led to the death of an operator. The lethal concentration of phosgene for one hour exposure (LC_{50}) is only five parts per million. Death is caused by pulmonary edema – i.e., drowning in one's own fluids – although, as this accident demonstrated, there can be some delay before the onset of symptoms.

The Belle plant used phosgene in its Small Lots Manufacturing (SLM) unit. Phosgene was supplied in 1-tonne cylinders by another company, VanDeMark, from Lockport, New York. Phosgene cylinders were stored in a shed with phosgene sensors installed in and around it. During routine changeovers of phosgene cylinders, the operator would wear only normal Personal Protective Equipment (PPE), i.e., hard hat, safety shoes, safety glasses, and flame-resistant clothing, together with a phosgene detector badge.

The incident occurred at about 1400 hours on January 23. An operator was in the phosgene shed when a stainless steel-braided PTFE hose used for transferring phosgene failed catastrophically. He was sprayed across the face and chest with about 1 kg of phosgene. It was estimated he received a fatal dose within less than one-tenth of a second. He was able to call for assistance using the phone in the phosgene shed.

He was taken to hospital, arriving within 35 minutes, and remained lucid and conscious. An X-ray showed no congestion of his lungs. However, almost 4 hours after the exposure, his condition deteriorated dramatically, and he died about 27 hours later at 2127 on January 24.

The hose that failed was made from PTFE tube surrounded by stainless steel braiding. The stainless steel braiding provides strength for internal pressure resistance, and also wear resistance for external abrasion. However, PTFE is very slightly permeable to phosgene. The failure occurred where a label had been attached to the hose with plastic tape. The tape had trapped some of the small amounts of phosgene that permeated through the PTFE, and the phosgene then decomposed to form hydrochloric acid which had attacked and corroded the stainless steel braid. Additionally, valves at either end of the hose were closed and liquid phosgene had remained trapped within the hose. Thermal expansion of the liquid phosgene then caused over-pressure and the hose failed, spraying and killing the operator.

The operators were supposed to change over the hoses every 30 days. However, the accident investigation showed that hoses had sometimes been left in place for as much as seven months. It was believed the failed hose had been in service for six months. Also, the operators seemed to be unaware why hoses were supposed to be changed every 30 days. This aspect was not communicated effectively.

In common with many operators of hazardous industrial plant worldwide, DuPont used SAP enterprise resource planning software. SAP is a large database system that, amongst other things, prints out work order cards as required for the scheduling of routine maintenance activities, such as the hose exchanges. In 2006, there had been a change to SAP data regarding the hose exchanges, and no more work order cards were issued by SAP for hose exchanges thereafter. DuPont were unable to find who had removed the SAP data on hoses, or why. There was no backup layer of protection such as weekly check-sheets, so thereafter it appears the hoses were only changed when someone saw fit to do it. The change to SAP data should have been recorded and reviewed through the DuPont Engineering Change (or MOC) process but this did not occur. It is possible or even likely that this change happened accidentally – so it was probably a latent error of commission – but there was not any backup system (checklists) to detect the latent error in the period of four years between its occurrence and the accident.

The use of automated work order cards is common throughout all hazardous industries. The Belle plant phosgene accident shows the danger of over-reliance on these systems; if a data error should occur, and an important routine maintenance activity gets accidentally dropped, how do engineers and managers know about it?

The US Chemical Safety and Hazard Investigation Board (the CSB) investigated these events at DuPont Belle and published their findings. Amongst their detailed findings, it was noted that DuPont Belle had had a loss of plant-specific knowledge because, over the period 2005–2009, there had been 85 retirements of people with an average of 30 years of service. There had also been 14 resignations and 14 transfers to other sites. This has led to a phenomenon described by CSB as "corporate memory fade."

There are three main generic learning points from the traumatic 33-hour episode at DuPont Belle in January 2010.

The first point is that a good safety record counts for nothing when things go wrong. The Belle plant was the top-rated DuPont plant for safety, in a company that was regarded worldwide as a leader in safety management – and yet, within 33 hours, they had three significant safety-related incidents, including one fatality. A reputation for safe operation can be ruined by one event, and there is no single easy way to assure safety. Safe operation of hazardous facilities requires thorough management processes and systems of course, but it also requires diligent, conscientious, motivated, knowledgeable personnel at all levels of the organization who worry about the safety of their plant. There is no shortcut on this.

The second point is about the use of SAP for the scheduling of routine activities, including important safety-related activities. The exam question for engineers and managers is this: How do you know that all the routine safety-related maintenance activities that are supposed to be carried out on your plant are actually taking place? How often is there a thorough review of scheduled activities? Also, do your staff know *why* they do all these maintenance activities? (See Fig. 6.2.)

The third point is about the management of aging workforces. Companies need to plan ahead and recognize the threat of "corporate memory fade" that may be posed by a skewed demographic profile in the workforce. This can happen, in particular, at remote sites with stable workforces that have been operating for a few decades – quite suddenly, over a period of a very few years, a large amount of corporate knowledge can vanish as people retire. The ways of mitigating this are not easy but include staggered retirement – offering early retirement to some, and offer delayed retirement with perhaps part-time working to others, combined with intensive recruitment, training and mentoring.

REFERENCES

[1] Investigation report, Dupont Methyl Chloride release, Oleum release and Phosgene release, January 22–23, 2010, Report no. 2010-6-I-WV, US Chemical Safety and Hazard Investigation Board, September 2011.

[2] Emergency Response to Chemical Accidents, World Health Organisation International Programme on Chemical Safety, Copenhagen, 1981. Section 3 Case studies on Seveso and Mississauga.

[3] Official Journal of the European Union, Directive 2012/18/EU of the European Parliament and of the Council of 44, July 2012 on the control of major accident hazards involving dangerous substances, amending and subsequently repealing Council Directive 96/82/EC.

Tragedies of the Commons 17

> *"We are locked into a system of 'fouling our own nest' so long as we behave only as independent, rational, free enterprises".*
> **Garrett Hardin**

Garrett Hardin's article in Science magazine in 1968, "The Tragedy of the Commons", took its title from the age-old problem of common (shared) grazing land: no one owns the land so it becomes over-grazed until it is all muddy and useless [1]. Each herdsman tries to obtain as much utility from the land as he can. He keeps adding one more animal, because if he does not, someone else will. Each herdsman becomes locked into a system, whether he likes it or not, that compels him to increase the number of his animals grazing on the common ground, because it is free, and if he does not use it someone else will take free grazing. In the end, the land becomes unusable by anyone. As Hardin puts it, "Freedom in a commons brings ruin to all."

Examples of Tragedies of the Commons, either real or incipient, are all around us. They range from the trivial to the genuinely tragic. Hardin quotes an example where a mayor decreed that all parking restrictions downtown were removed during the period before Christmas, as a seasonal gesture – and of course chaos ensued. Free parking for all led to roads becoming blocked and traffic jams.

Over-fishing of areas of sea is another example of a Tragedy of the Commons. Each fisherman has sought to maximize his catch by means of bigger nets, or nets with tighter mesh size, or nets that scraped along the ocean floor, or by using sonar, to seek out bigger and bigger catches even as fish stocks have dwindled. The Grand Banks off Newfoundland are the textbook example of this. The cod stocks seemed limitless, but the catch collapsed over a very few years between 1968 and 1974. Despite a moratorium, cod numbers have not since recovered.

An example of an incipient Tragedy of the Commons is the gradual failure of antibiotics to control bacterial infections. We all want access to antibiotics when we need them. Some countries keep control over their prescription, but in other countries people may buy antibiotics over the counter in any pharmacy. Also, some doctors have historically been too willing to prescribe antibiotics to insistent patients with non-bacterial infections. The net effect is that antibiotic-resistant strains of many common infections have developed, and there is now serious concern amongst many medical experts that it is only a matter of time before there is a pandemic of untreatable strains of bacterial infections such as gonorrhea or tuberculosis.

Problems of pollution are also Tragedies of the Commons. For air pollution, the air is the "commons". For each person (or country) the cost of discharging waste to

High Integrity Systems and Safety Management in Hazardous Industries. 978-0-12-801996-2

the atmosphere is less than the costs of purifying the waste before releasing it. And so, as Hardin noted, we become locked into a system of "fouling our own nest" so long as we behave only as independent, rational, free enterprises.

Tragedies of the Commons may also sometimes be versions of a "Prisoner's Dilemma". The classic Prisoner's Dilemma [2] goes like this: Two people are arrested for murder and are held in separate cells, unable to communicate with each other. Each knows if they both stay silent (or "cooperate"), they will only receive short sentences. However, if one should offer to provide evidence against the other (or "defect"), he will walk free and the other will be hanged. The best way forward for their mutual benefit is to cooperate – but they cannot communicate with each other. Does each trust the other sufficiently for both to remain silent?

Over-fishing and antibiotics abuse are both Prisoner's Dilemmas, with many participants. The best way forward is that everyone cooperates (to minimize fishing/ antibiotics use), but how can you be sure no one is defecting? When fishing quotas are set for trawlers, how can fishermen be sure that others are adhering to the limits? Similarly, if one country imposes tough rules on the prescription of antibiotics it will not be of any benefit so long as other countries still allow free access to antibiotics; new, resistant strains of bacteria will not respect national boundaries. To ensure cooperation in Prisoner's Dilemmas, all parties have to be confident that no one is defecting. There has to be enforcement.

Tragedies of the Commons are a different category of major accident from the others considered in this book. In other major accidents, the failures are mechanical failures, or system failures, or management failures, or human failures. By contrast, Tragedies of the Commons happen because of the small contributions of many, and hence they require organized and enforced collective action to prevent their occurrence.

THE GREAT STINK OF 1858

In mid-Victorian times, London still had very primitive sanitation. Fresh water supplies came from well water or direct from the River Thames or its tributaries. Sewage generally went to cesspits which could be emptied if the owners had enough money; otherwise the sewage might just be allowed to overflow. From 1815 it was also allowed just to route sewage into the Thames – the same river that was used for fresh water supplies. In addition, the advent of flush toilets increased the volume of water flowing into cesspits, causing yet more overflow.

Of course, cholera followed, but no one could at first understand what caused the disease – it was thought by many to be transmitted by the foul smell which was called "miasma". However, London physician John Snow demonstrated that there was a connection between cholera and sewage-contaminated water supplies when he traced an outbreak of cholera in Broad Street, Soho, in which 127 people died in August 1854, to a particular source of contaminated water.

In the summer of 1858, the smell from the Thames became so strong that it affected the Houses of Parliament – the year of the "Great Stink". Money and resources were made available and soon afterward Sir Joseph Bazalgette led the construction of two sewers that took sewage downriver and away from the city center.

The Great Stink was another example of a Tragedy of the Commons. In this case the river itself was the "commons". However, in this case, there was no Prisoner's Dilemma: the cause was overpopulation, ignorance and lack of investment.

THE GREAT LONDON SMOG OF 1952

Another Tragedy of the Commons affected London in 1952 when the Great Smog occurred [3]. This event had had a number of notable precursor events: London was long infamous for its "pea-soup" fogs. Charles Dickens wrote of London fogs in *Bleak House*, published in 1852, and Arthur Conan Doyle wrote of bad fogs in his Sherlock Holmes stories. Claude Monet painted the Houses of Parliament in a fog in 1904. The Gershwin brothers wrote "A Foggy Day in London Town" in 1937.

Smog occurs when there is a temperature inversion, i.e., when the air at ground level is colder and denser than the air above. This traps smoke and other pollutants at low level, preventing dispersion. If this happens coincident with fog – when there are moisture particles in the air – the moisture absorbs chemicals and particles from the smoke, and the moisture becomes highly irritant to the lungs.

Especially bad fogs that killed hundreds occurred in London during the winter of 1873–1874, but there was no real outcry of complaint. Some very weak smoke abatement measures were introduced in the 1875 Public Health Act, but there were many loopholes, so the legislation did not fix the problem. For example, no controls were placed on domestic dwellings, and mining and smelting operations were exempted. Even then, the penalties for the remainder were only small fines.

Four more smoke-fog events happened between 1880 and 1892, with each event killing hundreds of people.

The word "smog" was invented in 1905 by London physician Harold Des Voeux, although it did not enter common parlance until nearly half a century later when the Californians began to use it.

A further attempt at a smoke abatement law was abandoned because of the start of the First World War. Then in December 1922 there was further heavy fog, and the consequent deaths from bronchitis and public awareness led to the 1925 Smoke Abatement Act, but it was again a weak piece of legislation. No real improvements occurred.

In early December 1930, severe smog affected a part of the Meuse valley in Belgium that was heavily industrialized. A temperature inversion trapped the fog and smoke for five days. The pollution and smoke mixed with fog, and people coughed, vomited and gasped for breath. Several hundred became seriously ill and some 60 died, and the event produced headlines around the world. Eminent people denied that smog alone could have been responsible – poison gas and plague were suggested as possible causes. The geography of the narrow Meuse valley was cited as a reason why the same sort of thing could not happen in London. In London, meanwhile, nothing changed.

After the Second World War, more official reports were written while still nothing changed. Another severe smog in London in November 1948 killed at least 300.

The Great Smog itself began on Thursday December 4, 1952 and lasted until Tuesday December 9. Impenetrable fog descended over the whole of Greater London, at its worst covering 92 miles of the Thames estuary and shutting down the London Docks. At its worst, the density of the smog seems hard to imagine: people could step out of their houses and literally just *get lost*. Swans were reported far from the river, lost on dry land. At first, traffic jams overwhelmed the city, because drivers just could not see where they were going. Buses stopped running.

The smog was described as yellow, pungent and dirty. It made clothes turn gray. The temperature inversion was such that smoke did not rise from chimneys, it sank. Coal-fired steam trains added to the pollution.

The sky became dark as the sun disappeared into the murk. At its worst, on Sunday December 7, visibility was reduced to only 1m – the distance of an outstretched hand from one's face.

The Great Smog lasted for six days. During the 1930 Meuse smog, deaths had begun on the third day, but illness and deaths in the Great Smog began on only the second day, Friday December 5. Worst affected were the ill and frail, and those with bronchitis. The overall death rate increased by 50%. By Saturday, travel in London became extremely difficult indeed, not least because drivers became lost. Meanwhile, hospitals filled up with people with respiratory problems, and general practitioners were forced to work flat-out.

Off-duty ambulance drivers were called in to help cope with the number of emergency calls. However, ambulances had to have someone walking in front carrying a lighted flare, to provide light and local heat to clear a small hole in the smog. By Sunday, the supply of flares was running out.

Over the weekend, there was no wind, and the anticyclone holding the smog remained motionless. However, anyone taking a short climb to the top of Hampstead Heath rose out of the smog into clear sunshine. Down in the smog, however, almost nothing was moving, there was no wind, and there was hardly any noise. A deathly calm fell over the city. Pedestrians spoke of having been lost for hours: a landmark might be found, but within a few meters they would again be lost.

On Monday, newspapers looking for an angle reported imaginary crime waves (how could the criminals find their way?), but no one seemed to pay much attention to the rising toll of illness and death.

The smog lifted on Tuesday December 9 as the anticyclone moved eastwards.

The Press barely seemed to have noticed the sudden increase in death rate until the Minister of Health, Mr Ian McLeod, made an announcement in the House of Commons that there had been approximately 3000 more deaths than normal during the week ending December 13. However, the government of the day was at first inclined to do nothing. Harold MacMillan (later Prime Minister), who was at that time the Minister of Housing and Local Government, told the House of Commons in late January 1953 that he did not think further legislation was needed. Norman Dodds MP accused the government of apathy.

Gradually, further reports were published about the cost and health implications of the smog, including those from a pressure group called the National Smoke Abatement Society, and in May 1953 the government appointed an independent inquiry chaired by Sir Hugh Beaver. Within a few months Beaver published an interim report (December 1953), which helped keep the pressure up for reforms. Beaver's Interim Report identified that the rise in mortality had equated to an extra 4000 deaths over the duration of the smog and the ensuing days. (It has since been determined that even the figure of 4000 was an under-estimate; figures as high as 12,000 [4] are now thought of as a realistic total because of continued higher than normal mortality for several months after the smog cleared.) Beaver's final report was published in November 1954, recommending the elimination of black smoke from private dwellings, controls on industrial emissions, and an improved supply of smokeless fuels.

Despite these reports, the real desired outcome for the government was still to do nothing – to "wait out" the issue until it became un-newsworthy. However, a Private Member's Bill led by Gerald Nabarro MP was drafted in cooperation with the National Smoke Abatement Society and introduced to the House of Commons in January 1955. Nabarro ensured as much publicity as possible, and many MPs spoke in favor. In early February 1955, the new Minister responsible, Duncan Sandys, capitulated; although Nabarro's Bill was rejected, Sandys committed the Government to produce new statutory legislation. The Clean Air Act eventually became law on July 1956 and has prevented any subsequent recurrences of choking smogs.

The Great Smog remains one of the world's three biggest industrial accidents, by mortality. The other two are the Bhopal accident in Madhya Pradesh, India on December 2–3, 1984 and the Morvi dam failure, in Gujarat, India, on August 11, 1979. Mortality figures for both of these accidents are uncertain, but it is likely that each caused approximately 10,000 deaths.

WORLD POPULATION IN THE TWENTY-FIRST CENTURY

Overpopulation has been a concern since the time of British economist and philosopher Thomas Malthus, who wrote "An Essay on the Principle of Population", first published in 1798. He thought that overpopulation must eventually lead to increased starvation and disease until further growth was not possible. However, he did not foresee the huge improvements in food production and medical knowledge; the human population has increased from about one billion when Malthus first published his essay to seven billion in 2013.

The population growth rate has also accelerated since Malthus' time (Fig. 17.1). It took 123 years for the 1804 population of one billion to reach two billion, in 1927. It reached three billion in 1960, and we passed seven billion in 2012. However, the rate of increase may be falling slightly – whereas it only took 13 years for the last billion of population growth to occur, it may take 15 years for the next billion. Current estimates are that we shall reach eight billion in 2027.

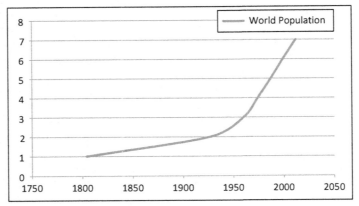

FIGURE 17.1

World population (in billions) from Malthus' time to the present day.

Furthermore, world population is highly focused in certain areas. The data in Table 17.1 (for 2009) show population data for nine countries or regions in the world, which together comprise two-thirds of the world's population. Population density is greatest in two main regions: South and East Asia (notably China, India, Indonesia, Pakistan, and Bangladesh) and Europe.

The issue here is that certain regions are particularly vulnerable to adverse environmental change. Where a region is already densely populated, any climate change, which might affect water or food availability, could become extremely serious. South and East Asia immediately look vulnerable; China, India, Pakistan, Bangladesh and

Table 17.1 The world's most populous regions

	Population (2009)	Percentage of world total	Density (per sq km)
China	1,334,230,000	19.6	141
India	1,172,800,000	17.2	382
Europe	830,400,000	12.2	82
USA	307,973,000	4.5	34
Indonesia	231,369,000	3.4	321
Brazil	192,064,000	2.8	23
Pakistan	167,999,000	2.5	229
Bangladesh	162,221,000	2.4	1034
Nigeria	154,729,000	2.3	175
Subtotal of top 9 regions/countries	4,553,785,000	66.9	
WORLD	6,810,000,000	100	47

Indonesia are all densely populated and together comprise 45% of the world's population. When other southern and eastern Asian countries are included, some 54% of the world's population resides in South and East Asia. Any significant environmental change in that area could pose enormous problems.

Forecasts for further population growth during the twenty-first century have large uncertainties. Highest growth is predicted in Africa, Asia, and South America. North America is predicted to grow more slowly, and Europe's population is expected to stabilize or fall slightly.

Looking ahead to the rest of the twenty-first century, at the lower end of the range of predicted growth (i.e., if education about and availability of birth control is widespread and successful), it is possible that world population might level off below nine billion. If this does not prove to be the case, high-end estimates are that world population may reach eleven billion by 2100.

Eventually, human population must be limited by war, famine, disease or contraception: there are no other options. If the limiting factor becomes "contraception", then we can consider that world population has achieved a "soft landing" – anything else is a "hard landing", a Tragedy of the Commons. Can the Earth sustain about 10 billion people? Almost certainly, the answer is "yes", if nothing else changes. We have the technology for clean water supply, we can sustain people with genetically modified food, and we can arguably produce enough energy. (Some people may not like the solutions, but they are possible.) Hence it is possible to envisage a soft landing where world population reaches a plateau, then perhaps begins to fall gradually...*if nothing else changes*. If that soft landing were to occur, we will have averted a Tragedy of the Commons because, in effect, enough people will have cooperated to contain their drive to reproduce in order to achieve the greater good for all. (Individuals or couples are not necessarily thinking in terms of the "greater good for all", of course. The individual decision to reproduce or not is based on a huge number of other factors including education, employment, and wealth.)

As an aside, it is worth considering whether unbridled population growth, in addition to being a potential Tragedy of the Commons, could also be an example of a Prisoner's Dilemma. This depends on whether, on the one hand, reproduction is just a matter of individual choice or, on the other hand, over-reproduction could be a form of strategy to achieve hegemony by specific cultures or religions. If most people on the planet become convinced of the need to reproduce sensibly, but some specific cultures or religions see over-reproduction as a means to become dominant, those cultures or religions are, in effect, "defectors" in a game of Prisoner's Dilemma, where the outcome might become war, famine or disease for all.

GLOBAL WARMING IN THE TWENTY-FIRST CENTURY

Global warming is the ultimate tragedy of the commons. First, its scale is bigger: we are in danger of changing our environment to such an extent that large parts of the world may become uninhabitable. Second, it is irrevocable: we cannot hope to revert to our previous unpolluted world by some quick agreed international consensus – we all need energy, and even if we did stop creating large quantities of greenhouse gases, it will be an extremely long time indeed before the world recovers to its prior conditions.

A huge amount has been written about global warming. Some are keen to see evidence of global warming in almost every daily weather forecast, but the effects will take many decades to become truly serious. Global warming most definitely requires sober reflection about its long-term effects, and not instant headlines. The next few paragraphs are a brief summary of the scientific consensus on this issue. (I have used the IPCC's 2007 report as prime reference [5]. The 2014 report from the IPCC by-and-large supports their earlier findings, supplemented with later more detailed research.)

Annual global emissions of carbon dioxide from fossil fuel combustion began to rise above their pre-Industrial Revolution levels soon after 1800. The rate of emissions accelerated following the Second World War, increasing about fivefold in the 60 years following. Although emissions from OECD nations have now more-or-less reached a plateau (albeit at a high *per capita* level), high levels of emissions growth are still expected for many years to come in non-OECD countries.

Correspondingly, atmospheric carbon dioxide levels (Fig. 17.2) have risen since the Industrial Revolution began. However, they too have also risen much more steeply since the Second World War.

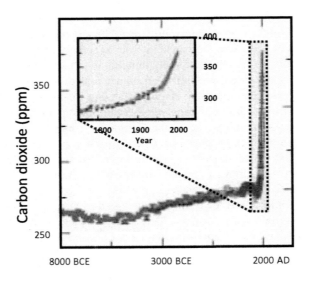

FIGURE 17.2

Atmospheric carbon dioxide levels from the end of the last Ice Age to the present day (Source: Climate Change 2007: Synthesis Report. Contribution of Working Groups I, II and III to the Fourth Assessment Report of the Intergovernmental Panel on Climate Change, Fig. 2.3. IPCC, Geneva, Switzerland.)

The correlation between carbon dioxide emissions and measured atmospheric levels seems obvious and undeniable. The United Nations Intergovernmental Panel on Climate Change (IPCC) published its fourth report in 2007 which reviewed the basis for climate change due to rising carbon dioxide levels and concluded that Global Warming is unequivocally taking place, and this is very likely due to anthropogenic greenhouse gases, in particular carbon dioxide. The data for measured climate change up until 1990 are presented in Fig. 17.3.

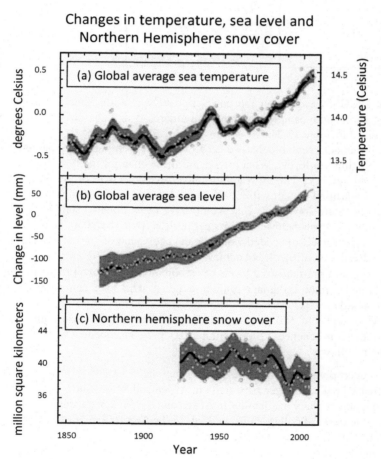

FIGURE 17.3

Measurements providing evidence of global warming (Source: Climate Change 2007: Synthesis Report. Contribution of Working Groups I, II and III to the Fourth Assessment Report of the Intergovernmental Panel on Climate Change, Fig. 1.1, IPCC, Geneva, Switzerland.)

(Note: This is the historical record, not projections. The 2014 IPCC report shows no change from these data.)

Table 17.2 The IPCC correlation between atmospheric CO_2, equilibrium temperature increase, and equilibrium sea level increase.

Range of equivalent atmospheric carbon dioxide concentrations (ppm)	Range of equilibrium increase in global average temperature (°C)	Equilibrium rise in sea level from thermal expansion only (i.e., neglecting any icecap melt) (m)
445–490	1.5–3.5	0.4–1.4
490–535	1.75–4.25	0.5–1.7
535–590	2.0–5.0	0.6–1.9
590–710	2.5–6.25	0.6–2.4
710–855	2.75–7.25	0.8–2.9
855–1130	3.5–8.5	1.0–3.7

The IPCC's views about the correlation between atmospheric carbon dioxide concentration and mean global temperature increase, and also sea level increase, are summarized in Table 17.2.

So, it seems undeniable that we face a future of rising mean global temperatures and also rising sea levels. What effects are these changes likely to have within the twenty-first century? The most authoritative review to date of the potential effects of global warming remains the 2006 Stern Review on the Economics of Climate Change [6]. Stern looked at what would have to be done to stabilize atmospheric carbon dioxide levels below 550 parts per million (ppm) by 2050. To achieve this, he estimated that the effort needed would be the equivalent of 1% of world GDP, which he considered to be difficult but achievable.

Returning to pre-industrial levels of carbon dioxide is not a practical option. The only available option is damage limitation, i.e., to what level can we limit the rise in carbon dioxide levels?

If we accept that 550 ppm is a realistic and achievable maximum level, what then are the consequences? Stern summarizes the consequences at 550 ppm carbon dioxide as follows.

- Falling crop yields in many developing regions
- Increased risk of hunger, especially in Africa and West Asia
- Reduction in water availability in Africa and the Mediterranean
- Possible onset of collapse of part or all of the Amazonian rainforest
- Large-scale species extinction
- Rising intensity of extreme weather events, including hurricane intensity in the USA
- Possible increased methane release and weakening of the Atlantic thermo-haline circulation
- Increased coastal flooding
- Reduction in mountain glaciers, causing more seasonal variation in river flows in India and China

- Possible irreversible damage to coral reefs
- Onset of irreversible melting of the Greenland ice-sheet
- There may be some improvement in crop yields in high-latitude countries

Just for clarity: this is the best outcome we can reasonably hope for if urgent measures are taken worldwide to limit the rise in carbon dioxide levels to 550 ppm. Stabilization (at any level) requires that global annual carbon dioxide emissions be brought down to *more than 80% below current levels* – yet all that has been achieved since the IPCC's fourth report in 2007 and the Stern report in 2006 is that global annual emissions have more-or-less leveled off. Furthermore, the principal reason for that leveling is because of the global financial recession since the 2008 crash, rather than concerted government actions to reduce emissions. Also, any reduction in carbon dioxide emissions by the "advanced" countries has been negated by increases amongst the rapidly industrializing countries such as China and India.

Carbon dioxide levels of 550 ppm are likely to cause a global average temperature increase of about 4°C. An interactive graphic published by the UK Government [7] indicates that local effects in southern and eastern Asia could be worse – an increase of 5 or 6°C, with consequent impact upon water availability and crop production, and coastal flooding.

In Europe, the mean temperature may rise by 5°C. Forest fires will become more common, and water availability and drought may become more problematic. Extreme weather will become more common.

- The Stern Report noted the consequences of an unusual heat wave in Europe during a 3-month period in the summer of 2003, during which there were exceptionally high temperatures, on average 2.3 °C hotter than the long-term average. Previously, a summer as hot as 2003 would be expected to occur once every 1000 years, but climate change has already probably doubled the chance of such a hot summer occurring. By the middle of the twenty-first century, summers as hot as 2003 are likely to be commonplace. Across Europe during the summer of 2003, the premature deaths of around 35,000 people occurred because of the effects of the heat, often through interactions with air pollution. In Paris alone, the excess mortality was about 15,000; the urban heat island effect kept nighttime temperatures high, and reduced people's tolerance for the heat the following day. In France, electricity became scarce because of a lack of river water needed to cool nuclear power plants. Farming, livestock and forestry suffered damages of $15 billion from the combined effects of drought, heat stress and fire.

Africa, currently much less densely populated than Europe or South and East Asia, will become the region of the world with fastest population growth in the twenty-first century. Climate change is expected to cause forest fires, crop yield reduction, and increased water shortages and drought.

The Americas may suffer crop reductions, forest fires, and water shortages.

Northern Siberia and Canada are expected to see the biggest increases in average temperatures, with rises in the range 8–12 °C. This will cause melting of permafrost, converting long-frozen land into soft bog – and releasing methane, which could accelerate global warming further.

All of the above will be happening at the same time as the world's population grows from its current seven billion up to nine, 10 or even 11 billion. It is the synergy of population growth and global warming that is the biggest threat for the twenty-first century; global warming is likely to force large-scale migrations across national boundaries, yet population growth is liable to make other countries even more unwilling, or simply unable, to accept massive migration. The consequences of this are difficult to foresee but frightening. How will Russia react if hundreds of millions are forced to migrate north into Siberia? What will happen to the 1.5 billion in the Indian sub-continent (India, Pakistan and Bangladesh) as their average temperatures increase by 5 °C?

Population growth and global warming, acting together, may yet be the greatest ever Tragedy of the Commons. We need renewable energy, nuclear energy, and carbon sequestration, and we need them as quickly as possible. The real worry is that concerted effort is needed now, yet some nations are still in denial.

REFERENCES

[1] G. Hardin, 'The Tragedy of the Commons, Science 162, 1968, 1243–1248.

[2] W. Poundstone, Prisoner's Dilemma, Anchor Books, 1992.

[3] W. Wise, R. McNally, For a fascinating description of the Great Smog and its consequences, see Killer Smog, 1968, reprinted by backinprint.com, 2001, Wise's account is thoroughly researched but reads like a thriller.

[4] M.L. Bell, D.L. Davis, T. Fletcher, A retrospective assessment of mortality from the London smog episode of 1952 the role of influenza and pollution, Environmental Health Perspectives, 112, 1, 2004, pp. 6–8.

[5] The IPCCs 2007 and 2014 reports on climate change can be downloaded at www.ipcc.ch.

[6] N. Stern, Stern Review on the Economics of Climate Change, HM Treasury, London, 2006.

[7] The impact of a global temperature rise of 4°C, HM Government, retrievable at http://webarchive.nationalarchives.gov.uk/20100623194820/.

Conclusions

18

> *"Hindsight is always twenty–twenty".*
> **Billy Wilder**

Major accidents have a seeming inevitability that can be quite depressing. Just when we think we have learned from past mistakes, we begin to repeat them, or else we discover new mistakes as technology pushes new boundaries. Recent new boundaries include, as we have seen, ultra-deepwater oil exploration, and the widespread use of computer technologies in high-integrity control and protection systems for aircraft and industrial plant. These provide major opportunities and challenges for engineers to ensure that the new technologies are deployed safely.

There has also been a trend to larger and larger plants, driven by economies of scale, which means that major accidents, when they occur, tend to be even bigger and more expensive. Nevertheless, society is dependent on the benefits that these technologies bring, so it is the job of engineers to ensure that the technologies are exploited safely.

This book has presented (in Part 1) a summary of some principal techniques for designing high-integrity instrumented systems to control and protect hazardous industrial plant, such as oil and gas exploration and production facilities, hydrocarbon processing plants, and nuclear power stations. These design techniques and management processes have then been illustrated by examples where mistakes were made.

There have been some remarkable successes in industrial safety. As discussed in Part 2, the threat posed by sudden catastrophic failures of large pressure vessels – which in the nineteenth century were commonplace – has been largely abated (*pace* SS Norway). This has been achieved by rigorous quality controls and in-service inspections and testing. These arrangements were driven originally by the requirements of insurance, and later supported by sound science to understand the causes of failures. The challenge for the twenty-first century is to ensure that we manage safely the new challenges posed by digital control and automation systems.

Part 3 gave an overview of the principal management processes used to ensure the safe operation of hazardous plant, illustrated with examples of safety management failures. Safety policy starts at the top of any company, and the strategic health, safety and environmental (HSE) policy must capture clearly each company's aspirations to be a safe and efficient operator. This begins with the recruitment, training and development of responsible and conscientious operating staff. There is a need for clear responsibilities, a sound safety culture, and an effective and honest approach to safety analysis. There is also a need for realistic simulator training, and for emergency exercises to enable effective crisis management. Active and robust problem reporting and corrective action tracking processes can be good indicators of a healthy safety culture.

To some degree, most accidents are "organizational accidents", since they will reveal underlying weaknesses of some sort in an organization's safety management arrangements. However, the term is usually reserved for accidents where the causes are really deep-rooted failings within the organization.

Most accidents have a human element and a technology element. Safety culture and good safety attitudes have to spread throughout an industry and throughout a company, at all levels. All staff have to be looking out for safety.

There is a need for plain speaking about the safety of hazardous industries, but this is sometimes made difficult by obfuscation, defensiveness, prejudices, and jargon. (This is particularly so where major new projects are being proposed, when both proponents and opponents can be guilty.) This book has tried to explain in straightforward terms the framework within which hazardous plant is designed and operated, in the hope that it will help encourage plain speaking.

A further objective of this book has been to attempt to foster greater understanding and awareness within the engineering community of the challenges faced by the various disciplines of our fellow-engineers. Designers often do not understand the challenges faced by operators or regulators (and vice versa) because each has little experience of the others' working environments. When design engineers get it wrong, it is generally the operators who have to cope with the resulting bad decisions.

Effective safety management requires oversight at multiple levels, from the micro to the macro. There has to be close consideration of the detailed implementation of daily work, but there also has to be oversight at system and plant level to ensure that overall technical and plant risk criteria are satisfied. Similarly, there has to be appropriate supervision at the working level to ensure employee and plant safety during normal day-to-day working, and there also have to be emergency procedures in place so that personnel safety is adequately addressed in potential major accidents.

Good safety management in hazardous industries requires a lot of different aspects to be just right, so that production can take place both efficiently and also in a safe and controlled way. It requires motivated, diligent, conscientious people who know that detail matters; it requires quality engineering; it requires procedures that are thorough and detailed, yet the procedures must not be so complicated that the business becomes unprofitable.

Accident precursors are everywhere, so personnel at all levels must be able to identify and react accordingly to those precursors.

SOME KEY THEMES IN SAFETY INSTRUMENTED SYSTEMS

Table 18.1 is a tabulation of the events, incidents and accidents relating to safety instrumented systems that were presented as case studies in Part 1 of this book. (The table also includes references to the I&C aspects of the Whatcom Park accident and

Table 18.1 I&C system failures case studies, categorized by significant themes

Themes case study	Chapter Ref.	Common-mode failure	Specification or coding failure	Engineering change failure	Regulatory issues	Accident investigation	HMI weakness	Cyber security	Operator error	FMEA failure	Aging I&C
Ariane 5	2	X	X								
Forsmark	2	X		X							
EPR I&C	2				X						
Paperless recorder	2		X					X			
Stuxnet	3							X			
APT1	3							X			
Birgenair 301	4						X				
Aeroperu 603	4					X	X				
Air France 447	4						X		X		
Qantas 72	5	X	X							X	
Uljin	5	X	X	X							
Kashiwaziki-Kariwa	5	X									X
North Sea pipe handling system	5		X								
Auto recalls	5										
Buncefield I&C	6		X	X							
Whatcom	12	X	X	X							
Macondo-*Deepwater Horizon* (BOP)	14 14									X	X

Note: Buncefield, Whatcom and Macondo-Deepwater Horizon also feature in Table 18.2.

Macondo-*Deepwater Horizon*, as described in Part 3.) They are categorized according to a series of key themes, which are discussed further below.

Common-mode failure, affecting redundant systems, can have a variety of causes. Case studies presented here included poor design (Ariane 5), poor plant modeling after an *engineering change* (Forsmark), unauthorized software changes (Uljin), and aging microprocessor-based equipment (Kashiwaziki-Kariwa).

Software system failures were considered in Chapter 5, and also were a significant factor in the Whatcom Park pipeline accident described in Chapter 12. These are an important area of risk as software-based control systems and protection systems have become ubiquitous. The Qantas 72 incident – which by good fortune killed no one – is an excellent example of *inadequate failure modes analysis*. Software systems can become so complex that it is very difficult to foresee all possible failure modes. Qantas 72 is also an excellent example of the challenges posed by using smart devices where the software has been produced by a third party. Finally, Qantas 72 is also an example of the additional difficulties that arise when there is not absolute separation between control and protection systems – as becomes necessary in aircraft systems, although this separation should always be maintained in hazardous process plants.

Software errors described in this book included deliberate acts (the paperless chart recorder), and poorly specified systems (the North Sea pipe handling system that re-set to a dangerous condition). There have also been numerous very expensive automobile recalls for software errors – however, we usually do not know whether or not these software errors led to accidents.

Engineering changes (or plant modifications) always have to be treated with the same degree of care and attention as the original design. Amongst all the many mistakes that led to the Whatcom accident (Chapter 12), the final mistake was an engineering change failure – there was poorly managed software development work on the SCADA system that controlled plant operations. Also, *temporary overrides*, where the potential exists for operators to place plant into a dangerous condition by ill-considered changes, need to be treated with the same degree of seriousness as permanent design changes. The Buncefield accident was at least partly attributable to this.

Safety regulators should be able to rise above day-to-day operational concerns and see the bigger picture. The sorry tale of the EPR I&C architecture – where regulators from three different European Union countries have each insisted on different architectures for a nuclear reactor design that is being marketed globally – does none of the regulators any credit. Different national safety regulators must try harder to talk to each other and agree consensus positions, instead of adopting a "we-know-best" approach.

Accident investigation of complex I&C system failures must try to get advice quickly to other operators. The Aeroperu 603 accident was a tragic example – advice from the ongoing investigations into the Birgenair 301 accident had not been disseminated quickly enough. Poorly designed *human–machine interface* (HMI), and in particular incomprehensible alarm messages, were the main cause of both accidents. This led to a *loss of situational awareness*, compounded by misplaced concerns that the entire digital system was suffering a complex failure.

Air France 447 remains an extraordinary accident. There were undoubtedly HMI and training issues, but the behavior of the co-pilot in the crucial four minutes was nonetheless bizarre. His action of holding the control stick back (despite numerous stall alarms, low indicated airspeed, and falling altitude, until it was too late for recovery) remains difficult to understand. Perhaps he was following his training for recovery from an *approach* to stall at *low* altitude – despite actually being *in* a stall at *high* altitude. *Operator error* was, in this case, the principal cause of the tragedy.

In 2007, the Baker Report into the Texas City accident dwelt on the topic of *process safety*, and how it was important that plant engineers and managers really understand the safety basis of their plant and manage the key safety aspects (and do not just use irrelevant performance indicators like personal injury rates). Another challenge should also now be put to plant engineers and managers, when so much of their hazardous operations are controlled by hidden software processes: Are there engineers on site who *really understand* the software that controls the operation of the hazardous plant, and the full implications of any software changes?

SOME KEY THEMES IN SAFETY MANAGEMENT

Table 18.2 presents a tabulation of the case studies used in Part 3 of this book ("Safety Management"), using themes which are discussed further below. Themes in this book have overlapped and merged, with some important themes being repeated. The following is an attempt to summarize the key themes from the examples presented.

At the highest level, accidents happen because of one or more of the following: Ignorance, stupidity, carelessness, hubris, complacency, and/or organizational neglect. The accidents described in this book all fall within one or more of these categories. However, in terms of knowing how an organization can avoid these characteristics, we need to drill down into more detailed issues and procedures.

A good *safety culture* within a profitable business is the sum of all the parts: good engineering design, good maintenance, well-trained and motivated personnel with questioning attitudes and a healthy concern about safety, good safety management processes, an organization that learns from its own and others' mistakes, good emergency planning, and clear leadership and accountability. Effective safety management in a profitable business requires the right balance to be struck between the following two constraints:

> If a business does not make profit, then before long there will be no business.
> "If you think safety is expensive, try having an accident" (Trevor Kletz).

Ultimately, any appearance of conflict between safety and profit can be resolved by thinking through the implications of *failing* to manage safety properly. The balance between safety, cost, and time pressures in any commercial hazardous installation is always difficult. Careful risk assessment is one part of the answer, a robust questioning safety culture is another, and a challenging safety regulator is a third part. The challenge of *balancing safety with costs and time pressures* has

Table 18.2 Safety management system failure case studies, categorized by key themes

Themes case study	Chapter Ref.	Poor safety culture	Cost or time pressures	Poor design engineering or safety justification	Inadequate independent assessment	Engineering change or temporary modifications	Work control arrangements	Conflicts of interest	Emergency planning and training	Land use planning	Weak regulation	Political or senior management involvement	Corporate memory fade
Buncefield	6	X		X					X	X			
SS Norway	7	X	X	X	X	X	X				X	X	
RAF Nimrod	9	X	X	X	X	X							
Port of Ramsgate	11			X	X								
Saudia 163	11	X							X				
Whatcom	12	X		X		X	X						
Equilon Anacortes	12	X	X	X			X	X					
Piper Alpha	13	X	X				X		X		X		
Mumbai High	13	X	X					X	X				
Texas City	14	X	X	X	X		X	X	X	X	X	X	
Macondo-Deepwater Horizon	14	X	X	X	X			X	X		X		
Chernobyl		X	X	X	X	X	X	X	X		X	X	
Fukushima		X		X	X				X		X		
DuPont Belle	16			X			X						X

Note: Buncefield, Whatcom and Macondo-Deepwater Horizon also feature in Table 18.1.

been a recurrent theme, in particular chapters 9, 12 and 14. For example, Chapter 9 discussed the case of the RAF Nimrod; it may have seemed incongruous at the time to those involved that an aircraft that had seen good, safe service for 30 years required a safety case to be produced at all. It may also have seemed hard for those responsible to imagine that there would need to be an expensive upgrade program when the aircraft only had a few years' service remaining. The cynicism of the contractors in carrying out a deficient safety review was discussed at length by Charles Haddon-Cave QC in his report. Likewise, the Equilon Anacortes accident (Chapter 12) showed the consequences of a hasty management decision, perhaps influenced by time and cost pressures. Texas City (Chapter 14) was at least partially attributable to cuts in maintenance budgets and staffing levels, for reasons of cost.

Hubris and complacency in organizations often go together, with hubris leading to complacency. An organization that runs hazardous processes cannot afford to think it has got a complete understanding and control of the safety aspects of its business; it must always be seeking to "try harder". The hubris that leads to a mind-set of "we know what we are doing" is followed quickly by complacency, and then it is only a matter of time before an accident follows. Several times in this book (notably Chapters 9, 12 and 14) there have been examples where organizations might be accused of hubris and complacency.

Organizations have to retain humility and seek to keep learning. In high-hazard industries, this has to include carefully considered arrangements for allowing employees to report problems without managers "shooting the messenger" (Chapter 10). Appendix 2 describes how one individual perceived strong resistance from his management when he raised a major safety issue; there need to be clear routes for dealing with employee's legitimate safety concerns. One part of a good safety culture is a "culture of encouragement" (as described in another context by the historian Paul Kennedy in his excellent book *Engineers of Victory*, Penguin, 2014).

A "learning organization" also has to review incidents within the organization and any events from elsewhere, to try to keep improving. It also has to keep abreast of technical developments. The development of pressure vessel safety is an historical example where a well-established nineteenth century technology was still subject to rapid and significant safety-related progress throughout most of the twentieth century (Chapter 7).

Organizational neglect or failure is an extreme form of poor safety culture, and more difficult to define, but with hindsight it is obvious. The Equilon pipeline accident in Whatcom Park (Chapter 12) and the BP Texas City refinery explosion (Chapter 14) were cases in point; there were multiple weaknesses in organizations where errors went unchecked, and where few seemed to worry about safety, or even to care. Safety management processes were in place but there seemed to be no local verification or enforcement. In the Texas City accident, an old refinery was suffering from long-term neglect. It was a small part of a very large organization, and it did not receive enough senior management attention and investment. It returned very low Lost Time Accident statistics and these were being used, quite wrongly, as a surrogate measure for overall plant safety – so everything was assumed to be alright. Open reporting of plant problems can be a good indicator of the safety culture of an organization (Chapter 10). At Texas City, there had also been significant *organizational changes*, the implications of which had not been adequately considered.

The worst sort of organizational failure occurred at Chernobyl (Chapter 15) because of *political interference* in the former USSR. In a centrally planned state, with top-down direction and little upwards feedback, an unsafe RBMK reactor design was built. The operators perhaps did not know that weaknesses in the reactor design could lead to its instability so, because of time pressures, they over-rode protection systems and disobeyed operating procedures until the reactor was in a dangerous state. Perhaps some sort of "need to know" secretive attitude meant that the operators did not actually even understand the full extent of the safety problems with the reactor design.

Some form of subtle, low-key and indirect political interference may also have been going on in developments that led to the Nimrod accident (Chapter 9). The RAF Nimrods were near the end of their operational lives when safety reviews were initiated on all aircraft. The translation by civil servants of politically driven financial constraints into detailed work plans meant that contractors were told to do a quick and cheap safety review on the Nimrod, and they were apparently expected not to find anything significantly wrong. As a result, major safety deficiencies were overlooked. One possible conclusion is this: You do not have to live in a dictatorship to feel political interference.

Politics and major accident safety are very difficult bedfellows. Politicians should not try to influence safety-related decision-making in any way but, when they do, strong senior managers have to be prepared to stand up to them. Budget cuts often imply program cuts; any attempt to cut the budget without either delaying the program or reducing the project scope may mean that safety is adversely affected.

An issue related to both political interference and safety culture is national culture. This was discussed with respect to Chernobyl and Fukushima, but it is also relevant elsewhere. Some national cultures hold respect for authority as a very important quality, and in some cases this may contribute to difficulty in the transmission of bad news upwards within the organization – the "bad news non-return valve." Where relevant, multinational organizations have to be aware of these issues.

Design engineering and safety justification are at the heart of all hazardous plant operations. If there are fundamental design weaknesses, then an accident is inevitable. The Port of Ramsgate accident (Chapter 11) and Chernobyl (Chapter 15) were both, in their own very different ways, examples where the causes of the accidents were fundamental design weaknesses. The protection against such accidents is ultimately vested in a combination of design verification, independent design reviews, and independent and robust safety regulation.

Ultimate safety systems are, by definition, only required *in extremis*. When they are required, they really *must* work – that is so obvious it should not have to be said. And yet, at Chernobyl (the reactor shutdown system), *Deepwater Horizon* (the blow out preventer) and Fukushima (the tsunami barrier), the ultimate safety system in each case was woefully inadequate and unable to prevent disaster when called for.

Wherever any design feature or system is identified as an ultimate safeguard, which by definition will only be needed rarely, then those features or systems must be subjected to rigorous and exhaustive design reviews, and must so far as is reasonably practicable be tested in a way that is representative of the hazards that are being protected against.

Experience, training and staff development are themes that were discussed several times, notably in chapters 4 and 11. An effective organization deals with gaps in knowledge of its staff through performance reviews, training programs, and ultimately redeployment for those who are assessed to be in the wrong job. Staff who are incapable of learning are a greater challenge – the best assurance against stupidity is to have careful selection and recruitment processes. The DuPont Belle accidents described in Chapter 16 demonstrated another area of difficulty – *corporate memory fade* – when large numbers of staff simultaneously approach retirement age.

The *suitability of people for operator roles* is a slightly different "take" on the issue of experience and training; no matter how much experience and training, some people are just not suitable to be operators of high-hazard equipment. The role is unusual and difficult. It requires a great deal of knowledge about the plant, a continuous awareness of the current plant state, an attention to detail, a commitment to follow procedures (while being circumspect for the possibility of errors in the procedures), and a tolerance of boredom, while simultaneously being ready to take control of a suddenly-developing emergency situation and make important decisions under pressure. Not everyone is suitable for this role – and yet those who are unsuitable may not be revealed to be so until a crisis arises.

Crisis management requires a cool head. In Chapter 13, we saw how the platform manager failed quickly enough to grasp the seriousness of the situation on Piper Alpha and take the drastic action – to order people to go down to the water – that was necessary to avoid everyone dying of smoke inhalation. Similarly, in Chapter 11, the captain failed to respond with enough urgency to the fire on board Saudia 163. Modern highly accurate and realistic simulators, and realistic emergency exercises, can now go a long way to help prepare operators for crises such as these.

At the other extreme, operators must resist urges for *impulsive action* when faced with a problem. The immediate cause of the Piper Alpha accident was precipitate, impulsive action to restore to service a pump that was under maintenance. Similarly, the Mumbai High accident occurred because of ill-considered action to help a crew member on the MSV *Samundra Suraksha* who had cut his fingers. "Group think" can also sometimes lead to bad, impulsive decision-making, such as the Equilon Anacortes accident. Also, the Chernobyl accident was partly attributable to operators who impulsively sought to bend the rules in order to complete a test. (A further, personal example of an ill-considered rush to action is described in Appendix 1.)

Hence, operations supervisors or managers have to have sound judgment. They must know when to step in, take control, and issue clear instructions to take action in an emergency – but equally they must know when circumstances are sufficiently abnormal that they have to hold back their staff (and themselves) and to consult others or otherwise consider actions carefully, before taking action. Operations supervisors or managers need to judge when situations require urgent action, yet also to spot when situations definitely do *not* call for urgent action. The selection and training of operations supervisors should aim to identify and develop this ability.

The training and development of *senior management* (either on the operational site or in head offices) need to ensure they have a good awareness and understanding of operational issues. Undoubtedly the preference in most organizations is to "grow their own management", i.e., to develop their staff so that they can, in time, take on senior management roles. However, this can lead to "in-breeding" where senior managers all have a similar view of how things should be done. Hence, it is usually considered a good thing in large organizations to have a minority of senior managers recruited from outside the business. This ensures that suitable challenges are made to the organization when the explanation offered for doing something wrong is, "But we've always done it that way". People recruited from outside an organization can sometimes see that arrangements are inadequate, where "home-grown" managers may not notice.

A further problem in global organizations is that it can become difficult to develop senior managers with a wide enough experience of the organization; one approach adopted to help solve this problem is, having selected suitable individuals for their future senior management potential, to give those individuals a series of short-term middle or senior appointments at different locations around the globe. While this approach will undoubtedly develop their management skills, it may have adverse effects on the locations where these people are posted. Short-term senior managers, running BP Texas City as "development roles", were probably a factor in its problems (Chapter 14).

The importance of *safety audits*, *independent safety assessment*, and *independent safety regulation* has been mentioned many times in this book. It is worth recapping on the differences between these three activities:

Safety audits are audits of particular safety management arrangements at (normally) a specific plant or location, against internal (company-specific) or external (national or international) standards. There will normally be an internal sponsor of the audit (i.e., someone who is an internal advocate of the need for the audit), although funding may be internal or external, depending on circumstances.

Independent assessment is normally a document review exercise to review design engineering or management aspects. As its name implies, the assessor will not in any way be associated with the people or organization being assessed.

Safety regulators are normally government employees (or contractors working for the government) whose job is to ensure that government regulations are being satisfied, and to initiate enforcement action if deficiencies are found. Operators always have somewhat awkward relationships with safety regulators (no matter what they might say!), because the regulators always demand improvement, even when reviewing already well-developed arrangements. Regulators tell the operators they are not doing enough about safety when operators may think they are already doing too much. A tension between the operators and the regulators is an essential part of their working relationship; if the relationship becomes too cosy, the regulators are not doing their job properly (so-called "regulatory capture"). The regulator also has an important role in ensuring that operators are challenged to keep up-to-date in areas of rapidly developing technology.

Safety regulators are unequivocally essential to maintain the balance between safe operation and profit. Without a robust regulator, there is always a risk that the short-term demands of shareholders and/or senior management may force inappropriate cuts in operational budgets. We can all feel safer because of safety regulators – no matter (within reason) how much they annoy the operators! Indeed, much of the benefits of safety regulation come from the regulators' low-key advocacy and awareness-building of the importance of good safety management practices, both within industry and also in the news media. The key challenge for a good regulator is always to show good judgment in choosing when to employ their full regulatory legal powers (the regulatory "big stick"), and when to work more in a "coaching" mode.

Partial blame for several accidents described in this book have afterward been ascribed to a safety regulator that was not sufficiently engaged or independent, or where the safety regulator has had a conflict of interest – typically, promoting an industry while also regulating its safety. Such major accidents include Macondo-*Deepwater Horizon*, Chernobyl, and Fukushima. In those three accidents, three different countries had each to learn, the hard way, the same lesson about why safety regulators have to be strong and independent.

Similarly, *conflicts of interest* can occur when the same individual is responsible for both safety decisions and production decisions. This seems to have occurred at Macondo-*Deepwater Horizon* (Chapter 14) and also at Chernobyl (Chapter 15), where individuals were put into (or else they assumed) the position of judging whether a dangerous short cut could be taken. Risks were then taken to try to minimize losses, with devastating results.

Safety management processes have been a consistent theme throughout much of this book. The importance of good safety management processes cannot be overstated – good processes are the bedrock of safety management. Chapter 10 was devoted to an overview of some of the more important safety management processes required in the operation of hazardous plants. Also, the DuPont Belle phosgene accident (Chapter 16) demonstrated the risks from over-dependence on enterprise management software systems for scheduling safety-related maintenance activities – how can the integrity of the databases be assured? How do engineers and managers know that all the intended maintenance activities are actually being carried out? How would they find out if an item just "disappeared" from the database?

The need for good, well-rehearsed *emergency planning* arrangements was also demonstrated in the Piper Alpha accident (Chapter 13) and, more recently, the Fukushima accident has demonstrated that the scope of emergency planning sometimes needs to include consideration of the breakdown of civil infrastructure (Chapter 15).

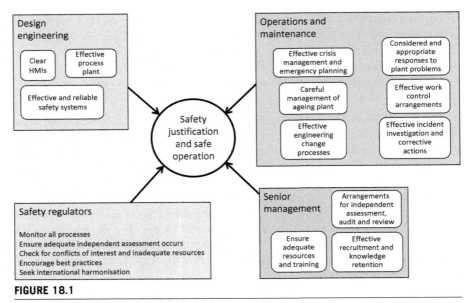

FIGURE 18.1

Responsibilities for safe operation.

One theme that has been touched on is the importance of *land use planning* around hazardous facilities to ensure that residential areas, schools and hospitals are not built next to existing hazardous plants. For example, this has been a significant issue in the Buncefield accident (Chapter 6).

Fig. 18.1 attempts to summarize all the sections in one picture. This presents a grossly simplified model for the division of responsibilities between design engineers, operations and maintenance, senior management, and safety regulators – the reality is, inevitably, more complicated – but perhaps it clarifies where the major divisions should lie.

Finally, Chapter 17 addressed a completely different category of accident – Tragedies of the Commons. In these accidents, everyone and no one is to blame, and the only hope for preventing them is that political leaders can act in good time. It is beginning to look unlikely that world leaders will take strong enough action, quickly enough, to stop the major global impact of the combined effects of overpopulation and global warming during the twenty-first century. I worry for future generations.

FINAL WORDS

Managing the safety of hazardous industrial processes is multi-faceted. The management of such enterprises requires: understanding of engineering design, maintenance, and operations; robust procedures; a sound safety culture (in all its many subtle aspects); well-trained and experienced staff; adequate financial resources; an engaged and competent supply chain; good arrangements for independent review, audit and regulation; and, when things go wrong, well-rehearsed and realistic emergency planning.

Case studies given in this book have mostly been of accidents in the last 20 years, with a few exceptions such as Saudia 163, Chernobyl and Piper Alpha. There are many other notable accidents from the 1970s and 1980s in particular, each with their own lessons, including:

- The Flixborough explosion (UK, 1974)
- Los Alfaques campsite fire (Spain, 1978)
- The Three Mile Island reactor accident (Pennsylvania, 1979)
- *Alexander Kielland* rig capsize (Norway, 1980)
- The San Juanico gas explosions (Mexico City, 1984)

Another whole area I left out was ammonium nitrate explosions. This hazard was one that we seemed to have learned about in the twentieth century, with major accidents at Oppau, Germany in 1923, and in Texas City in 1947. However, there have been at least two major ammonium nitrate explosions causing multiple fatalities in the twenty-first century: the Toulouse AZF explosion (France, 2001) and the West explosion (Texas, 2013). We seem to keep forgetting about the hazardous nature of ammonium nitrate, perhaps because it is such a commonplace fertilizer.

Lastly, one other accident that might have been included was the sinking of Petrobras P-36, a semi-submersible platform, off Brazil in 2001. P-36 was a relatively new platform that had been built at a cost of 350 million dollars. At the time of the accident, it was the world's biggest semi-submersible oil platform. Explosions caused by overpressure and then leakage of hydrocarbon vapor caused flooding, and the platform eventually sank five days later. This accident was an extreme example of the dangers of corporate hubris. Prior to the accident, an un-named executive in Petrobras had allegedly given an infamous speech on quality assurance. This was later widely distributed as a slide presentation that superimposed his words against images of the sinking P-36. The un-named executive said that the company had "...established new global benchmarks for the generation of exceptional shareholder wealth through an aggressive and innovative program of cost cutting on its P-36 production facility. Conventional constraints have been successfully challenged and replaced with new paradigms appropriate to the globalized corporate marketplace. The project successfully rejected the established constricting and negative influences of prescriptive engineering, onerous quality requirements, and outdated concepts of inspection and client control. Elimination of these unnecessary straitjackets has empowered the project's suppliers and contractors to propose highly economical solutions, with the win-win bonus of enhanced profitability margins for themselves. The P-36 platform shows the shape of things to come in the unregulated global market economy of the twenty-first century." Eleven people died in the accident, and the platform was lost in 4000 feet deepwater.

Those who cannot remember the past are condemned to repeat it.

FURTHER READING

I could present a large list of books that have been written about safety, accidents and their causes, and the technical assessment of hazardous plant, but such lists are available elsewhere (notably an extremely wide-ranging list in Volume 3 of Frank Lees' authoritative work on technical safety, listed below). Accident and other reports mentioned earlier in the text are not re-stated. This list is therefore a highly personal selection of a few particularly useful and relevant books and reports, which have not been previously mentioned.

TECHNICAL SAFETY ASSESSMENT AND ACCIDENT CASE STUDIES

FP Lees, Butterworth-Heinemann, Loss prevention in the process industries, second ed., vols. 1–3, 1996

J. Spouge, A guide to quantitative risk assessment for offshore installations, CMPT, 1999

JR Thomson, Engineering safety assessment, Longman, 1987 (available for download at www.safetyinengineering.com)

W Wong, How did that happen? Engineering safety and reliability, Professional Engineering Publishing, 2002

T. Kletz, What went wrong: case histories of process plant disasters and how they could have been avoided, Butterworth-Heinemann, 2009

VC Marshall, How lethal are explosions and toxic escapes? The Chemical Engineer, August 1977, pp. 573–577.

C. Matthews, A practical guide to engineering failure investigation, Professional Engineering Publishing, 1998

PL Clemens, Human factors and operator errors, Jacobs Sverdrup, 2002

H. Petroski, To engineer is human: the role of failure in successful design, Vintage, 1992

V. Bignell, J. Fortune, Understanding systems failures, Manchester University Press, 1984

J. Reason, Managing the risks of organizational accidents, Ashgate, 1997

N. Storey, Safety-critical computer systems, Prentice Hall, 1996

Guidelines for safe and reliable instrumented protective systems, Centre for Chemical Protective Safety, Wiley-Interscience, 2007

OTHER BOOKS AND REPORTS

H. Devold, Oil and gas production handbook, ABB, 2006

Population: One planet, too many people? IMechE, 2011

N. N. Taleb The Black Swan, Penguin, 2007

DJ Bennet, JR Thomson, Elements of nuclear power, third ed., Longman, 1989 (available for download at www.safetyinengineering.com)

REGULATORY GUIDES AND STANDARDS

HSG 65, Successful health and safety management, HSE books, 1997

Reducing Risks, Protecting People HSE Books, 2001

The tolerability of risk from nuclear power stations, HMSO, 1992

Guidelines for the development and application of Health, Safety and Environmental Management Systems, report 6.36/210, E&P Forum, 1994

IEC 61508, Functional safety of electrical/electronic/programable electronic safety-related systems, second ed., 2010

IEC 61882, Hazard and operability studies (HAZOP studies) Application Guide, 2001

Appendix 1
Experience and Judgment

All accident descriptions and reports risk over-simplifying the actual circumstances. The accident case studies in this book sometimes do not do full justice to the complexity of the actual events. This Appendix presents a first-hand detailed account of a minor event in 1982 when I accidentally caused a nuclear power station to shutdown. Even this simple incident had complex root causes.

There is always a danger in discussing major accidents that one can sound like a "Monday morning quarterback" – someone who is very good at saying what others should have done during times of difficulty. For this reason, I am always a little suspicious of those who write about safety and accidents but have never actually worked in front-line operational roles. This is not to say their views are invalid, but rather to suggest that perhaps they may not completely understand the full picture.

Plant operators work under pressures and constraints that most of us do not face. Shift work can be exhausting and difficult for family life. At 3 am, there is no one else to ask for advice when something goes wrong, so one either uses one's initiative and works through problems as they emerge, or one does nothing for the remainder of the nightshift. The latter option will be unlikely to get much praise from day staff when they arrive at about 8 am – so one generally uses one's initiative and works through the problem.

At times in this book, I may appear to have been too sympathetic to the operators – for example, Colin Seaton on Piper Alpha, or Captain Khowhyter on Saudia 163. My reason for being sympathetic is that I have been a shift manager of a nuclear power station, and I know how easy it is to make errors of judgment when things go wrong.

When I was a 26-year-old assistant shift manager in 1982, I caused a nuclear power station to shutdown. It was my mistake. It was a routine reactor "trip"; no one was harmed. However, the reactor trip was my fault alone, because I acted impulsively. I thought I knew what I was doing, and I was keen to prove myself to others.

"What's happened? You look as if you've aged 10 years". These were the words of my wife when I arrived home after work on Friday 9 March 1982.

"I tripped the reactor by mistake". The day had been one that I have never forgotten – a day that was humbling and humiliating, a day that I wished I could begin all over again so that, next time, I could get it right.

It was also a day when I think I learned more about how accidents actually happen than many people will learn in a lifetime. That may sound a bit arrogant but it is just a statement of reality – few people experience the sort of things that happened to me that day. Things had gone wrong, though no one was hurt, but it was my fault. Hence it is not a statement of arrogance but of self-knowledge and humility. I had screwed up in a way that few people get the opportunity to experience.

I had completed my PhD in process engineering in 1979. I decided that the Fast Reactor program looked interesting. Oil prices were sky-high because of the Iranian revolution. The Three Mile Island nuclear accident in Pennsylvania had just happened and nuclear power looked challenging. The UK Government was talking about a major nuclear build program. Above all, fast reactors offered the possibility of energy for centuries, at a time when oil was expected to run out within decades and coal miners had already brought down one government. We had an energy crisis and maybe fast reactors were the answer.

Caithness has a stark and unique beauty if seen on its rare sunny days. For the last 20 miles to Thurso, the road goes across the Cassie Myre, which means "the Causeway across the Bog", in an area which is known as the Flow Country. Caithness, in contrast to the rest of northern Scotland, is mostly flat and the Flow Country is the largest expanse of blanket bog in Europe. The Cassie Myre road itself is surprisingly quiet – there are few cars or buildings for many, many miles. There are no trees, just mile after mile of peat bog.

Eventually, the road comes alongside the River Thurso and a few miles later you enter the town of Thurso, which somewhat has the feel of a frontier town. When asked "What is Thurso like?" by people who have never been there, I usually say "Well, its a 7-h seven hour round-trip to the nearest Marks and Spencer's, which is in Inverness". For city dwellers, this helps put it in perspective. If you are living in Caithness, despite being part of mainland Britain, you might as well be living on a separate island.

In Caithness, the winters are long and dark and the wind howls for what seems like months without end, while the rain crosses horizontally. By contrast, the summers can be sublime, if very brief and unpredictable. Thurso is, at 59 degrees north, the northernmost town on the British mainland. In midsummer, although the sun does set briefly, it doesn't get dark. One of the golf clubs even holds a midsummer competition which tees off at midnight.

Fast reactors are cooled by liquid sodium and fueled by plutonium and uranium. These facts alone made Dounreay attractive to me: the engineering challenge was wonderful. Also, fast reactors are in principle able to extract hundreds of times more energy from their fuel than other types of reactors so, with the various oil crises of the 1970s just behind us, they seemed like a good idea.

Work at Dounreay site had begun in 1955. By 1979, the site was huge and employed some 2500 people. The site was divided between the fuel cycle area where used fuel was reprocessed, and the Prototype Fast Reactor (or PFR) itself which could generate enough electricity for a city. There was also a smaller fast reactor, the original Dounreay Fast Reactor (with its iconic "golf ball" containment building) that had shutdown in 1977.

The economy of Caithness was completely dominated by Dounreay, with 2500 employees out of a total population of about 20,000.

A Ministry of Defence site next door (and completely separate) housed a prototype reactor for the Royal Navy submarines, and seemed to be very secretive. Also, a couple of miles further down the road, there was an ultra-long wave US Navy communications station with a huge aerial several hundred feet high which (so it was said) could send messages to submarines even when they were submerged – but no one was really sure. In 1979, the Cold War had been going for thirty years, and lots of things were still kept secret.

I was initially appointed as a junior engineer in the operations department, looking after final preparation of new fuel before loading to the reactor. A first impression was that all the staff seemed old. When I started work at Dounreay I was 24, and most of the rest of the staff at PFR seemed to be over 40. PFR had been operating since 1974, and its construction had begun in the late 1960s. In addition, many of the staff had been transferred to PFR when the old Dounreay Fast Reactor had closed. Hence most of the staff had been at Dounreay for a long time and there had not been many recent new recruits.

There seemed to be a significant division between most engineering staff and the very senior management of PFR. These were a few people – only two or three – who seemed to have been parachuted in from Head Office at Harwell in Oxfordshire, and could best be described as Gentlemen Physicists. They generally came from an impeccable Oxbridge academic background with some achievement in theoretical physics behind them. They really liked talking about academic aspects of nuclear reactors, but always looked distinctly uncomfortable when the conversation became more practical. The problem here was that PFR was primarily a working power station with all the problems that operating power stations have, and no amount of excellence in theoretical physics was going to help a person understand about turbine bearing problems or seaweed blocking the seawater pump-house. Thus there was a divide between the Gentlemen Physicists (who were nominally in charge, but whose main role seemed to be to explain to Head Office what was going on) and the plant engineers (whose role was to try to make the power station work).

In 1979, the culture in an engineering activity like PFR was still extremely heavily male-dominated. Female technical staff were the exception. Out of perhaps 100 engineers and scientists at PFR, I recall only two women. Also, at that time, every face was white.

Another early impression was the pernicious effect of secrecy. Everything that happened at Dounreay was covered by the Official Secrets Act, and all documents were marked "Confidential". This seemed very strange since this was a civilian project with no military significance. The only information that could possibly be of interest to the Russians might conceivably be the size of the plutonium stocks.

Another early impression of the local culture was the engineering culture at PFR itself. The atmosphere could only be described as neurotic, high blood pressure, and high-energy. There was a daily morning meeting at 0900 which was chaired by the Operations Manager at times when the station was generating, or by the Maintenance Manager at times when it was shutdown. The Gentlemen Physicists did not attend the morning meetings, which were strictly for the plant engineers. These meetings were often bad-tempered affairs with lots of shouting and swearing.

The reason for this collective neurotic and high-pressure behavior was that the PFR power station simply *was not working very well*. The nuclear reactor worked brilliantly, achieving very good availability, but it was regularly let down by the boilers and the turbine plant, the "conventional" plant that should have been the easy part.

There were three underlying reasons for this.

The first reason was that this Government-funded project had cut corners to save money – power stations depend on everything working, so any normal power station would have two Condensate Extraction Pumps to pump water out of the turbine condenser, two Heater Drains Pumps to pump water through the feed heaters, and two Main Boiler Feed Pumps to drive water into the boilers. Each of the pumps could do the job by itself, but a redundant spare would normally be provided to ensure overall plant reliability. However, at PFR we only had one of each. This will have saved a lot of money during construction, but it had been a false economy, since one failure meant the whole power station stopped making electricity.

The second underlying reason was that the seawater pump-house, where seawater was pumped up to the turbine condenser for cooling the steam before returning liquid water to the boilers, had been built down at the seashore in what the locals called a "goe", a gulley which was also a natural growth and collection point for seaweed. During spring tides, when the tidal range is biggest as the moon and the sun are in alignment, low tides are at their lowest. At these times, the pumps in the pump-house could suck in massive rafts of seaweed, many tens of tonnes, which would block the pump-house completely and force a shutdown of the power station. This must have happened, on average, about twice per year.

The third and final underlying reason was that there was a flaw in the design of the boilers, or "evaporators", which were unique to PFR – and therefore every bit as experimental in their design as the reactor itself was. The evaporators were heated by liquid sodium which was pumped by the Secondary Sodium Pumps around narrow diameter steel boiler tubes containing boiling water at very high pressure. The boiler tubes were "U" shaped and the water was pumped through them by the boiler circulating pumps. The boiler tubes were welded onto a thick steel tube plate, which was above the liquid sodium, and the space between the liquid sodium and the tube plate was filled with inert argon gas. The problem with this design was that the welds between the boiler tubes and the tube plate had not been adequately heat-treated, which made them susceptible to cracking, and retrospective heat-treatment in situ was not possible. Of course, if the weld cracked, high-pressure steam entered the argon gas above the liquid sodium – and, from school chemistry, sodium and water react very vigorously indeed.

This third reason, then, became the major Achilles' heel of PFR. Steam leaks into the gas space above the sodium happened repeatedly over many years, more or less from the start of power operation in 1975 until a major project was instigated to fix the problem finally in the mid-1980s. Steam leaks could be detected at a very early stage by special instrumentation which detected hydrogen in the argon gas space above the sodium, so the reactor could be shutdown before serious damage was done, or before the emergency sodium and steam dumping safety systems were actuated.

Once shutdown, the boilers would be inspected to find out which tube (of hundreds) had leaked, and the leaking tube would be plugged.

The three underlying problems described above caused PFR, while I worked there, to be extremely unreliable. The problems also led to highly strung engineers having many excitable morning meetings where strong swear words were freely exchanged.

The management of the power station – plant engineers and Gentlemen Physicists alike – became completely pre-occupied with the huge technical challenges facing PFR. I suppose pressure from Head Office to achieve reliable generation must have been intense. Threats from Government, where budgets were being cut across all activities, must have been very real, so the Gentlemen Physicists will have been under no doubt what was expected. Power station performance is measured in terms of load factor, which is the ratio of achieved electricity output to the maximum potential output at 100% generation. PFR should have been achieving load factors of at least 60%, but in practice it was achieving less than 20%.

In any large organization, there are many issues other than technical ones that top management must keep an eye on. One is succession management. There always have to be young people being trained and available to take over if older people resign or become ill.

PFR was run 24/7 by the shift teams. There were six shift teams; four were normally on shift rota, with two teams doing "three months' days", i.e. a spell of three months doing training and otherwise helping out on day shift, while also being available for holiday or sickness cover for those on shift.

A military approach had been adopted, probably because of the Government-led nature of the enterprise. Each shift comprised a number of "general workers" (unskilled labor, mostly locally recruited), a number of technicians for operations activities and maintenance (although most maintenance was done during daytime), two maintenance foremen, two operations foremen, a "desk operator" (who sat in the control room and carried out plant maneuvres), two Assistant Shift Managers, and a Shift Manager. Of these, only the Shift Manager was a graduate professional engineer. The Assistant Shift Managers would normally be qualified to National Certificate level, and would have been "promoted from the ranks", i.e. they were ex-foremen. The desk operator was at that time from a non-technical background – quite frankly, I never understood how they were selected and trained; later on, after I left, this was changed into an engineering role.

Holiday and sickness cover was from the shift teams on "three months' days" There was no procedure for temporarily upgrading from Assistant Shift Manager to a Shift Manager. The military-style hierarchy was absolute; the Shift Manager was an officer, but the Assistant Shift Manager was only an NCO.

In late 1980 or thereabouts, two of the shift managers (out of six) resigned to take up jobs outside the UKAEA, and no new shift managers were in training. This must actually have led to a crisis, although I was not immediately aware of it. Until new shift managers could be trained, this meant the remaining four had to work shifts continuously (without getting periods of three months' days) and, furthermore, if one of them wanted to take a holiday it necessitated the others working 12-hour shifts to cover for him.

Normally at nuclear power stations, a shift manager's period of training and development will last several years. Suddenly, in early 1981, PFR had to find new shift managers as soon as possible.

My immediate boss said I should apply. He said senior management must be in a bit of a panic, they had dropped the ball on succession management. I had only been at PFR some 18 months, and I was still not quite 26 years old. I checked out the existing shift managers; their average age was 42.

Suddenly the opportunity existed for a fast track to a job with kudos and responsibility, while earning good money. Members of the six shift management teams (i.e. the shift managers, assistant shift managers and foremen) received very respectable shift allowances, some 50% above regular salaries. For that money, they were expected to work a 4-week shift cycle consisting of seven night shifts, followed by two days off (which were really required to catch up on sleep and re-adjust your body clock), then seven back shifts (4pm till midnight), two days off, then six day shifts, followed by a long weekend of four days off. In other words, one weekend off in every four – a complete change of lifestyle. And every six months, if all the six shift teams were filled with their full complement, each person would have a period of three months of days doing training and special project work at the direction of the Operations Manager, in between covering as required for other shift staff who were sick or wanted to take holidays.

In short, when you began shift work, you sold your soul. You committed to a monthly cycle of antisocial hours with poor sleep. You also entered the "shift trap" – the money was so good that it was very difficult to stop doing it.

Six of the younger engineers (which in practice meant every engineer or physicist at PFR under 30 years old) applied for the training posts. An interview panel was arranged and within 2 weeks we were all invited to interview.

When the day of the interview arrived, I was quietly confident.

The interview panel consisted of one of the Gentleman Physicists and one of the senior plant engineers, with another person from Human Resources to ensure fair play, and someone from Head Office who did not say much. The Gentleman Physicist chairing the meeting was a guy called Tony Judd, a lovely guy whom I subsequently held in highest regard. He was an Oxford graduate and had published a lot of theoretical studies on possible reactor accidents. The senior plant engineer was Ed Adam. Ed was very tall, powerfully built, with black hair and thick glasses. He had been a champion swimmer in his youth. Ed had an ability to stare unblinkingly eyeball-to-eyeball during conversation which I always found a little disconcerting. Also, when sitting down, his knee twitched continually – it bounced up and down maybe a hundred times a minute. He was an incredibly hard worker who seemed to almost live at the power station. He also seemed unflappable; his manner was always the same, no matter what problems arose. I never heard him raise his voice, which was abnormal for PFR.

It was made clear that the objective of the interview was not to test the extent of my knowledge, which they accepted would be incomplete at this stage, but to test my suitability to eventually become a shift manager. Ed made it clear that we were talking of a timescale of perhaps two years, during which the successful candidates (probably two out of the six) would be put through accelerated training and would be expected to be present at all major plant activities such as start-ups and shutdowns. They would then work for a while as Assistant Shift Managers, before further assessment and eventual promotion to Shift Manager.

I remember the key moment, the crux, of the interview. Tony Judd asked me, "There is a low sodium level reactor trip on the primary sodium pump intakes. Why do you think we have that feature?"

This was referring to instruments which measured the level of sodium coolant in the reactor vessel. Above the sodium, there was inert argon gas. If the level was too low, the instrumentation would initiate an automatic reactor shutdown, known as a "trip".

I mumbled a reply about how we would not want the pumps to suck in any of the argon cover gas.

"Why not?"

"Well, if argon was sucked into the reactor core it would reduce heat transfer from the fuel, so the fuel might get too hot."

"OK, so supposing that was happening – we were getting argon sucked into the core – as an operator, how would you know?"

At this point my brain was whirring – I had never thought about that before. There was a silence which seemed very long while I tried to think how to answer.

Ed interrupted. "Maybe that is a bit advanced…"

I said "No, no, I'll have a go, I just haven't thought about that before". Pause. "I don't think you would detect any significant difference in reactor coolant outlet temperature. The instrument response time will be too slow". Pause. "However, I know the sodium coolant absorbs some neutrons, because that is how radioactive sodium-24 forms in the primary coolant. So, if bubbles of argon were passing up through the reactor core, the absence of sodium would mean that the neutron flux would increase a bit and we would see some noise on the signals from the neutron detectors".

My answer was perfect. And I had worked it out under pressure in front of an interview panel. I got the job.

Nuclear power stations (or refineries, or oil platforms, or aircraft, or any other large artifact of technology) have a huge amount of equipment. At PFR, as well as the reactor plant and its associated refueling equipment, there were boilers and pumps, the turbo-generator and condenser, the condensate and feed heating system, and the seawater cooling system. There were the electrical systems, switchgear, transformers, electrical protection systems, and electrical breakers. There were also the reactor shutdown systems and emergency cooling systems.

PFR was a very advanced nuclear power station for its time. Its control room was completely computerized – the power station was run by means of television screens driven by two early 1970s vintage mainframe computers. Almost everything the operators knew about the plant came from the computer systems.

In addition to all the plant and equipment, there were all the procedures. There were start-up and shutdown procedures. There were safety rules for working with high-voltage electricity, for working at height, for working in confined spaces, for building scaffolding, for welding, for doing radiography. There were emergency planning arrangements. And then there were all the Human Resources issues associated with managing a team of about 35 staff on every shift – what was done if someone went sick, what were the disciplinary arrangements, etc.

A shift manager did not have to know the details of all of these things, but he had to know enough – he had to be able to spot trouble from far away, he had to know what was good and what was wrong. He had to be able to set expectations for others. He had to be able to smell a rat. Some of the shift managers seemed to live on their nerves, but some others seemed to rise above it and keep focussed on what was current, yet be able to change direction instantly if circumstances changed, while remaining calm. (Much later, I have heard this described as "hands-on detachment", which neatly summarizes the paradox.)

I would have to learn about all of the above – enough to pass several interviews in front of people who had been there very much longer than me and who knew much more than I could possible learn in a couple of years. Having a PhD does not help you with this stuff. I have often wondered if the Gentlemen Physicists thought that a PhD might be able to assimilate all the training more quickly. (The Gentlemen Physicists had, of course, never had to learn how to operate a power station.) A PhD working in shift operations would certainly be a novelty, and maybe they thought I was the solution to their succession management problem. I was aware of an unspoken expectation that I should make rapid progress through my training.

However, you can accelerate training, but you cannot accelerate experience.

On Friday 9th March 1982, I had been working on shift as an Assistant Shift Manager on the "steam side" – i.e. the turbo-generator and boiler side of the plant – for maybe six months. I had still only been working at PFR for about 2.5 years, which I am sure most power station engineers would say is not enough time to be in a shift operations management role.

While training, my shift manager was Paul, who was a very laid back Lancastrian, quite short with thinning fair hair, in his early forties. I liked Paul because he was not permanently in a state of panic like some of the other shift managers. He had a good sense of irony, too – he was able to see how the high-pressure situation at PFR was making people behave oddly.

We were on a day shift. Day shifts were hard work – all the non-shift engineers would be in and out of the control room badgering the shift team for information, and there would be meetings to attend in the administration building. There was lots of hassle.

At about 9am, Paul called me to the control room.

"Jim, we've got a strange alarm that has just come up. It says the isolating valve for the level control valve for feed water heater 4 isn't fully open as it should be to allow the control valve to operate properly. This has happened before. What will

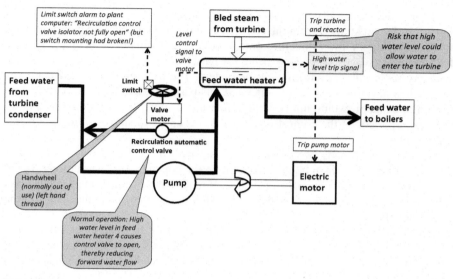

FIGURE A.1

Direct contact (DC) heater water level control.

probably have happened is that the hand-wheel on the isolator will have vibrated down and away from its limit switch. Can you go down to the turbine hall basement and just wind the hand-wheel back up to its fully open position, please?"

"OK, I'll get a radio." I collected a radio from the office – but I omitted to check the battery, which should be done every time. When the batteries ran out of charge, the radios could die quite suddenly.

I was quite pleased to be asked to do something like this – it seemed to suggest that, after a few months, Paul was beginning to trust me.

The feed water heaters (Fig. A.1) are there to pre-heat the water from the turbine condenser, before going back to the boilers. It is all to do with the thermodynamics of the Rankine steam cycle – feed water heaters make the overall efficiency better. The more heat that can be transferred between the steam bled from the turbine and the feed water, the better. Hence, in the 1960s, designers had moved to a design of feed water heater called a Direct Contact (DC) heater, where the steam from the turbine condenses directly onto the surface of the feed water, in a large cylindrical vessel, in order to maximize the heat transfer and the overall efficiency. The potential drawback here is the risk that, if the water level in the heater vessel rises too high, water might go back up the bled steam pipe and into the steam turbine. The steam turbine rotates at 3000 revolutions per minute, so water entering the turbine will cause a great deal of damage – the turbine would be written off, the cost of repair would be immense, and it would take many, many months to replace it. For this reason, instruments are fitted to detect high water level in the DC heater and send signals to the turbine to shutdown (trip). When the turbine trips, then the reactor trips too.

From Paul's diagnosis, the problem was not a big deal, and there was no real hurry, but I went straight from the control room down to the turbine hall basement. It was a few floors lower than the control room, a few minutes walk away. Turbine halls are noisy places, so I was wearing my hard hat and goggles and ear defenders, with my radio plugged into my ear defenders. The lighting in the basement was poor. I had never before actually operated this particular hand-wheel, but I had no reason to expect it would not be straightforward. I found the valve and radioed to the desk operator in the control room that I was about to start winding the hand-wheel open.

At this point I made two elementary mistakes. I did not check that Paul's preliminary diagnosis, which he had given me in the control room, was correct. In the poor lighting I did not notice the real cause of the problem – the limit switch bracket had corroded and the switch had fallen off. Hence the alarm was false, because the switch was faulty.

Furthermore, I did not know that the hand-wheel on this particular valve had a left-hand thread; it is almost universal that you turn a hand-wheel, a nut, or a bath-tap clockwise to tighten and anti-clockwise to loosen. I proceeded to turn the hand-wheel anti-clockwise to move it in the "open" direction, but actually I was closing it…

And also, just about at this point, the battery on my radio died…

So, I kept winding the hand-wheel anti-clockwise, thinking I was opening it. What I was actually doing was forcing the isolating valve into the closed position. This prevented the level control valve from doing its job and forced too much water forward into feed water heater 4.

After maybe a minute, I heard something crackly on the radio as if the desk operator was trying to speak to me. "Jim, what…crackle, crackle…"

I kept winding the hand-wheel and then, quite suddenly, I heard some bangs as various valves closed and electrical breakers opened, and the 3000 revolutions per minute tone of the huge 250 MW spinning main turbine suddenly lowered in pitch as the turbine began to slow down. With a sinking feeling, I realized the whole plant had tripped.

Afterward, it was like everyone was being too polite to tell me what they thought.

Paul and I had a quick de-brief before several day staff came to the control room to find out what had happened. My sense of shame was made worse when we discovered the shock of the plant trip had caused another steam leak in the steam generator, so the plant foreman and I had to deal with that for a couple of hours.

No one reprimanded me or swore at me, or even really spoke to me. I guess they could sense how I felt and they just gave me plenty of space. Eventually, in the afternoon, someone came from Technical Department to ask me what had happened so that he could write an incident report.

Nobody was hurt, and no damage was done apart from the steam leak in the steam generator. But I knew the plant had lost a lot of revenue because I had screwed up. (It normally takes about two days to return a nuclear power station from shutdown to full power – that is, two days' lost electricity generation. However, the steam generator leak would mean that we would be shutdown for quite a bit longer than that.)

Later on, all the monthly report said was this: *"Initially an alarm, later found to be incorrect, indicated that the DC heater regulator valve had closed and action in response to this led to problems in level control in DC heater 4 which led to a turbine and reactor trip."* Monthly reports can be very good at glossing over the details.

When there is an incident or accident, engineers do a root cause analysis (Chapter 10). Ultimately, a root cause analysis is trying to answer the question, "If you had a time machine, how far back in time would you have had to go to stop this event from happening?"

The immediate causes of this incident were threefold:

- **(i)** The false alarm was not investigated properly by the operator (me) (an error of omission).
- **(ii)** Incorrect action was taken by the operator (me) because the diagnosis of the cause of the alarm was incorrect (an error of commission).
- **(iii)** The operator (me) did not know that the recirculation isolation valve had a left-hand thread (a cognitive error).

There were also four contributory factors:

- **(i)** The operator (me) had not ensured the battery on his radio was adequately charged (an error of omission).
- **(ii)** The lighting in the turbine hall basement was poor (a design or maintenance issue).
- **(iii)** Maintenance had failed to identify the corrosion of the bracket holding the limit switch.
- **(iv)** The designer had specified a left-hand thread on the hand-wheel.

The key cognitive error was that I did not know that the recirculation isolation valve had a left-hand thread. If I had known to turn the hand-wheel clockwise, I would have immediately realized that the valve was actually already open, which would have made me look harder and think again – and then I would have noticed the corroded limit switch bracket. I showed impulsiveness to fix a problem that had not yet been adequately diagnosed. A more mature head than mine would have considered the problem much more carefully before jumping to a diagnosis and a solution.

One root cause would have to be that the operator (me) was insufficiently experienced for the job he was doing. Really, everything above makes it quite clear that it was unequivocally my fault.

However, a contributory factor might be that the plant senior management had not been looking after succession planning and, to recover the position, they were trying to push young engineers into positions of responsibility before they were ready. Or does this sound like self-justification? Am I just trying to blame others for my mistakes, even decades later?

I made a series of small mistakes in quick succession. Experience means recognizing when not to be in a hurry, when to say, "I need to ask someone about this, to think about it". But experience can also mean recognizing the easy quick fix, when something can be done easily and quickly and thereby avoid making a minor issue into a problem. (It would have been a great feeling to have gone back to the control room saying, "There's nothing wrong, it is just the limit switch bracket has corroded. I'll raise a job card.")

Maybe I am beating myself up. My pride was hurt – and it still is, whenever I think about this. I had been a fool, and I had rushed in.

The opportunities to make mistakes in power stations are innumerable, because of the enormous range of equipment that has to be learned and understood. From my current, late-career perspective, the management problem is to ensure somehow that mistakes can be made safely, without blame, so that lessons can be learned without consequence.

So, the exam question for senior managers is this: How do you *know* when people have enough experience? You could of course keep extending the training period, introducing longer and longer periods that young staff remain in junior roles; stifling their sense of worth, and stifling initiative, while those at the other end of their careers remain in post longer, becoming less dynamic and showing less novel thinking. If you do that, you drive up costs, and the good young guys may become bored and leave. In succession management, how do you carry out fast-track promotion without risking mistakes?

Another contributory factor was the shift manning arrangements. There was only one graduate engineer on each shift team, and it was deemed unacceptable to temporarily upgrade an assistant shift manager if the shift manager was on holiday or was sick. Hence, holiday and sick leave shift cover for the shift managers had to come from the "three months' days" shift managers. There were only six shift manager posts in total, so the whole system was very vulnerable to resignations – which is what had led to me being fast-tracked through shift operations training. There needed to be a better system that could allow for resignations and long-term sickness to occur without immediately causing a crisis.

In summary, then, the organizational contributory factors were as follows.

1. There was inadequate succession planning.
2. The shift team arrangements were inflexible and could lead to a crisis if shift managers resigned or became long-term sick.

And so I learned the hard way how accidents happen (even though this was not an accident, it was really just a minor incident). In short, accidents happen because of one or more of the following: Ignorance, stupidity, hubris, complacency, and/or organizational neglect.

The difference between ignorance and stupidity is important. A smart person can be ignorant because he has not been taught something. A stupid person may not understand something even after he has been taught it. So, ignorance is a training issue, but stupidity is a recruitment issue.

Lastly, organizational complacency can be one cause of organizational neglect, although there are others. Also, at PFR, perhaps organizational overload had led to neglect, which manifested itself as poor succession management.

I did become a shift manager at PFR, in December 1982. Soon, however, the dullness and routine of shift life and my inability to sleep well when working on night shifts made me restless for change. I left Caithness in 1984. I never worked in Operations again – all my later work has been in design engineering, project management, safety management, engineering management, and consultancy.

The Prototype Fast Reactor was closed in 1994, and the UKs fast reactor program was abandoned. The Dounreay site is still being decommissioned.

At time of writing (2014), France is considering a new fast reactor program, under the banner of a "Generation IV" advanced reactor design, a new BN-800 880 MW fast reactor is undergoing commissioning tests at Beloyarsk in Russia, and a 500 MW fast reactor is being commissioned at Kalpakkam in India.

Appendix 2
Roger Boisjoly, the *Challenger* Accident, and Whistle-Blowing

In October and November 2000, I had the great privilege of attending a 5-week management seminar, "Senior Nuclear Plant Management", at the Institute for Nuclear Power Operations (INPO) in Atlanta, Georgia. This seminar consisted of 14 experienced people from nuclear power companies – 12 from the USA, one from France, and me.

This seminar was specifically not a technical training course. Instead, we discussed how best to manage nuclear power plants in terms of safety, economy and efficiency. We also discussed the difficulties and challenges faced by nuclear plant managers, and some of the ethical dilemmas. We visited three operating nuclear plants and had discussions with their operators. Invited speakers came to talk to us.

One of the invited speakers was Roger Boisjoly[1], who spent a whole morning with us, telling us his version of events leading up to the *Challenger* accident in 1986, and answering our questions. After he left, we spent the afternoon reviewing what he had said and its relevance to the nuclear industry. Roger had worked as a mechanical engineer for Morton Thiokol, in Utah, USA, from 1980 to 1986. Among other things, Morton Thiokol supplied the Solid Rocket Boosters (SRBs) for the Space Shuttle program. Roger had worked as a project engineer with responsibility for the rubber O-ring seals that made a gas-tight connection between the sections of solid fuel in the long SRBs.

On 28th January 1986, the Space Shuttle *Challenger* exploded 73 seconds after takeoff, killing all seven crew. I do not intend to review the accident here because it is very well documented in many other places (in particular, [1, 2]). In short, however, the launch happened in extremely cold weather; the rubber O-ring seals in the Solid Rocket Boosters lost their elasticity when cold; this led to hot rocket exhaust blowing past and destroying one of the O-ring joints, and the hot exhaust impinged on the main hydrogen tank, which failed and exploded.

I want to describe my impressions of Roger Boisjoly, his role in what happened, and how he interacted with his own management. Roger spoke extremely passionately indeed about his involvement. I felt that it mattered to him to persuade us of his version of events, and maybe to atone for the guilt he felt for those events. He seemed to be haunted by what had happened; he spoke with a deep conviction that he had been part of a dreadful accident that was in some way shameful to him, to his company Morton Thiokol, to NASA, and to the United States. He had only been an extra in a much larger drama, but he acted as if he were carrying the guilt for the whole nation.

---

[1]Roger Boisjoly died in 2012.

Roger sat down as he spoke to us. He had thinning grey hair and blue, slightly bloodshot eyes which darted around, seldom looking directly at anyone in the room. I got the impression he had told the story often before, but it still seemed important to him to keep re-telling. His voice was soft, but it could suddenly rise when he was making an important point. There was a sort of missionary zeal about him.

From the start, the use of SRBs in the Space Shuttle had been a compromise, he said. The initial concept had been for a two-stage, wholly-reusable craft, but the development costs were going to be too great, so a decision was made that only the Orbiter would be fully reusable. Instead, lift-off and initial boost would use SRBs, which were a development of boosters used in other projects.

"The entire Shuttle program was run on a shoestring", he said. After the fourth flight, the shuttle was declared operational, which meant there were hardly any funds left for Research and Development (R&D). In the 1970s, NASA said the Shuttle would achieve 60 flights per year. By the mid-1980s, this had been reduced to 24 flights per year, but NASA had indicated to Morton Thiokol that 18 flights per years was a realistic target. The Shuttle had been advertised as a routine way of traveling into space, but in reality it was still very much an advanced prototype.

Roger said, "The O-ring problem was well-known and had been documented as far back as 1977". The second launch in 1981, and again the fifteenth launch in January 1985, had shown very significant O-ring erosion which was discovered in post-flight examination of the recovered boosters. NASA was informed but, according to Roger, their response was "We don't want to know that".

During the April 1985 launch, 80% of one O-ring seal was found to have eroded, and NASA became concerned. A memo was written by Roger to NASA on 2nd July 1985 which said the seals were vulnerable below 50°F (10°C). He described 33 possible ways of fixing the problem in another inter-office memo dated 31st July 1985, although he said, "That memo was killed by my boss". The memo had said amongst other things:

> It is my honest and very real fear that if we do not take immediate action to dedicate a team to solve the problem...then we stand in jeopardy of losing a flight.

An unofficial O-Ring Task Force, run by Roger, was formed in Morton Thiokol in August 1985 to address the problem, and NASA was informed.

"The culture in Morton Thiokol was inward-looking, and career progression was based on nepotism". He said there was a complacent, non-challenging attitude; Morton Thiokol had been making boosters like the SRBs for 20 years and they knew best. He said quality assurance was poor, and "there was 1950s' equipment and a 1960s' mindset". He also said if any member of staff made a complaint, it was put on his file – he described this as a "punishment system". He described the culture within Morton Thiokol as one of "punishment and intimidation – a blame culture".

He also said the Shuttle SRBs provided about fifty per cent of Morton Thiokol's profits – so the Shuttle program was crucial to the company.

He said, "The O-Ring Task Force had no resources, no authority, and no power". He said that NASA peer engineers sat in on the Task Force meetings and were sympathetic but no tangible support was forthcoming. Nevertheless, he thought a short-term fix would be developed, and he never thought about going outside the organization. He described the activities of the Task Force as "trying to do R&D in a production environment", i.e. he still retained his operational responsibilities.

In the autumn of 1985, there were a series of escalating memos on the topic within Morton Thiokol, and some progress towards design changes was made, for example towards thicker O-ring seals. However, no actual design changes were made before a cold-weather launch was proposed in January 1986. The Task Force was asked its opinion and a launch postponement was recommended. This got NASA's attention and a Flight Readiness Review teleconference was arranged between NASA and Morton Thiokol, to be held on 27th January 1986 – the day before the scheduled launch date. Morton Thiokol managers attended the meeting with Roger. It was clear that NASA was not willing to postpone the launch, and the teleconference became heated. NASA attacked Morton Thiokol for providing late information.

George Hardy, who was NASA's Deputy Director of Science and Engineering, said, "I am appalled by your recommendation" (for launch postponement). Larry Mulloy, NASA's manager for the SRB project, made it clear he was not willing to delay the launch.

At this point, Roger said the mute button on the phone was pressed by a senior manager in the Morton Thiokol conference room. For several minutes, the Morton Thiokol team discussed among themselves what they should say next.

"I was screaming at them, 'Look at the photos!', Roger said, referring to photos showing O-rings damaged during the April 1985 launch.

Morton Thiokol Senior Vice-President Jerry Mason said "Thiokol has to make a management decision. Am I the only one who wants to fly?"

In the end, of course, Morton Thiokol gave their go-ahead to NASA for the launch. The approval was signed by Joe Kilminster, Vice-President of Space Booster Programs, who according to Roger was not actually an engineer. Notably, Kilminster's deputy Al McDonald refused to sign.

NASA accepted this "recommendation" without further query.

After the explosion during takeoff the following day, Roger said NASA and Morton Thiokol managers tried to distort the truth. He said he and Al McDonald were relegated to secondary positions, and that he suffered severe mood swings and depression. After he spoke to the Commission of Inquiry in July 1986, he said he was shunned by his colleagues and branded as a whistle-blower, while some people who had supported the launch were still in post.

He resigned from Morton Thiokol in July 1986. He said he was diagnosed with post-traumatic stress disorder – he had clearly had some kind of nervous breakdown after the accident. He filed a lawsuit against Morton Thiokol but it was dismissed.

Roger ended his presentation with his thoughts on what employees should do in an ethical dilemma. He summarized his views as "Exit, voice or loyalty'; i.e. if you wish to remain silent you must leave the company, or you can stay in the company and try to be heard, or you can stay in the company and not rock the boat. "Management must recognize that what they want to hear is not always what they need to know, and subordinates must recognize that what they want to tell is always what the managers need to know".

He concluded that NASA would have launched on 28th January 1986 no matter what Morton Thiokol had said.

The Presidential Commission report made some wide-ranging recommendations, some of which were as follows:

- Redesign of the solid rocket booster
- Set up an independent oversight group
- Review the shuttle management structure, put astronauts into management, and establish a Shuttle Safety Panel
- Carry out a critical item review and hazard analysis
- Review the safety organization within NASA

The report of the challenger accident investigation to the House of Representatives [2] concluded that the failure of the O-ring seal joint was due to a faulty design, which neither NASA nor Thiokol had fully understood prior to the accident. Also, the test and certification programs for the joint were inadequate; neither NASA nor Thiokol had responded adequately to available warnings about the defective joint design. The safety, reliability, and quality assurance programs within NASA were grossly inadequate. NASA was advised to review its risk management activities to define a complete risk management program.

Former astronaut Frank Borman described the shuttle as "a hand-made piece of experimental gear".

Listening to Roger Boisjoly was one of the most interesting mornings I have ever spent.

The first thing to say is, quite unequivocally, that he was of course absolutely right – the O-ring seals were unsafe.

My problem with Roger, however, was his personality. He will have been a management nightmare. I imagine he was the sort of employee who will have demanded huge amounts of the managers' already busy schedule. He was emotionally extremely committed to his job, but he seemed to lack judgment. One of the most telling parts for me was when he said he had drafted the memo on 31st July 1985 listing 33 options for fixing the problem; this indicates to me that all he had done was carry out some unfiltered brainstorming of the problem, with no real intelligence being applied regarding the practicality of the options. If I had been his manager I would have said, "OK, so what am I supposed to do with this? Come back with a recommendation". His manager could not have gone to NASA with that memo – Morton Thiokol would have looked ridiculous and incompetent.

Roger's personality seemed to lack the essential engineering ability to make pragmatic judgments. No research and development budget would ever have satisfied him because he would always have wanted to explore yet more possibilities instead of making a judgment and taking a decision. I imagine he will have been the sort of engineer that frustrated his managers by always exceeding budgets and not achieving deadlines.

I imagine also that he may have had a record of crying "Wolf" to his managers, seeing many things that were wrong or dangerous in the program that turned out, on closer inspection, to be actually alright. I suspect this will have desensitized his managers. "It's only Roger", they may have said in private. His manner, his self-righteousness, may have antagonized his managers. *The problem of course was that, this time, there really was a wolf.*

Roger had said that a NASA engineer had also written recommending against the launch on 28th January 1986, so he was not completely alone, no matter how it may have felt to him at the time. The plain fact is that the O-ring seal problem *was ultimately NASA's problem*, and if NASA was not prepared to listen, then Morton Thiokol was talking to the wrong people in NASA, at the wrong level. NASA's attitude was completely wrong – they wanted Morton Thiokol to provide proof that it was *unsafe* to launch whereas they should have been asking for justification why it was *safe* to launch. Hence, Morton Thiokol was in a no-win position; if their managers had refused to give their blessing to the launch, their company would have been in serious trouble with their client, and there would have been uproar in the media about "NASA incompetence" – after all, was not the Space Shuttle now proven technology, just a bus into space, as NASA's Public Relations seemingly wanted people to believe?

Morton Thiokol should have recommended launch postponement, of course, but the consequences for the company would undoubtedly have been serious because of the bad attitude that existed in NASA toward safety at that time. Instead of thanking Morton Thiokol for their concerns and their openness, the company would have been blamed by NASA for the public relations difficulties that NASA would have faced after (yet another) launch postponement.

However, Roger was claiming that everyone in Morton Thiokol was indifferent to safety, that the management structure was awful, that the company was out-of-date in its design and production practices, and that the wrong people were promoted – in short, almost everything in the company was wrong and he was the only person who could see this. I began to see he was the sort of employee that was never happy. I expect he was in his manager's office on a daily basis, complaining repeatedly and at length about all the things that were not to his liking. I expect he was very good at saying what other people should be doing, but maybe he was less good at prioritizing and delivering his own responsibilities. Also, his communication skills were poor – he became too emotional about problems. When he said, "I was screaming at them, 'Look at the photos!', I remember thinking, No, that was the wrong way to go about it – you were alienating the people whose support you wanted.

So, if we go back in a time machine to the teleconference on 27th January 1986, what should Morton Thiokol have done? They should, *of course*, have refused to approve the launch.

What more should they have done *before* 27th January? Many months earlier, Morton Thiokol senior management should have contacted top NASA managers with their concerns. They must have had very high-level contacts in NASA – so they should have used those contacts to go over the heads of the NASA people making launch decisions, and expressed their safety concerns to someone senior enough to recognize the full potential horrendous implications of a cold-weather launch, someone who could recognize that, yes, launch postponement would be a PR disaster but there were more important things to consider. It is on occasions such as this – making difficult decisions and taking the public criticism – that senior managers earn their salaries.

The real root cause was NASA's unwillingness to deal with safety concerns. NASA was under public pressure to meet its launch program, and in turn this public pressure was generated by political pressure on the NASA budget. Hence, meeting project deadlines became more important than ensuring safety. This attitude was ultimately responsible for causing the accident, and (indirectly) it also forced Roger into some sort of nervous breakdown. The sad truth is that, in 1986, NASA's safety culture and organization was irredeemable; it required an accident to change it.

How should Morton Thiokol managers have dealt with Roger Boisjoly? Once the problem had been identified in July-August 1985, a Morton Thiokol manager (and not Roger himself) should have been chairing the "unofficial Task Force" so that the problem would remain visible to (and owned by) management. They should have recognized that Roger was not the right person to manage the responsibility of such a high-profile problem, but kept him as a useful and important team member, who could undoubtedly make valuable contributions, but who needed management support and guidance.

However, that is not a generic solution to the problem of "employee concerns" (or, as it more usually known, "whistle-blowing"). All companies working with serious safety issues need to have a "safety net" process to catch employees' safety concerns; there has to be some route that enables employees to send messages to the very top of the company organization or even to their clients (in this case, NASA). So, another lesson here is having an effective means of listening to, and responding to, employees' safety concerns. All hazardous, high-technology industries have this problem from time-to-time; employees who are convinced that some aspects of the company's operations are being managed in an unsafe way, and who deserve to be listened to, but who find that their line managers will not listen.

The worst-case scenario for both the company and the employee is if the employee goes to the news media. In that event, the company receives bad publicity, and the whistle-blowing employee may suffer later retaliation from either management or fellow-employees.

Of course, some of these employee safety concerns may be unfounded – in which case the relevant employee deserves a good explanation as to why this is so. However, many concerns will be valid, and it only takes one real safety issue to cause an accident. The important thing is to deal with employee safety concerns in a transparent way.

There are at least four complementary ways of dealing with this. The first way is to have an independent Director or Vice-President responsible exclusively for safety who reports directly at CEO-level and who should be accessible to employees. The second is a means for employees to report their safety concerns anonymously, for example by putting their written concerns in a drop-box on site. The third in unionized companies is for employees to take their concerns to their union representatives. The fourth is that there is a wholly independent government safety regulator who is visible to, and approachable by, the workforce. Often, all four are used simultaneously.

REFERENCES

[1] The report of the Presidential Commission (The Rogers Report) is available at http://history.nasa.gov/rogersrep/genindex.htm.
[2] The full report to Congress of the investigation of the challenger accident is available at http://www.klabs.org/richcontent/Reports/Failure_Reports/challenger/congress/64_420a.pdf.

Index